JACARANDA

SENIOR GEOGRAPHY 1
FOR QUEENSLAND

UNITS 1 & 2 | FOURTH EDITION

BILL DODD

MICK LAW

IAIN MEYER

PHIL O'BRIEN

LESSA GORE-BROWN

KIMBERLEY REEVE

jacaranda
A Wiley Brand

Fourth edition published 2024 by
John Wiley & Sons Australia, Ltd
Level 4, 600 Bourke Street, Melbourne, Vic 3000

First edition published 2000
Second edition published 2008
Third edition published 2019

Typeset in 10.5/13 pt TimesLTStd

© Bill Dodd, Mick Law, Iain Meyer, Phil O'Brien 2000, 2008, 2019, 2020, 2024

The moral rights of the authors have been asserted.

ISBN: 978-1-394-27000-2

Front and back cover images: © Greg Brave / Adobe Stock; © Vector light Studio / Adobe Stock; © Coral_Brunner / Adobe Stock

Typeset in India by diacriTech

A catalogue record for this book is available from the National Library of Australia

Printed in Singapore
M130213_260824

The publisher of this series acknowledges and pays its respects to Aboriginal Peoples and Torres Strait Islander Peoples as the traditional custodians of the land on which this resource was produced.

This suite of resources may include references to (including names, images, footage or voices of) people of Aboriginal and/or Torres Strait Islander heritage who are deceased. These images and references have been included to help Australian students from all cultural backgrounds develop a better understanding of Aboriginal and Torres Strait Islander Peoples' history, culture and lived experience.

It is strongly recommended that teachers examine resources on topics related to Aboriginal and/or Torres Strait Islander Cultures and Peoples to assess their suitability for their own specific class and school context. It is also recommended that teachers know and follow the guidelines laid down by the relevant educational authorities and local Elders or community advisors regarding content about all First Nations Peoples.

All activities in this resource have been written with the safety of both teacher and student in mind. Some, however, involve physical activity or the use of equipment or tools. **All due care should be taken when performing such activities.** To the maximum extent permitted by law, the author and publisher disclaim all responsibility and liability for any injury or loss that may be sustained when completing activities described in this resource.

The publisher acknowledges ongoing discussions related to gender-based population data. At the time of publishing, there was insufficient data available to allow for the meaningful analysis of trends and patterns to broaden our discussion of demographics beyond male and female gender identification.

Contents

UNIT 1 RESPONDING TO RISK AND VULNERABILITY IN HAZARD ZONES

TOPIC 1 NATURAL HAZARD ZONES

TOPIC 2 ECOLOGICAL HAZARD ZONES

About this resource

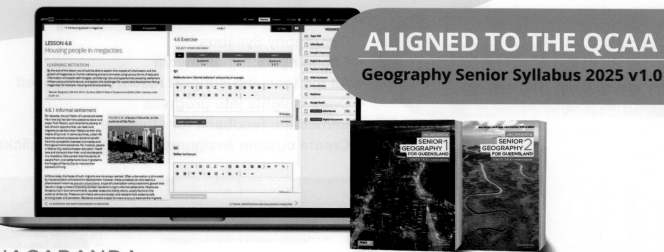

ALIGNED TO THE QCAA
Geography Senior Syllabus 2025 v1.0

JACARANDA
SENIOR GEOGRAPHY 1
FOR QUEENSLAND | UNITS 1 & 2 FOURTH EDITION

Developed by teachers for students

Tried, tested and trusted. The fourth edition of the *Jacaranda Senior Geography 1 for Queensland* series is the only resource written specifically for the revised Queensland Senior Geography Syllabus, with a toolkit of resources to provide a skills-rich, inquiry-based approach to engage students of all abilities.

Because both *what* and *how* students learn matter

Learning is personal

Whether students need a challenge or a helping hand, you'll find what you need to support your students as they develop their geographical skills and thinking.

Whether in class or studying at home, students can progress and achieve success! Skills-rich, inquiry-based and supported by real-world case studies, students are exposed to a variety of contemporary problems and challenges. Automarked differentiated question sets with detailed solutions, video lessons and interactivites provide a range of learning experiences.

Learning is effortful

Learning happens when students push themselves. With learnON, Australia's most powerful online learning platform, students can challenge themselves, build confidence and ultimately achieve success.

Learning is rewarding

Through real-time results data, students can track and monitor their own progress and easily identify areas of strength and weakness.

And for teachers, Learning Analytics provide valuable insights to support student growth and drive informed intervention strategies.

Learn online with Australia's most

Everything you need for each of your lessons in one simple view

- Aligned to the revised Queensland Senior Geography Syllabus
- Engaging, rich multimedia
- All the teaching-support resources you need
- Deep insights into progress
- Immediate feedback for students
- Create custom assignments in just a few clicks.

Practical teaching advice and ideas for each lesson provided in teachON

Links to the revised Queensland Senior Geography Syllabus

Reading content and rich media including videos, interactivities and audio files.

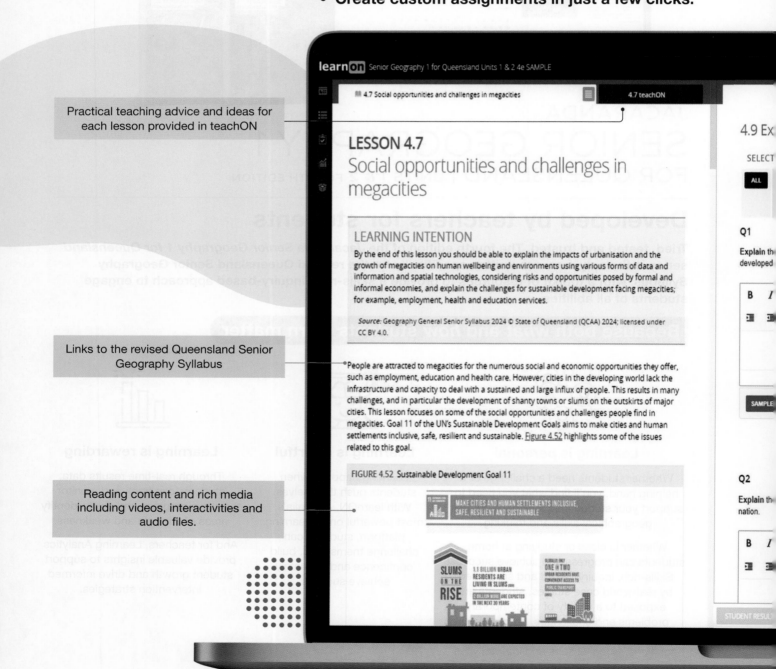

learnOn Senior Geography 1 for Queensland Units 1 & 2 4e SAMPLE

4.7 Social opportunities and challenges in megacities   4.7 teachON

LESSON 4.7
Social opportunities and challenges in megacities

LEARNING INTENTION

By the end of this lesson you should be able to explain the impacts of urbanisation and the growth of megacities on human wellbeing and environments using various forms of data and information and spatial technologies, considering risks and opportunities posed by formal and informal economies, and explain the challenges for sustainable development facing megacities; for example, employment, health and education services.

Source: Geography General Senior Syllabus 2024 © State of Queensland (QCAA) 2024; licensed under CC BY 4.0.

People are attracted to megacities for the numerous social and economic opportunities they offer, such as employment, education and health care. However, cities in the developing world lack the infrastructure and capacity to deal with a sustained and large influx of people. This results in many challenges, and in particular the development of shanty towns or slums on the outskirts of major cities. This lesson focuses on some of the social opportunities and challenges people find in megacities. Goal 11 of the UN's Sustainable Development Goals aims to make cities and human settlements inclusive, safe, resilient and sustainable. Figure 4.52 highlights some of the issues related to this goal.

FIGURE 4.52 Sustainable Development Goal 11

MAKE CITIES AND HUMAN SETTLEMENTS INCLUSIVE, SAFE, RESILIENT AND SUSTAINABLE

SLUMS ON THE RISE

1.1 BILLION URBAN RESIDENTS ARE LIVING IN SLUMS

3 BILLION MORE ARE EXPECTED IN THE NEXT 30 YEARS

GLOBALLY, ONLY ONE IN TWO URBAN RESIDENTS HAVE CONVENIENT ACCESS TO PUBLIC TRANSPORT

4.9 Ex

SELECT

ALL

Q1

Explain th
developed

B   I

SAMPLE

Q2

Explain th
nation.

B   I

STUDENT RESUL

powerful learning tool, learnON

Differentiated question sets

Teacher and student views

Textbook questions

eWorkbook

Answers and sample responses

Digital documents

Video eLessons

Interactivities

Enhanced teaching support resources

Interactive questions with immediate feedback

Get the most from your online resources

Online, these new editions are the complete package

Trusted Jacaranda theory, plus tools to support teaching and make learning more engaging, personalised and visible.

Each lesson is linked to the content and skills from the revised Queensland Senior Geography Syllabus

Key terms glossary to help develop and support effective communication of concepts and skills

onResources link to targeted digital resources including video eLessons and weblinks.

Tables and images break down content, allowing students to understand complex concepts.

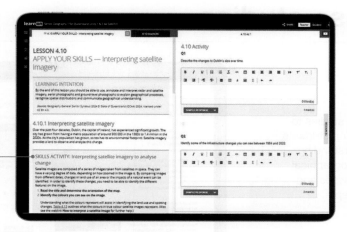

Skills activity lessons to develop geographical skills and understanding

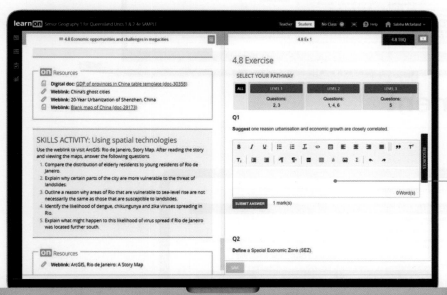

Online and offline differentiated question sets, with immediate feedback, help students to overcome misconceptions as they occur and to challenge themselves at their own level.

Online and offline exam questions with immediate feedback are available in each subtopic.

Case studies

Case studies to develop students' understanding of geographical processes and concepts.

Skill development

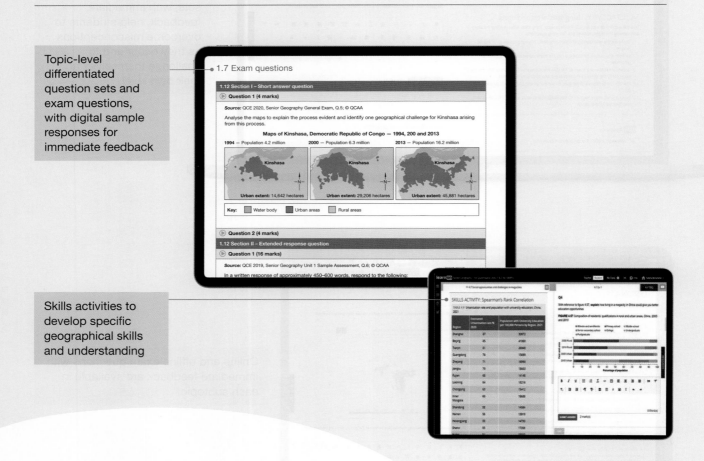

Topic-level differentiated question sets and exam questions, with digital sample responses for immediate feedback

Skills activities to develop specific geographical skills and understanding

A wealth of teacher resources

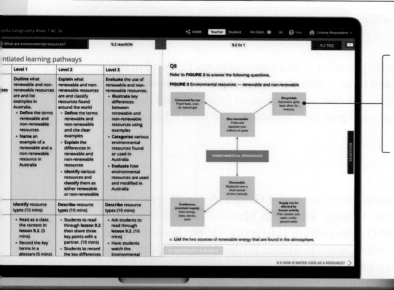

Enhanced teaching-support resources for every lesson, including:
- work programs and curriculum grids
- practical teaching advice

Customise and assign

An inbuilt testmaker enables you to create custom assignments and tests from the complete bank of thousands of questions for immediate, spaced and mixed practice.

Reports and results

Data analytics and instant reports provide data-driven insights into progress and performance within each lesson and across the entire course.

Show students (and their parents or carers) their own assessment data in fine detail. You can filter their results to identify areas of strength and weakness.

Meet our author team

Bill Dodd

Bill Dodd is a highly experienced and enthusiastic geography teacher with an extensive background in text writing and field studies. Although having been a HOD and former state panelist, his prime passion is delivering engaging lessons in the classroom and providing stimulating outdoor experiences to inspire future custodians of our beautiful planet. He believes there has never been a more urgent time for students to have a knowledge and appreciation of their world and its geography.

Dr Philip O'Brien

Dr Philip O'Brien is an experienced Secondary school teacher who has taught Geography at all levels in the United Kingdom and Queensland including Brisbane Boys' College. Philip has Masters' degrees from University of Queensland and Griffith University and a PhD from Griffith University. He has been involved in writing all editions of Jacaranda's Senior Geography textbooks and has written textbooks for Western Australian and Victorian Geography syllabuses. He has also written a number of articles for 'Geography Review', a magazine designed for A-level students in the United Kingdom.

Mick Law

Mick Law is a passionate educator and geographer working with digital tools whenever possible. Mick has been involved in geospatial education at the school level for over 15 years. As a classroom teacher, Mick developed geospatial programs at two Queensland schools before managing ESRI Australia's GIS in-schools program. Mick then established his own geospatial educational consultancy, Contour Education. Mick now works for Queensland's Department of Natural Resources, Mines and Energy advocating for geospatial technologies and developing resources for teachers and students to bring these exciting tools into Queensland classrooms.

Mick is active in the geospatial industry and is a former Executive Committee member and journal editor of the Geography Teachers' Association Queensland (GTAQ) and current chair of the Destination Spatial Queensland (DSQ) Executive.

Iain Meyer

Iain Meyer was educated in England and graduated from London University in Geography and Economics. Has taught geography for over 40 years in the UK and Australia and has held the position of Head of Geography at Melbourne Grammar School.

Iain has co-authored Geography textbooks for senior school students in the UK and Australia. He has, along with co-author Phil O'Brien, been a regular contributor to Geography Review, a quarterly magazine published in the UK for senior school geography students. He has been Chief Examiner for the Oxford and Cambridge Schools Examination Board and a member of the QCAA District Review Panel. In 1985, he was awarded a BP Education fellowship at Oxford University.

Still passionate about geography and the intrinsic value of the geographical perspective, Iain's current interest is in the application of visible thinking in Geography, as pioneered Ron Ritchhart and Project Zero team at Harvard Graduate School of Education.

Lessa Gore-Brown

Lessa Gore-Brown is an experienced educator with a background in field studies and a passion for Geography. She has taught across UK, Australian, and IB exam boards, covering a wide variety of Geography topics and syllabi. For the past seven years, she has been teaching in Queensland schools. Lessa is keen on using innovative methods such as GIS and drones in teaching and fieldwork. She has held curriculum leadership roles, developed educational resources, and has experience in script marking and assessment moderation. She has most recently been involved in updating the Senior Geography resources in line with the revised QCAA syllabus.

Kimberly Reeve

Kimberly Reeve is an experienced educator teaching Geography both in Australia and the UK. She has a passion for humanities and its place in the world and has taught both Geography, Religious Studies and History. She has been an external examiner in the UK at both GCSE and A-Level since 2005 and, after moving to Queensland in 2015, has been involved with QCAA assessment and moderation. Kimberley is currently Assistant Head of Humanities and has previously held a Head of Geography position in both the UK and Australia. She is currently responsible for the planning, resourcing, and teaching of Senior geography.

Acknowledgements

The authors and publisher would like to thank the following copyright holders, organisations and individuals for their assistance and for permission to reproduce copyright material in this book.

The full list of acknowledgements can be found here:

www.jacaranda.com.au/acknowledgements/#2024

Every effort has been made to trace the ownership of copyright material. Information that will enable the publisher to rectify any error or omission in subsequent reprints will be welcome. In such cases, please contact the Permissions Section of John Wiley & Sons Australia, Ltd.

1 Natural hazard zones

UNIT 1 TOPIC 1

SUBJECT MATTER

In chapter 1, students:
- **explain** the geographical processes that result in geological, geomorphic and atmospheric hazard zones
- **recognise** the spatial patterns of hazard zones and the implications for people and places
- **investigate** natural and anthropogenic factors to identify why some places are more at risk than others from specific types of natural hazards and disasters.

Students conduct a case study of two locations to:
- understand vulnerability and risk in hazard zones for places in less developed and more developed countries
- understand why people in particular places have a greater or lesser capacity to respond to natural hazards and natural disasters.

LESSON SEQUENCE

Fully worked solutions for this topic are available in the Resources section at www.jacplus.com.au.

LESSON
1.1 Overview

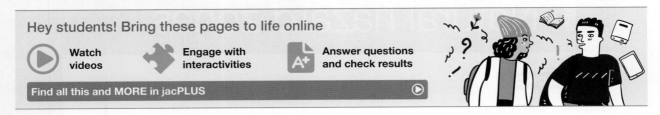

1.1.1 Introduction

In this topic, you will examine different types of natural hazards and the zones in which they are most likely to occur. These include **atmospheric hazards**, such as severe storms and cyclones; **geomorphic hazards**, such as landslides and mudslides; and **geological hazards**, such as earthquakes and volcanic eruptions.

You will also learn about the processes and patterns of natural hazards, and why they are sources of risk. By analysing data and information, you will assess why some hazards seem more common, predictable or frequent, while others seem to occur almost randomly.

Finally, you will apply your understanding of natural hazards to examine their potential impacts and how different communities might be able to minimise the damaging effects they have on people, property and the environment.

atmospheric hazards a potentially damaging natural event generated in the troposphere, such as a severe storm, tropical cyclone (typhoons and hurricanes), tornado, blizzard and wind storm

geomorphic hazards a potentially damaging event on the Earth's surface — such as an avalanche, landslide or mudslide — that is often caused by a combination of natural and human processes

geological hazards a potentially damaging natural event occurring in the Earth's crust, such as a volcanic eruption, earthquake or tsunami

FIGURE 1.1 Geological hazards such as volcanic eruptions can create hazard zones in their surrounding areas due to lava, ash fall, heat, pyroclastic clouds and toxic gases.

1.1.2 Syllabus links

Syllabus links	Lessons
○ Explain, using a range of representations such as maps, conceptual models, block diagrams, and cross-sections, how natural hazards are the result of processes that occur within the earth (geological) on the surface of the earth (geomorphic), in the atmosphere (atmospheric) or a combination of these and may result in natural hazards.	**1.2** **1.6** **1.7** **1.9–1.14**
○ Recognise hazard zones, represent these on a map (using spatial technologies) and analyse spatial distribution of the hazard to describe geographic patterns and identify the implications for people and environments.	**1.2** **1.4**
○ Explain how the severity of the impacts of natural hazards is influenced by factors such as • speed of onset • magnitude • frequency • duration • sequencing and seasonality of events (that is, random or regular).	**1.4**
○ Explain how climate change may affect the severity and incidence of some natural hazards and increase risk.	**1.4**
○ Explain the concepts of vulnerability and risk as applied to natural hazard management.	**1.3** **1.5**

○ Conduct a case study to investigate one natural hazard that has affected a place in a less developed country and a place in a more developed country. As part of this case study, students must
• manipulate, adapt and transform data, using spatial technologies and information and communication technologies, to represent and describe the nature, extent and characteristics of the hazard zone for the case study locations
• analyse threat and vulnerability data and information (including representations in maps) for each place to explain how these factors contribute to risk for the case study locations including
  • exposure to the hazard (threat)
  • population density and settlement patterns
  • topographic features such as hydrology and elevation
  • level of economic development
  • degree of preparedness
  • location, speed and effectiveness of emergency responses
• analyse the use of technologies (including spatial) in responding to natural hazards
• apply geographical understanding from their analysis to explain the impacts for communities
• propose preparedness, mitigation, prevention and adaptation strategies to manage risk, referring to practices from various agencies such as NGOs, governments and individuals
• communicate understanding using appropriate forms of geographical communication.
	1.9–1.14

KEY QUESTIONS

1. What is a natural hazard?
2. Where do natural hazard zones occur and why?
3. What are atmospheric, geomorphic and geological hazards?
4. What factors affect the severity of impact of a natural hazard?
5. What factors affect a community's response to a natural hazard?
6. What factors affect a community's vulnerability to the risks of a natural hazard?
7. How are people in developed and developing communities affected differently by natural hazards?
8. How do people in developed and developing communities respond to natural hazards?

LESSON
1.2 Natural hazards

LEARNING INTENTION

By the end of this lesson you should be able to define hazards, disasters and hazard zones, and understand how humans use risk management to prevent or lessen the impact of hazards.

Source: Adapted from Geography General Senior Syllabus 2024 © State of Queensland (QCAA) 2024; licensed under CC BY 4.0.

1.2.1 Hazards and disasters

A **natural hazard** is any extreme geophysical event that has the potential to cause harm to people, other living things, property and the environment. They can occur in the Earth's crust, on the surface of the Earth or in the atmosphere, and are created by powerful forces that generate high levels of destructive energy. Because of the dynamic nature of the Earth, natural hazards occur almost everywhere and affect all parts of the **biophysical environment**: natural, managed and built (see figure 1.2).

A natural event becomes a hazard when its **magnitude** (size), speed of onset, duration or frequency create serious risk to people and has the potential to result in considerable damage. These hazards create **risk** (exposure to some form of dangerous situation). When individuals and communities are at risk, they have to assess how to manage and lessen the effects of that risk for their communities and the local area.

Natural hazards can cause death or injury to people, and damage buildings, property, infrastructure, crops and farmland. When a hazard is responsible for many deaths and the loss of homes, and will cost huge sums of money for repairs and compensation, it is called a **natural disaster**. Natural disasters may involve extensive disruption that requires a long-term recovery plan.

natural hazard an extreme event occurring either in the lithosphere or in the atmosphere. It can be highly destructive and cause considerable harm to living things and property. Examples include tropical cyclones, tornadoes, earthquakes and volcanoes.

biophysical environment both living (biotic) and non-living (abiotic) surroundings of an organism or population, made up of the elements of the atmosphere, hydrosphere, lithosphere and biosphere

magnitude a measure of size; for example, earthquakes are measured according to magnitude on the Richter Scale

risk the potential for something to go wrong. This is a subjective assessment about actions that may be predictable or unforeseen.

natural disaster a large natural event, such as a cyclone, flood, earthquake or landslide, that causes considerable loss of life, damage to property and infrastructure, and/or destroys sections of the environment

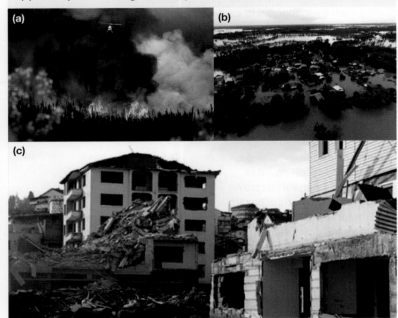

FIGURE 1.2 (a) Wildfires in Canada, (b) flooding in Thailand, (c) earthquake damage in Türkiye

In December 2023, Queensland experienced the wettest cyclone in Australia's recorded history when Cyclone Jasper swept through Far North Queensland. Some areas reported over 2000 mm of rain in just five days as a result of the cyclone. This resulted in major flooding events across the region. Over 1000 people in the Cairns region called for help from the Queensland State Emergency Service.

Although the death count was low, with one confirmed fatality, the impact on the local economy was huge. Experts estimated the cyclone was responsible for $1 billion in damage, while the tourism industry was hit hard and required a tourism recovery plan, costing $5 million.

FIGURE 1.3 Flooding in Cairns after Cyclone Jasper

1.2.2 Types of natural hazards

The Earth is made up of four interconnected systems, which form the biophysical environment: three non-living and one living, as shown in table 1.1 and figure 1.4.

TABLE 1.1 The Earth's four interconnected systems

Living system	Non-living systems		
Biosphere	**Lithosphere**	**Atmosphere**	**Hydrosphere**
The Earth's living things — plants, animals and organisms (also called the ecosphere)	The Earth's core, mantle and crust (also known as the geosphere)	The mix of gases surrounding the Earth	The Earth's water, such as oceans, rivers, lakes and glaciers

Hazards can occur in any of the Earth's systems. Because the systems are connected, energy is easily transferred within and between them. For example, solar energy reaches the Earth and warms the land (lithosphere) and water (hydrosphere), and each of these can warm the atmosphere; seismic wave energy from tectonic plate movement and earthquakes can cause landslides and buildings to collapse, and an earthquake below the ocean can lead to a tsunami.

Because of the way they form and how and where they happen, natural hazards are grouped into categories. Some hazards fit into several categories because they are caused by a combination of processes.

Table 1.2 shows the different categories of natural hazards, and some real-world examples of these.

FIGURE 1.4 The Earth's physical systems

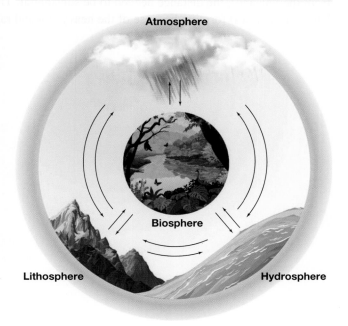

Atmosphere

Biosphere

Lithosphere

Hydrosphere

TABLE 1.2 Types of hazards

Type of hazard	Description	Examples	Real-world example
Atmospheric	Hazards that occur in the atmosphere or because of atmospheric conditions	Storms, cyclones, tornadoes, bushfires and droughts	2019–20 Black Summer bushfires, south eastern Australia
Geological	Natural events that occur in the Earth's crust	Volcanic eruptions, earthquakes	2023 earthquake, Türkiye–Syria
Geomorphic	Natural events that occur on the Earth's surface	Avalanches, landslides, mudslides	2024 landslide, Maco, Philippines
Climatological	Events caused by irregular climatic conditions	Bushfire, drought, heat waves, cold snaps	2023 heatwaves, Europe
Hydrological	Extreme events with a substantial water component	Floods, storm surges, tsunamis	2011 tsunami, Tōhoku, Japan

1.2.3 Hazard zones

The term **hazard zone** is used to identify areas at risk of being affected by a specific hazard or hazards, and to indicate which areas are at greater or lesser risk. For example, areas that have been flooded or are at high risk of flooding due to the location of drains will be identified and outlined in a local council's urban flood map. This kind of hazard zone map is then used in urban land-use planning or by insurance companies when they are calculating premiums for clients.

Hazards and risks can also change over time. For example, hazard zones are declared around active volcanoes when the risk of eruption is assessed to be at a high level. When Mt Agung erupted in Bali, Indonesia, in 2017, people were evacuated from the area surrounding the base of the volcano to a distance of 12 kilometres — any closer was considered a high-risk zone for ash and rock fallout. Because these volcanoes have a history of releasing toxic fumes and **pyroclastic clouds** (hot clouds of gas and debris from the volcano), the distance needed to be substantial. There was also the risk of lahars (mudflows) forming because of the heavy tropical rainfall.

climatological hazard a hazard that occurs due to the climatic conditions of an area, such as bushfires, droughts and heatwaves

hydrological hazard an extreme event with a high-water component, such as flash flooding, cyclones, ice melt, storm surges and tsunamis

hazard zone an area that may be affected by a natural hazard; for example, areas vulnerable to flooding based on past events or areas likely to be affected by pyroclastic flows from a volcano

pyroclastic clouds rapidly moving currents of hot air, gases and ash that run from the crater down the sides of a volcano. They are extremely lethal due to their high speed and lack of sound.

1.2 Exercise

1.2 Exercise

Learning pathways

■ LEVEL 1	■ LEVEL 2	■ LEVEL 3
1, 2	3, 4	5, 6

Explain and comprehend

1. Refer to figure 1.5.
   a. **Identify** the type of natural hazard that is shown in figure 1.5.
   b. List the natural and built features in the image.
   c. **Describe** the appearance of the water over the road. **Suggest** why it may not appear to be moving or flowing.

FIGURE 1.5 Road covered by floodwater

2. **Explain** why a flooded road would be regarded as a hazard.

Analyse and apply

3. **Determine** the effects a flood might have on local traffic flow.
4. **Predict** what possible scenarios might unfold if a person attempted to drive through floodwaters.

Propose and communicate

5. If you were a member of a local emergency response team, what actions would you **recommend** to reduce the risks flooding poses to commuters?
6. Consider a new bridge is constructed to allow vehicles to bypass land regularly inundated with floodwaters. It cost $12 million and has a daily traffic flow of 5000 vehicles. **Evaluate** the extent to which constructing the new bridge was justified.

Sample responses are available in your digital formats.

LESSON
1.3 Natural hazards in Australia

1.3.1 Variation in types of natural hazards in Australia

Australia is considered a naturally hazardous country because its risk level is relatively high, especially for atmospheric hazards. However, it is a very large country with a small population. Consequently, few natural hazards in Australia become catastrophic natural disasters on the scale of the 2004 earthquake and tsunami in the Indian Ocean, which killed more than 230 000 people, or the earthquake that killed approximately 220 000 people in Haiti in 2010 (see figure 1.6).

FIGURE 1.6 Australia rarely experiences natural disasters on the scale of (a) the 2004 Indian Ocean earthquake and tsunami, or (b) the 2010 Haitian earthquake.

In Australia, considerable variation exists in the types of natural hazards that occur between and within states (see figures 1.8 and 1.9). The perception of natural hazards in Australia is often less acute than in other countries. This is mostly due to:

- Australia's landscape is vast and varied, which can diffuse the impact of such events.
- Events such as droughts are often slow onset and are not recognised as immediate threats.
- The increasing unpredictability of weather patterns, particularly in coastal areas, can make perception of hazard unreliable.

EXAMPLE: Newcastle earthquake, 1989

Before 1989, few people in New South Wales would have considered earthquakes a risk. At the time, the Building Code of Australia, which is designed to safeguard people against major structural failure and loss of life, classified Newcastle's buildings as having a low earthquake risk. Consequently, specific building design for protection against earthquakes was not considered necessary.

Given this, the impact of the earthquake in December 1989 was significant because of the low levels of preparedness and preparation for such an event.

FIGURE 1.7 Damage from the 1989 Newcastle earthquake

FIGURE 1.8 Distribution of Australia's geomorphic and geological hazards

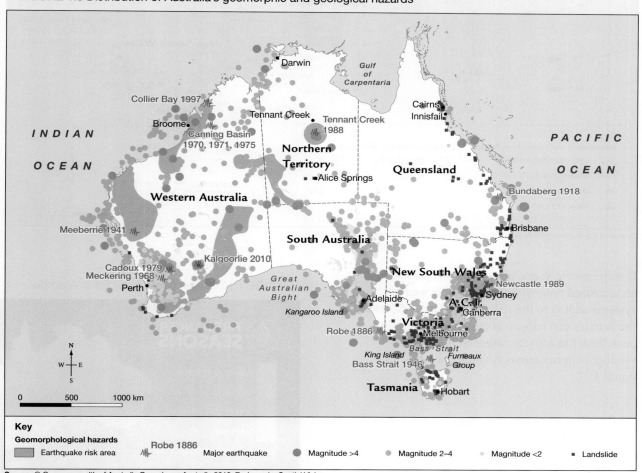

Source: © Commonwealth of Australia Geoscience Australia 2018. Redrawn by Spatial Vision.

FIGURE 1.9 Distribution of selected atmospheric, climatological and hydrological hazards in Australia

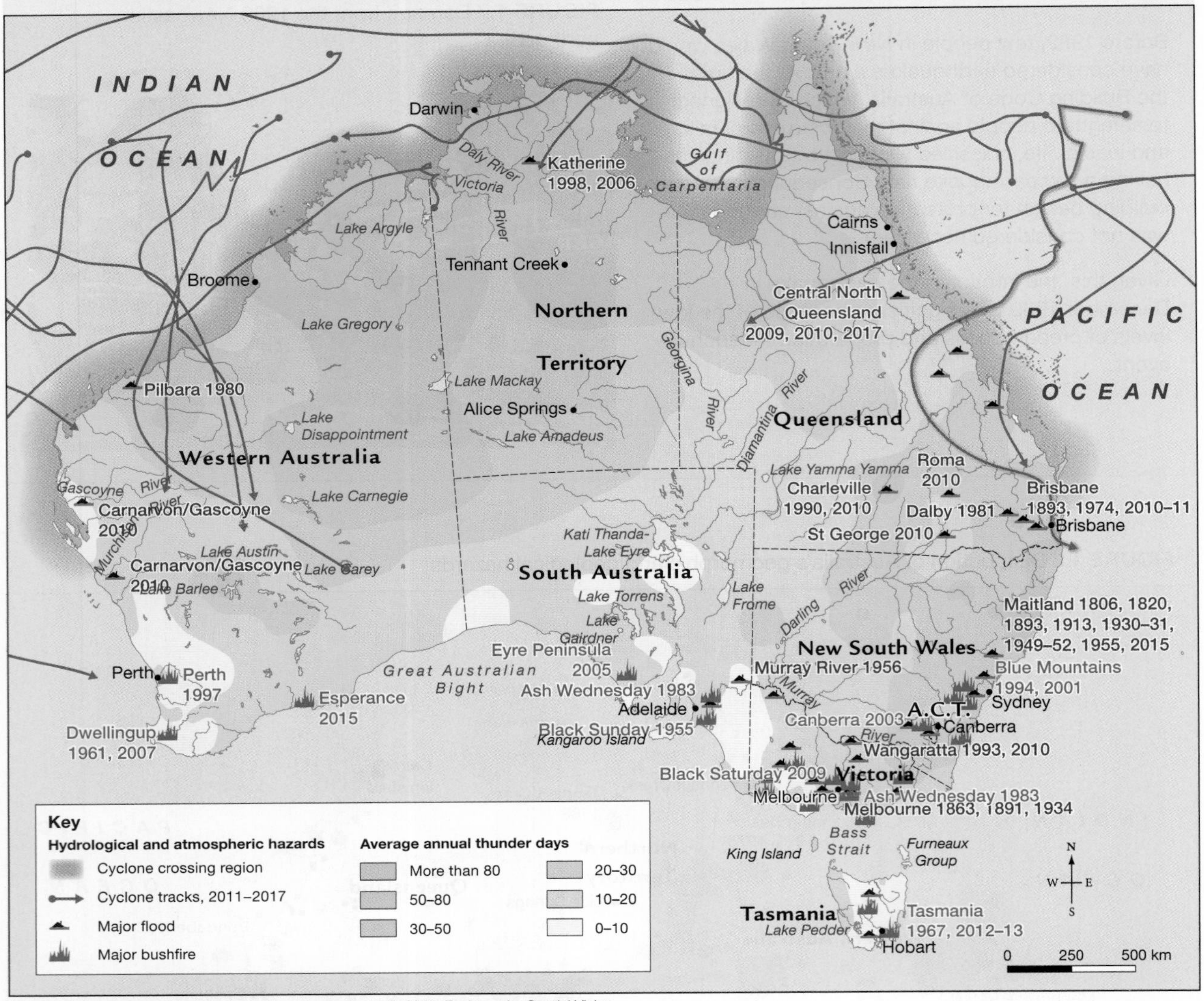

INDIAN OCEAN

Darwin

Daly River
Victoria River

Katherine
1998, 2006

Gulf of Carpentaria

Lake Argyle

Tennant Creek

Broome

Lake Gregory

Northern Territory

Cairns
Innisfail

Central North Queensland
2009, 2010, 2017

PACIFIC OCEAN

Pilbara 1980

Lake Mackay

Lake Disappointment

Alice Springs

Lake Amadeus

Queensland

Western Australia

Lake Carnegie

Gascoyne River
Carnarvon/Gascoyne 2010

Murchison River

Lake Austin
Carnarvon/Gascoyne 2010

Lake Barlee

Lake Carey

Georgina River

Diamantina River

Lake Yamma Yamma

Charleville
1990, 2010

Roma 2010

Dalby 1981

St George 2010

Brisbane
1893, 1974, 2010–11
Brisbane

Kati Thanda-Lake Eyre

South Australia

Lake Torrens

Lake Gairdner

Lake Frome

Darling River

Maitland 1806, 1820, 1893, 1913, 1930–31, 1949–52, 1955, 2015

New South Wales
Murray River 1956

Blue Mountains
1994, 2001
Sydney

Eyre Peninsula
2005

Great Australian Bight

Ash Wednesday 1983

Adelaide

Black Sunday 1955

Kangaroo Island

Perth
Perth 1997

Esperance 2015

Dwellingup 1961, 2007

Murray River

Canberra 2003

A.C.T.
Canberra

Wangaratta 1993, 2010

Black Saturday 2009

Victoria

Melbourne

Ash Wednesday 1983
Melbourne 1863, 1891, 1934

King Island

Bass Strait

Furneaux Group

N
W   E
S

Tasmania
Lake Pedder

Hobart

Tasmania
1967, 2012–13

Key

Hydrological and atmospheric hazards

Cyclone crossing region

Cyclone tracks, 2011–2017

Major flood

Major bushfire

Average annual thunder days

More than 80

50–80

30–50

20–30

10–20

0–10

0   250   500 km

Source: © Commonwealth of Australia Geoscience Australia 2018. Redrawn by Spatial Vision.

In more recent years, fire season has presented an increased risk, with unpredictable weather systems adding to the risk. While dangerous bush fire activity can happen at any time, generally, peak bush fire activity varies with seasonal weather patterns across Australia, as shown in figure 1.10.

FIGURE 1.10 Danger seasons for fire in Australia

FIRE DANGER SEASONS

Winter and spring

Spring

Spring and summer

Summer

Summer and autumn

FIGURE 1.11 Bushfire in southern New South Wales

SKILLS ACTIVITY

Use the **ArcGIS Australia's population distribution** weblink to complete the following.

1. Add a new layer to the map to show recent cyclones.
2. Describe the spatial association between Australia's population and its cyclones.
3. Explain possible reasons for this spatial association.
4. Research Australia's ten most economically damaging floods of the last 20 years.
5. Use the sketch function on ArcGIS to label these floods.
6. Describe the spatial association between these floods and Australia's population.
7. Explain possible reasons for this spatial association.

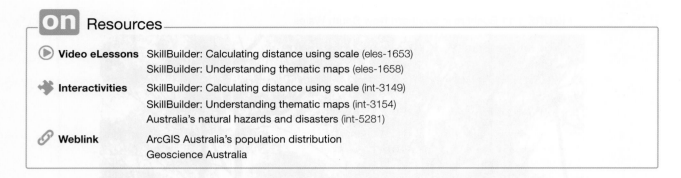

on Resources

▶ **Video eLessons** SkillBuilder: Calculating distance using scale (eles-1653)

SkillBuilder: Understanding thematic maps (eles-1658)

✦ **Interactivities** SkillBuilder: Calculating distance using scale (int-3149)

SkillBuilder: Understanding thematic maps (int-3154)

Australia's natural hazards and disasters (int-5281)

🔗 **Weblink** ArcGIS Australia's population distribution

Geoscience Australia

1.3 Exercise

1.3 Exercise

Learning pathways

■ LEVEL 1	■ LEVEL 2	■ LEVEL 3
1, 2, 3	4, 5	6

These questions are even better in jacPLUS!
- Receive immediate feedback
- Access sample responses
- Track results and progress

Find all this and MORE in jacPLUS ▶

Explain and comprehend

1. **List** the natural hazards that are identified in figures 1.8 and 1.9.
2. Considering the range of natural hazards in Australia, **identify** which hazards might be the most closely linked to the following features.
   a. Long periods of low rainfall
   b. Movement of the lithosphere
   c. Very high temperatures
   d. A severe tropical storm that forms over the ocean and brings torrential rain and destructive winds
   e. Destructive winds with rain and hail
3. Consider figures 1.9 and 1.10. **State** which parts of Australia are most adversely affected by bushfires. During what time of the year?
4. **Describe** the challenges that hazards might create for people in your area.

Propose and communicate

5. **Suggest** which state is the most hazardous or least hazardous place to live in Australia. **Justify** your answer.
6. **Hypothesise** the new challenges you might face if you moved from Darwin to a bushland property in central Victoria. What information might you need to minimise any risk to your family?

Sample responses are available in your digital formats.

LESSON
1.4 The impact of natural hazards

LEARNING INTENTION

By the end of this lesson you should be able to explain how the severity of the impacts of natural hazards is influenced by a number of factors and explain the concept of vulnerability as applied to natural hazard management and explain how climate change may affect the severity and incidence of some natural hazards and increase risk.

1.4.1 Variables affecting the impact of a hazard

The impact of a hazard on a specific area and its people, both in the long and short term, depends on a range of factors. These factors are shown in figure 1.12.

These factors can also be broken down into separate categories:

- the hazard itself
- the location's exposure to the hazard
- the location's vulnerability to the impacts of the hazard.

This lesson explains the factors affecting the impact of hazards, identifying why some hazards prove to be more disruptive than others to the physical environment and the people who live there.

FIGURE 1.12 Factors affecting the impact of hazards

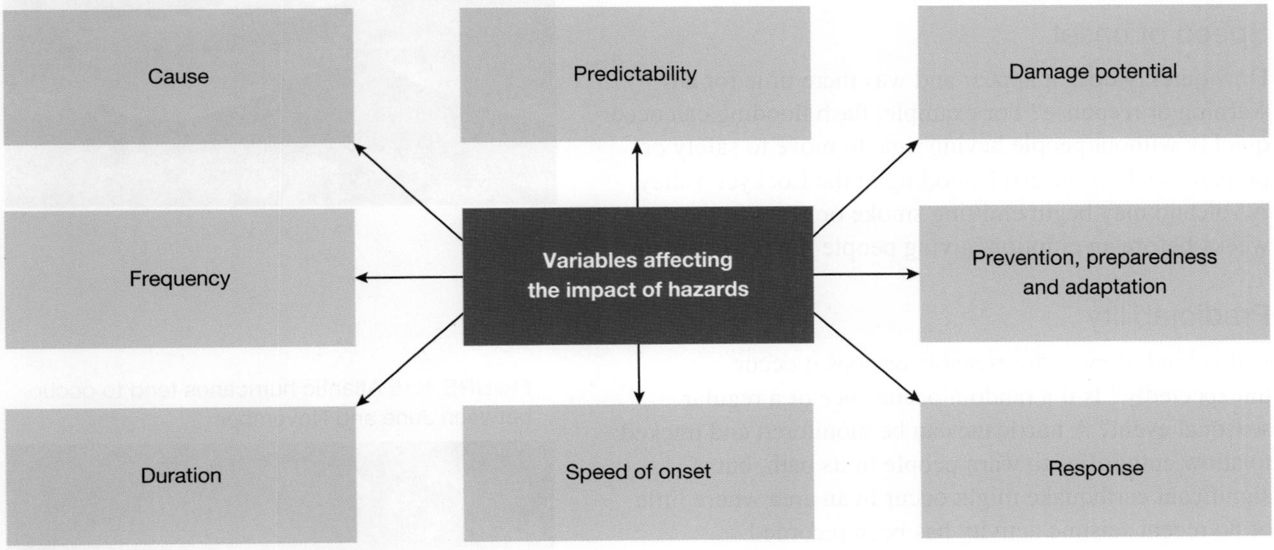
Cause
Predictability
Damage potential
Frequency
Variables affecting the impact of hazards
Prevention, preparedness and adaptation
Duration
Speed of onset
Response

Cause

What is the origin of the hazard? For example, a landslide (figure 1.13) might be triggered by deforestation of slopes (human causes) whereas a flood might be caused by torrential rain after a storm or tropical cyclone (natural causes). Other hazards may be triggered by a combination of causes.

Frequency

How often does it happen? Some hazards are seasonal, such as bushfires or cyclones, while others can occur at any time or without warning, such as earthquakes or a tsunami. If hazards occur with greater frequency, this leaves less time for rebuilding and risk management strategies to be put in place. If hazards occur infrequently, people may not be prepared sufficiently for an event to occur.

Duration

How long does it last? A severe storm may only last for an hour or so, while a drought can go on for months or years. Coping with the impact of an event over a long period of time will stretch the available economic and social resources, and might mean that people affected need to leave the area permanently, or that the land may no longer be safe for human habitation. A long-running natural disaster will also affect the wellbeing of the people in the area.

Speed of onset

How quickly does it appear and was there time for any warning or response? For example, flash flooding can occur quickly without people having time to move to safety or prepare, such as the 2011 flooding in the Lockyer Valley. A volcano may begin emitting smoke or gases in the days or weeks before an eruption, giving people time to evacuate.

Predictability

Is this kind of event foreseeable or does it occur unexpectedly? Is it a random occurrence or a regular seasonal event? A hurricane can be monitored and tracked to allow authorities to warn people in its path, but a significant earthquake might occur in an area where little or no recent seismic activity has been recorded.

Prevention, preparedness and adaptation

How much control do people have over the impacts and outcomes? Can they prevent a hazard from occurring or prepare and adapt to increase their chances of surviving a hazard? The threat in some bushfire hazard zones can be mitigated with controlled burns (figure 1.16) and careful land management, but this does not always lower the risk

FIGURE 1.13 A landslide may be triggered by human causes.

FIGURE 1.14 An increase in seismic activity was recorded at Mount Agung, Bali, Indonesia, before it erupted in 2017–2019.

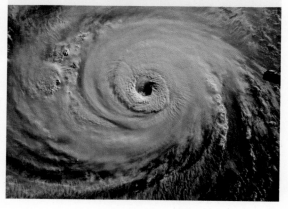

FIGURE 1.15 Atlantic hurricanes tend to occur between June and November.

when other factors occur, such as very high temperatures, winds, lightning and arson. People might prepare for the risk of cyclones by shuttering or boarding up windows as a cyclone approaches, choosing to move away during high risk periods or building cyclone shelters. Authorities might construct sea walls to prevent storm surges flooding coastal communities.

FIGURE 1.16 Controlled burns can reduce the risk of future fire events.

Damage potential and magnitude

How large or intense was the hazard? What is the potential for loss of life or large-scale damage to infrastructure and the environment? The size or magnitude of a hazard may not equate to greater damage. For example, a weaker cyclone hitting a heavily populated urban area may cause more damage to buildings and infrastructure than a stronger cyclone that makes landfall in an unpopulated area.

Ability to respond

Can people respond quickly, or must they wait for assistance due to safety concerns or inaccessibility? A landslide in a remote mountain area may cut off all access for emergency crews or make existing access dangerous because of unstable ground.

1.4.2 Factors affecting vulnerability to a hazard

Even though hazards can be quite different in their structures, the way they form and the way they disrupt an area, some common factors still influence people's vulnerability to their impact.

Physical factors

Factors in the physical environment, such as the weather, the season (summer or winter) or terrain, can affect how people cope in the short term when a hazard occurs.

EXAMPLE: Cyclone Debbie, Queensland 2017

When Cyclone Debbie struck north Queensland in April 2017, relentless torrential rain made it difficult for people to put temporary covers on unroofed and damaged houses for several days after the initial gales.

Flooded roads also made it impossible for emergency workers such as police, paramedics or electricity line engineers to reach those in need.

FIGURE 1.17 Flooded road caused by Cyclone Debbie

Economic factors

Preparedness, mitigation, prevention and adaptation strategies can be expensive to implement, so a community's level of economic development can affect the impact of a natural hazard. Countries with limited financial resources also have a greater chance of fatalities from hazards because they lack the money to provide the required emergency aid quickly. This emergency aid might include well-resourced emergency response teams, medical supplies and healthcare workers, shelter for survivors, and fresh food or water.

When hazards such as earthquakes occur in less developed countries, affected residents must often wait for overseas aid. For example, after the Papua New Guinea earthquake in 2018, much of the rescue and recovery work was organised by an oil drilling company that was in the area because the government did not have the resources to do so. Infrastructure in less economically developed countries is also less likely to be able to withstand severe natural disasters. Building collapses and disrupted transport routes can result in increased casualties due to delays in assistance.

EXAMPLE: Impacts of Cyclone Idai in Mozambique

Cyclone Idai struck eastern Africa in March 2019 and left a trail of destruction. One of the hardest hit countries was Mozambique, where over 600 people died. Mozambique, like many other less economically developed countries, faces significant challenges in infrastructure, health care, and emergency response capabilities. When Cyclone Idai made landfall, the destruction it caused included flooding, landslides and damage to infrastructure.

One of the immediate challenges was the inundation of entire communities, which left thousands of people stranded and in need of rescue. However, Mozambique's limited resources and damaged infrastructure made it difficult to reach those affected quickly. Roads were washed away, bridges had collapsed and communication networks were disrupted, hampering the ability to coordinate rescue and relief efforts.

FIGURE 1.18 Poor transport routes made reaching many of those in need in Mozambique more difficult.

SKILLS ACTIVITY

FIGURE 1.19 Ten deadliest natural disasters in 2023

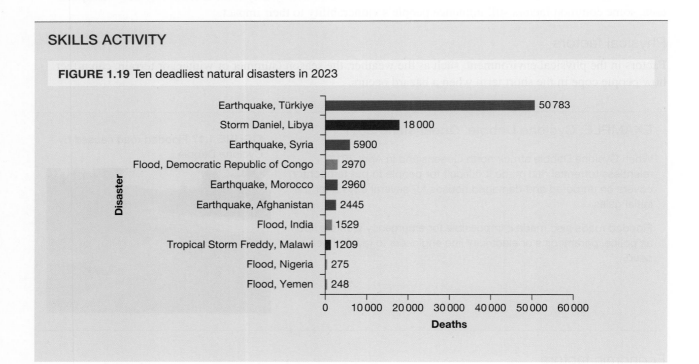

1. Based on the data in figure 1.19, identify what type of hazard caused the most fatalities in 2023.
2. Research online the natural disasters listed in figure 1.19. Using spatial technology or a print map of the world, shade the areas affected by these disasters, using colour to show the different types of hazard.
3. Based on your map data, do certain types of hazards seem to occur in specific regions or areas? Write a short paragraph to describe the geographic patterns you can identify.
4. Create a table showing the natural disasters that occurred around the world during the last year. Sort the data into a table showing the types of hazard, the location of the event and the number of fatalities. Does the data support your answer to question 3? Explain in a paragraph whether your assessment is supported by recent catastrophic hazards.

Social and political factors

After a natural disaster, the social structures of a community and country also become an important part of the recovery. A highly functioning society might benefit in the following ways:

- An immediate and positive response from internal government agencies (political and military), such as declaring a state of emergency. This helps to ensure that the rescue and recovery process runs smoothly and efficiently.
- Efficient and generous aid efforts, because these can be affected by the ways in which government bodies, community organisations and the media mobilise non-affected people to help sufferers. Aid efforts in a well-functioning society could include generous donations of medical items, fresh water and food, building supplies, clothing and money.
- Higher or more resilient morale, which can help with rescue and rebuilding efforts to continue, especially when victims have lost loved ones or are left homeless.

 Resources

 Weblink Vulnerability to extreme weather events

1.4.3 Climate change and natural hazards

Within the Earth's 4.5 billion years, its atmospheric, lithospheric and oceanic systems have experienced periods of great stability, as well as periods of major fluctuations. Prior to the arrival of humans, natural hazards occurred within these systems due to natural processes, such as volcanic activity and shifts in solar radiation. However, in the past couple of centuries, human activities have accelerated changes in Earth's natural systems, particularly through the release of greenhouse gases such as carbon dioxide and methane. These activities have magnified the impacts of climate change, leading to an increase in the frequency and intensity of certain **extreme weather events**.

Scientific principles, particularly those of thermodynamics, help explain some of these changes. Warmer air has the capacity to hold more moisture, which can lead to more intense rainfall, especially in coastal and wet regions where condensation is prevalent. However, the effects of climate change on precipitation patterns are complex and can vary by region. Research suggests that while some areas may experience more intense rainfall, others, particularly arid and desert regions, may face prolonged periods of drought due to changes in atmospheric circulation patterns and precipitation trends (see figure 1.20).

> **extreme weather event** a weather event that is rare at a particular place and/or time of year, with unusual characteristics in terms of magnitude, location, timing or extent

FIGURE 1.20 (a) Some places are experiencing more extreme rainfall events as a result of climate change, while (b) others are experiencing more extreme droughts.

Most scientists believe that increases in CO_2 levels and subsequent general warming of the atmosphere (global warming) contributes to climate change and consequently affects natural hazards. The most obvious indicators of climate change are:

- rising global temperatures due to the atmosphere retaining more heat
- more frequent and extreme droughts
- more frequent and damaging **wildfires**
- more severe and destructive tropical cyclones and hurricanes
- more frequent and destructive tornadoes
- rapid melting of glaciers, sea ice and ice caps
- melting of **permafrost** in tundra regions
- gradual sea level rising that adversely affects estuaries and low-lying coastal plains
- weakening of the **polar vortex** causing prolonged icy periods and intrusions of warm air into parts of the northern hemisphere.

wildfire an uncontrolled fire that spreads through vegetation

permafrost soil, rock or sediment that remains frozen for two or more years, commonly found in polar regions and high mountain areas

polar vortex a large area of low pressure and cold air that typically resides over the polar regions during the winter months, but can occasionally shift southward, bringing frigid temperatures and winter weather to lower latitudes

Unfortunately, the widespread awareness of global warming and its possible effects on climate only really occurred in the 1980s, when marked increases in temperatures were recorded. Today, meteorological science and its implications for hazard management is widely accepted. Whether for safety or insurance, governments, industry, institutions and the public are being forced to include the effects of global warming in their everyday risk management. Clearly, these strategies are having a positive effect, with statistics now showing that although natural hazards are increasing, people are taking improved steps, where possible, to mitigate harm, and so reducing fatalities and injuries.

The impacts of climate change are also affected somewhat by increasing population and urban expansion. As the human–nature interface expands, communities are becoming more vulnerable to the effects of natural hazards, especially in terms of costs of allocation and distribution of emergency response personnel and resources. This is particularly evident in Australia's fire-prone and flood-prone regions.

A considerable quantity of data indicates that overall fatalities from natural hazards have been decreasing, as shown in figure 1.21, which tracks deaths from disasters since 1900. This decrease is even though these hazards continue to occur throughout the world (see figure 1.22). This raises certain questions:

- Do most fatalities occur from occasional super-catastrophic events such as the Indian Ocean tsunami (in 2004, when about 230 000 people died) or the Haiti earthquake (in 2010, when at least 220 000 were killed and 300 000 were injured)?
- Are more communities better prepared in both mitigation and response strategies than in previous years?

FIGURE 1.21 Annual number of deaths from disasters (decadal average)

Disasters include all geophysical, meteorological and climate events including earthquakes, volcanic activity, landslides, drought, wildfires, storms, and flooding. Decadal figures are measured as the annual average over the subsequent ten-year period.

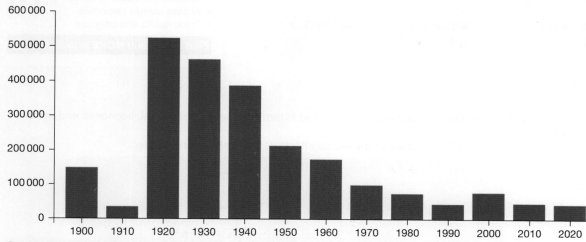

FIGURE 1.22 Global reported natural disasters by type, 1970 to 2024

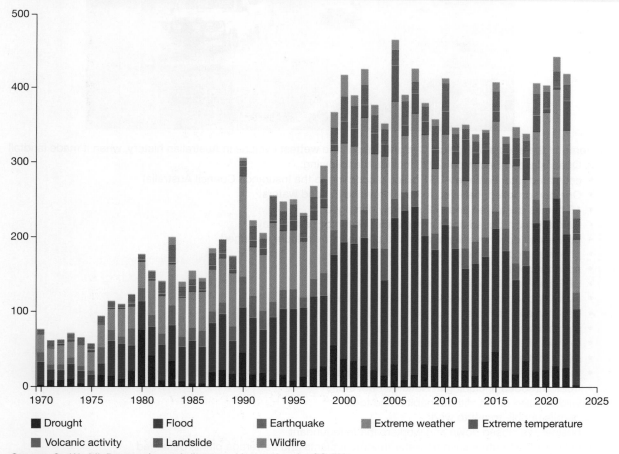

Legend: Drought ■ Flood ■ Earthquake ■ Extreme weather ■ Extreme temperature ■ Volcanic activity ■ Landslide ■ Wildfire

Data source: EM-DAT, CRED/UCLouvain (2024).
Note: Graph largely reflects increases in data reporting, and should not be used to assess the total number of events.

1.4 Exercise

1.4 Exercise

Learning pathways

■ LEVEL 1	■ LEVEL 2	■ LEVEL 3
2, 3	1, 4	5, 6

These questions are even better in jacPLUS!
- Receive immediate feedback
- Access sample responses
- Track results and progress

Find all this and MORE in jacPLUS ▶

Explain and comprehend

1. The severity of impacts of a hazard is often measured in terms of the environmental, economic and social effects.

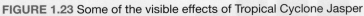

FIGURE 1.23 Some of the visible effects of Tropical Cyclone Jasper

Some of the effects of Tropical Cyclone Jasper, the wettest cyclone in Australian history, when it made landfall in Queensland in December 2023 include the following:
- estimated total damage AU$1 billion (according to the Insurance Council Australia)
- One person killed, believed to have drowned in flood waters
- 18 homes destroyed and 168 severely damaged
- Cairns Airport closed and tarmac under water for several days
- powerlines down over large area and phone infrastructure damaged
- phone and communication towers damaged
- schools closed
- SES staff unable to help people
- roads flooded or damaged and bridges washed away
- beaches and marinas damaged
- medical staff unable to get to hospitals
- tourism sites and resorts closed for weeks
- many farms damaged (e.g. pastures buried under silt)
- fruit crops such as mangoes, citrus (limes), papayas, bananas and avocado crops lost
- trees shredded, including in National Park areas
- wildlife killed or left without habitat and food
- people unable to get to work
- businesses closed down and workers made redundant
- critical infrastructure such as water treatment plants and sewerage plants damaged
- supermarkets and shops unable to be resupplied.

a. **Create** a table to categorise each impact as an environmental, an economic or a social factor. Use a table similar to the one provided.

Environment effects	Economic effects	Social effects

b. Using these ideas, write a paragraph about each type of effect, giving reasons for why they might have occurred and outlining some of the secondary and tertiary effects. For example, environmental effects included destructive winds over 180 km/h that shredded and uprooted trees in the coastal national parks, forcing cassowaries out of the rainforests onto roadways in search of food. This also resulted in a number being struck by cars or chased by domestic dogs. Powerful storm surges inundated coastal areas with saltwater, resulting in …

2. **Identify** five types of natural hazards.
3. **Explain** the meaning of the term 'decadal average' from figure 1.21.
4. Refer to figure 1.21.
   a. **Estimate** the number of casualties for the worst decade shown.
   b. **Calculate** the difference in fatalities between the worst decade and the second-worst decade in figure 1.21.
   c. **Explain** the trend shown in figure 1.21.

Analyse and apply

5. Refer to figure 1.22.
   a. **Identify** which natural disasters occurred most frequently between 1970 and 2024.
   b. **Explain** the limitation of the data in figure 1.22.
   c. **Consider** reasons for the levelling of data in figure 1.22 over the last 20 to 30 years.
   d. If floods and drought are increasing the most, **debate** a possible connection between this trend and climate change.

Propose and communicate

6. **Consider** the trend in figure 1.21. **Hypothesise** why overall fatalities have been declining as the types of natural disasters are increasing, particularly atmospheric hazards.

Sample responses are available in your digital formats.

LESSON
1.5 Assessing and responding to natural hazards

1.5.1 Risk assessment

To manage a risk and reduce the possibility of harm, planners assess likely hazards and the potential worst-case scenario. A tool known as a risk assessment allows planners to conduct a step-by-step analysis of these problems. Once the hazards have been identified, actions can be put in place to improve a community's preparedness, prevent and/or mitigate risk, and help to build adaptation strategies so that people can remain safe. This process is called **risk management**.

A risk assessment for a natural hazard involves the following steps.

1. Understand the possible effects of the hazard, largely based on previous events and data.
2. Examine the physical features and topographic area where the hazard may occur.
3. Appreciate the type and distribution of human features, infrastructure and logistics in the area.
4. Consider the **demographic profile** of the area and ability of residents to respond to challenging situations.
5. Understand the role and availability of media, communications, emergency services and support teams that may be accessed.

A risk may be mitigated by a reduction in the size of any of the following three main variables:
- type of hazard
- elements exposed (for example, people and buildings)
- vulnerability.

Consider figure 1.24, based on Crichton's Risk Triangle. According to David Crichton, people have little or no control over a hazard's type, frequency and magnitude, but they can do something about exposure and vulnerability, particularly in places where improved technologies and communications allow residents to be better informed.

Exposure refers to the things likely to be affected by a hazard, including people, crops, livestock, buildings and infrastructure. It also refers to intangible assets with economic value, such as work or communications. For example, if a tropical cyclone was heading towards a settlement, the residents and homes would be exposed to substantial risk. **Vulnerability** measures the degree of risk according to its location, amount of preparedness, and counter-response resources available.

risk management strategies and actions to reduce or mitigate risk based on the known consequences of encountering a hazard

demographic profile a detailed description of a group of people, including information such as their ages, genders, ethnicities, incomes, education levels, jobs, family sizes and where they live

exposure the degree or likelihood of a place, person or thing being affected by a hazard, in terms of risk assessment

vulnerability the degree of risk faced by a place, person or thing, based on an approaching hazard's potential impact, the place's degree of preparedness, and the resources available to respond

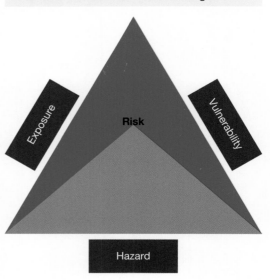

FIGURE 1.24 Crichton's Risk Triangle

In the risk management model shown in figure 1.24, the large blue area assumes the initial risk with each of the three variables contributing equally to the risk. However, if steps are taken to reduce both exposure (for example, by not living too close to a volcano) and vulnerability (for example, closely monitoring and recording volcanoes for signs of activity), the smaller, green triangle shows that the overall risk has been lowered.

Researching events of the past also gives people an increased probability of being able to forecast what might happen in the future. People can't prevent natural hazards from happening, but communities can reduce the risk and manage the effects and response.

FIGURE 1.25 People living near the Tajogaite volcano in the Canary Islands have high exposure to volcanic eruptions.

In most cases, proactive effort and readiness often prevent a natural hazard from becoming a disaster. This is best achieved through education, technologies and funding.

EXAMPLE: Preparing for earthquakes in Japan

Today, Japan is known for its ability to cope with powerful earthquakes. Innovative technology, cutting-edge engineering and response training are among the many mitigation strategies used in Japan. These allow the country to withstand huge earthquakes that would devastate other parts of the world. This hasn't always been the case, though. Learnings from past events have played a critical role in modern-day Japan's resilience to such hazards. On 1 September 1923, a huge earthquake caused death and destruction across Tokyo. Today, the first day of September is known as Disaster Prevention Day: a day on which the Japanese commemorate the events of 1923 and prepare for future hazard events.

FIGURE 1.26 Destruction in central Tokyo after a huge earthquake on 1 September 1923

Estimating a level of risk about a specific event (for example, an approaching cyclone) is difficult because precise locations and/or time of impact cannot always be known. However, risk models combining information about past events, including frequency and intensity, are now being developed to help experts predict possible hazard scenarios.

Crichton's model can be summarised as:

$$\text{Risk} = \text{Hazard type} \times \text{Exposure} \times \text{Vulnerability}$$

These models are now used as a guide for emergency services to prepare for a range of effects, including response strategies and damage estimates. The scale and frequency of various natural disasters now form an integral part of any risk assessment and equivalent insurance considerations.

Equally important is that efforts are made to ensure critical infrastructure sites such as power stations, water treatment systems, sewage disposal plants and telecommunication networks are less vulnerable to natural hazards. When minimal disruption occurs to infrastructure, recovery is more rapid and less costly.

CASE STUDY: Analysing vulnerability and exposure to earthquakes in Japan and Nepal

Quick facts: Japan and Nepal

Earthquakes annually:
- **Japan** up to 1500
- **Nepal** over 60

Human Development Index (HDI):
- **Japan** 0.95 (19th in the world)
- **Nepal** 0.6 (146th in the world)

GDP per capita:
- **Japan** US$54 184
- **Nepal** US$4934

Risk is assessed as a function of hazard, exposure and vulnerability. All three variables are required to co-exist in the same place to assess the presence of a risk.

Both Japan and Nepal experience strong earthquakes regularly. If a major earthquake were to hit each country, both of a similar magnitude, Crichton's Risk Triangle can be used to assess the risk to people and economies in both places.

Figures 1.27 and 1.28 show how the risk triangle changes for these two countries.

FIGURE 1.27 Japan earthquake risk triangle

Japan:
- Buildings constructed using earthquake-energy absorbing features, including laminated rubber to reduce seismic motions, dampers, springs, and ball bearings to reduce exposure to damage.
- Transport infrastructure constructed to highest standards.
- High-quality medical and rescue infrastructure.
- People have regular evacuation drills in office buildings, shopping centres and schools.

As a result of these measures, Japan's vulnerability to earthquakes is significantly reduced. The hazard itself remains significant and the exposure to it is also relatively high – the people of Japan are still in a part of the world that will expose them to earthquakes. Overall, however, risk is lowered.

FIGURE 1.28 Nepal earthquake risk triangle

Nepal:
- Buildings generally constructed using basic materials due to availability and affordability.
- Transport infrastructure is basic, particularly in rural parts of the country. This can leave parts of the country inaccessible and isolated.
- Health care and medical clinics are limited.
- Rescue infrastructure is limited.

As a result of these factors, Nepal's vulnerability to earthquakes remains high. Although Nepal's exposure to the hazard is a little lower than Japan's, as they occur less frequently in Nepal, the country's vulnerability when they do occur means the risk remains high.

1.5.2 Managing impact

While scientific technology, such as satellites and weather instruments, enables us to measure changes to the Earth's surface, oceans and atmosphere, it is the accurate analysis of data and clear communication to the public that are most crucial for shaping the level of impact. Precise and current information passed on to emergency personnel, community decision-makers and the media also means that people have time to prepare, respond or evacuate before the hazard occurs, therefore reducing the population's vulnerability to the hazard. Governments, councils and emergency teams need to be aware of the extent of the hazard zone, and the event's potential severity and degree of impact if they're to prepare the best response in terms of safety and mitigation of damage. This process is often referred to as disaster management.

FIGURE 1.29 Emergency response teams help reduce vulnerability to a hazard.

DID YOU KNOW?

The island of Taiwan is also subject to frequent powerful earthquakes. In April 2024 a magnitude 7.4 struck the Longitudinal Valley region of the island. Straddling two converging plates (the Philippines Plate and the Eurasian Plate), Taiwan is wedged in a tectonic vice. Regular rehearsal drills and an almost instant warning system greatly contributed to a low death toll in the 2024 earthquake.

SKILLS ACTIVITY

FIGURE 1.30 Australian natural disaster fatalities (fire), 1900–2010

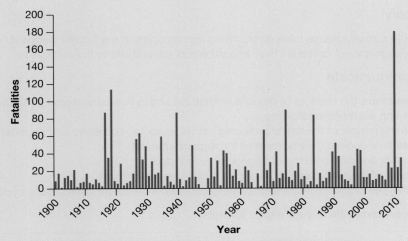

Source: Used with permission from the Bushfire and Natural Hazards CRC

1. Examine figure 1.30 and identify the number of fatalities for the years in which each of these natural disasters occurred:
   - Tasmanian bushfires (1967)
   - Ash Wednesday fires (1983)
   - Black Saturday fires (2009).

   Research each of these events and write a short paragraph about each explaining some of the variables that contributed to the impact of each of the fires, including suggestions for why these events may have had a greater impact than other fires.

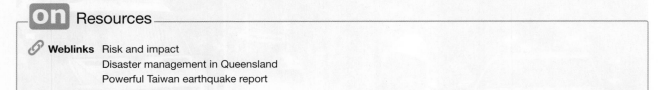

Resources

🔗 **Weblinks**   Risk and impact
                 Disaster management in Queensland
                 Powerful Taiwan earthquake report

1.5 Exercise

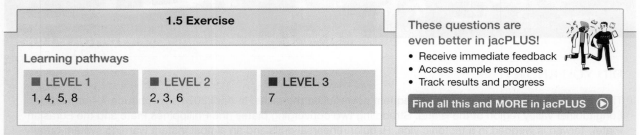

1.5 Exercise

Learning pathways

■ LEVEL 1	■ LEVEL 2	■ LEVEL 3
1, 4, 5, 8	2, 3, 6	7

These questions are even better in jacPLUS!
- Receive immediate feedback
- Access sample responses
- Track results and progress

Find all this and MORE in jacPLUS ▶

Explain and comprehend

1. **Explain** how technology may be used to mitigate the impacts of a bushfire hazard. Give three specific examples to support your explanation.
2. **Explain** how risk can be mitigated by lowering exposure and vulnerability to bushfires.
3. **Describe** the purpose of a risk assessment and the types of information this assessment typically includes.
4. **Explain** briefly why it is often difficult to estimate the level of risk of an approaching hazard.
5. **Identify** three emergency response roles adopted by people in the community.

Analyse and apply

6. What impact could climate change have on bushfires and droughts in the future? **Suggest** how communities in areas of high risk might act to reduce their exposure and vulnerability to these hazards.

Propose and communicate

7. **Research** and **compare** the impacts of drought in Australia and in the Sahel region of Africa. Write a paragraph explaining each of the following.
   a. The impact on the people of the regions affected (consider social, economic and physical impacts)
   b. The differences in economic development of the regions affected
   c. How governments and non-government agencies responded
   d. The effectiveness of the responses
8. **Design** a list with three proactive steps a community could take before an approaching cyclone hits.

Sample responses are available in your digital formats.

LESSON
1.6 Atmospheric hazards

1.6.1 Understanding atmospheric hazards

Atmospheric hazards are extreme weather-related events that happen in the lower levels of the atmosphere (the troposphere). They are all part of the Earth's climate system. The most common hazards are severe storms (thunderstorms), blizzards, snowstorms, sandstorms, tropical cyclones (hurricanes and typhoons) and tornadoes. Floods can also occur as a combination of atmospheric and **geomorphic** processes. Many of these hazards arrive and occur quickly during a short period; for example, over a few hours or a few days. Other longer term atmospheric hazards that are linked to natural cycles or the climate are dry spells and drought. Some atmospheric hazards also occur from human causes, such as air pollution from dust, chemical vapours, industrial fumes, fogs and smog. Because these have an adverse **toxicological** effect on living things and can impede breathing, they are considered hazards.

geomorphic related to the formation of the Earth's surface and its changes

toxicological related to the negative impacts of chemical substances

FIGURE 1.31 Atmospheric hazards are often closely connected to extreme weather conditions

(a)

(b)

(c)

(d)

1.6.2 Processes that create atmospheric hazards

The atmosphere is a clear layer of gases surrounding the planet. It keeps animals and plants alive and protects them from extreme cold. Within the atmosphere are numerous circulations of air, energy and water — a complex system powered by energy from the sun. This system can also create the conditions for atmospheric hazards to occur, as follows.

1. Solar radiation is absorbed by the Earth's surface and the atmosphere, initiating the planet's energy budget. This absorption leads to uneven heating across the globe due to the curvature of the Earth and the varying composition of the atmosphere.

2. Uneven heating results in distinct temperature variations across latitudes. Equatorial regions receive more solar energy, leading to higher temperatures, while polar regions receive less, resulting in cooler climates. This temperature gradient drives atmospheric circulation and weather patterns.
3. The distribution of land and sea plays a crucial role in modulating temperature. Landmasses heat up and cool down more rapidly than oceans due to differences in heat capacity. As a result, coastal regions often experience more moderate temperatures compared to inland areas, influencing local weather patterns.
4. Ocean currents and wind patterns redistribute heat from the equator towards the poles and from the poles towards the equator. This process, known as atmospheric and oceanic circulation, helps regulate global climate patterns.
5. Variations in temperature lead to differences in atmospheric pressure, driving the formation of weather systems. Warm air rises in regions of low pressure, such as near the equator, leading to the formation of cyclones and other storms. Conversely, regions of high pressure, such as subtropical high-pressure belts, influence weather patterns by promoting sinking air and stable conditions.

These factors are also illustrated in figure 1.32.

FIGURE 1.32 Factors that create atmospheric hazards

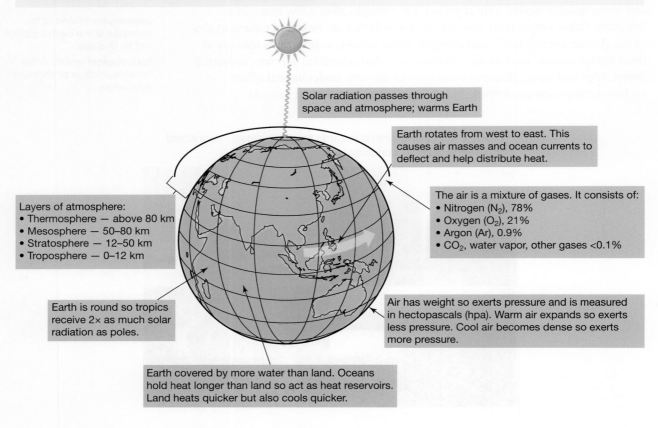

Solar radiation passes through space and atmosphere; warms Earth

Earth rotates from west to east. This causes air masses and ocean currents to deflect and help distribute heat.

The air is a mixture of gases. It consists of:
- Nitrogen (N_2), 78%
- Oxygen (O_2), 21%
- Argon (Ar), 0.9%
- CO_2, water vapor, other gases <0.1%

Layers of atmosphere:
- Thermosphere — above 80 km
- Mesosphere — 50–80 km
- Stratosphere — 12–50 km
- Troposphere — 0–12 km

Earth is round so tropics receive 2× as much solar radiation as poles.

Air has weight so exerts pressure and is measured in hectopascals (hpa). Warm air expands so exerts less pressure. Cool air becomes dense so exerts more pressure.

Earth covered by more water than land. Oceans hold heat longer than land so act as heat reservoirs. Land heats quicker but also cools quicker.

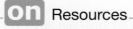 Resources

Video eLesson SkillBuilder: Reading a weather map (eles-1637)

Interactivity SkillBuilder: Reading a weather map (int-3133)

1.6 Exercise

1.6 Exercise

Learning pathways

■ LEVEL 1	■ LEVEL 2	■ LEVEL 3
1, 2, 3	4, 5	6, 7

Explain and comprehend

1. **Identify** the three TRUE statements from the following.
   a. Atmospheric hazards are extreme weather events.
   b. Weather takes place in the troposphere.
   c. The two most common gases in the atmosphere are oxygen and carbon dioxide.
   d. Air masses move around the planet from west to east.
2. Highlight the hazards in the list provided that are weather related.

   | Tornadoes | Heat waves | Tropical cyclones | Tsunamis | Drought |

3. Complete the sentences correctly.
   Many hazards arrive quickly and occur for short periods (hours or days) such as _____. However, others like _____, occur slowly and last for long periods (weeks or months).
4. **Identify** which weather hazards are most common in the area where you live.

Analyse and apply

5. **Identify** when the last weather hazard occurred in your area and how it affected your region.
6. Provide two reasons different parts of the Earth are cooler or warmer than others.
7. **Consider** the reason a high-pressure air cell may pass over Western Australia before it reaches the east coast of Australia.

Sample responses are available in your digital formats.

LESSON
1.7 Atmospheric hazards — thunderstorms

LEARNING INTENTION

By the end of this lesson you should be able to explain, using a range of representations such as maps, conceptual models, block diagrams and cross sections, how natural hazards are the result of processes that occur in the atmosphere (atmospheric).

Source: Adapted from Geography General Senior Syllabus 2024 © State of Queensland (QCAA) 2024; licensed under CC BY 4.0.

1.7.1 Thunderstorms

One of the most common types of atmospheric hazard is the thunderstorm. Thunderstorms occur all around the world, with up to 2000 happening at any one time. Thunderstorms form when warm moist air rises high into the atmosphere due to relatively hot weather and unstable air. Water vapour condenses forming huge cumulonimbus clouds.

In warmer climates, thunderstorm cells are related to low air pressure, where strong winds and updrafts carry moisture up to 20 kilometres into the sky. This rapid thermal updraft allows huge volumes of water to remain suspended in the sky until it eventually falls as rain. This process is shown in figure 1.33. Thunderstorms may only last for an hour or so but can cause havoc when they do occur due to their enormous release of energy.

Thunderstorms also develop in cooler areas. When a mass of cold air, along a cold front, forces warm air to rise, large cumulonimbus clouds form and a thunderstorm eventuates (see figure 1.35).

Thunderstorms form when the following occurs:
- Warm, moist air rises into cooler regions of the atmosphere.
- As the warm air rises, it cools down, and the moisture condenses into water droplets, forming towering cumulonimbus clouds.
- Eventually, precipitation occurs, and the cycle of rising and falling air, known as a convection cell, leads to the thunderstorm's development.

The main hazards associated with severe thunderstorms are torrential rain and associated flooding, hail, destructive winds and lightning strikes.

Types of thunderstorms

There are three main types of thunderstorm, each with its own distinctive features. These are:
- a single-cell storm — this type is limited to a single heavy downpour. This then breaks up quickly as cool downdrafts of wind smother the original warm air. A single-cell storm may only last an hour or so and will seldom produce severe weather.
- the multicell thunderstorm — this type is most common and larger, often consisting of successive storms in sequence. Because it is larger and stronger, a multicell thunderstorm produces severe weather with heavy rain, hail and wind gusts.
- the supercell — this is a very large and dangerous storm with a continuous powerful updraft that seems to control the surrounding atmosphere. It has a dominant cloud shape that reaches high into the troposphere and a dark, threatening appearance. A supercell may last for many hours and is capable of very heavy rain, severe hail and destructive winds.

FIGURE 1.33 Formation of thunderstorms by warm convection currents (tropics)

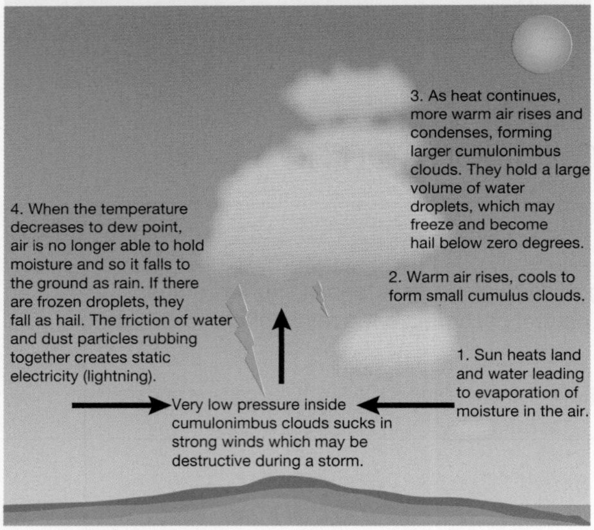

3. As heat continues, more warm air rises and condenses, forming larger cumulonimbus clouds. They hold a large volume of water droplets, which may freeze and become hail below zero degrees.

4. When the temperature decreases to dew point, air is no longer able to hold moisture and so it falls to the ground as rain. If there are frozen droplets, they fall as hail. The friction of water and dust particles rubbing together creates static electricity (lightning).

2. Warm air rises, cools to form small cumulus clouds.

1. Sun heats land and water leading to evaporation of moisture in the air.

Very low pressure inside cumulonimbus clouds sucks in strong winds which may be destructive during a storm.

Source: Bill Dodd

FIGURE 1.34 A cold front (the blue lines) is the boundary between warm air and relatively cooler air.

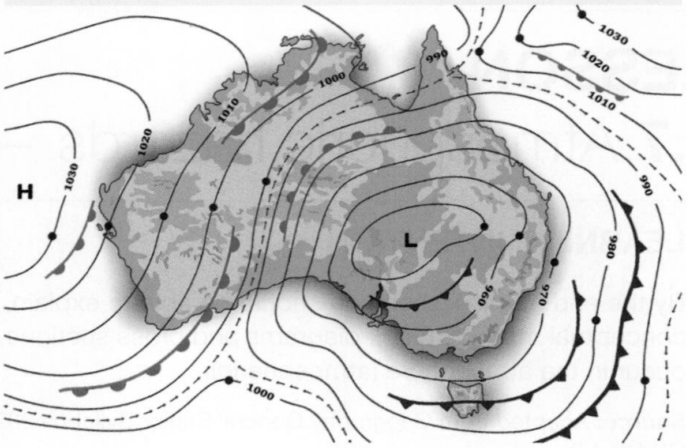

FIGURE 1.35 Formation of thunderstorms from a cold front (temperate regions)

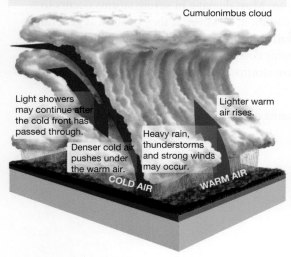

Cumulonimbus cloud

Light showers may continue after the cold front has passed through.

Lighter warm air rises.

Denser cold air pushes under the warm air.

Heavy rain, thunderstorms and strong winds may occur.

COLD AIR

WARM AIR

FIGURE 1.36 A storm cell moves across the Brisbane suburbs.

Source: Bill Dodd

DID YOU KNOW?

In recent years, the term 'rain bomb' has been used to describe torrential downpours associated with severe storms. A rain bomb occurs outside the tropics in cool weather and is the result of a cool column of sinking air caused by a rapid drop in air pressure. Rain bombs cause considerable damage due to destructive winds and flash flooding. For example, South East Queensland and northern New South Wales experienced a powerful rain bomb in February 2022.

Preparing for a thunderstorm

Here are some basic steps that people in storm areas should consider if a storm is approaching or forecast.
1. Keep your yard and exposed areas (such as a balcony or patio) free of loose items, especially outdoor furniture.
2. Keep roof gutters, downpipes and drains clean to prevent blockages.
3. Ensure garden trees and branches are trimmed to avoid them falling on your home.
4. Fix or cover any roof damage (broken or missing tiles) to prevent water leaks.
5. Have an emergency plan or kit ready in case of loss of power or hail damage.
6. Listen to local media for weather warnings and updates.
7. Check your insurance policy is up to date.
8. Ensure pets are secured.

Once a storm arrives, you can only stay inside and remain safe. Storms are very loud due to the powerful wind gusts, torrential rain, hail and thunder. Once the storm has passed, you can assess any damage.

Other than injury, the most common forms of damage are:
* fallen trees and branches
* flash flooding
* damaged windows, roofs or other items from hail
* fallen power lines.

Because thunderstorms are the most frequent natural hazard in Australia, their effects are unsurprisingly very costly. For example, the Christmas Day and Boxing Day storms that hit South East Queensland in 2023 were reported to have cost insurance companies around AU$1.4 billion. Most damage in severe storms occurs to house roofs, other household items and cars. Fatalities are not common; however, if they occur it is generally from falling trees or people drowning at road crossings or in drains. Insurance experts all agree on one thing — the onset of climate change will increase both insurance and repair costs. As shown in figure 1.37, the vast majority of costs from natural disasters come from floods, severe storms and tropical cyclones.

FIGURE 1.37 Present value of economic costs and the components of costs under low emissions scenario by type of natural disaster, $billion

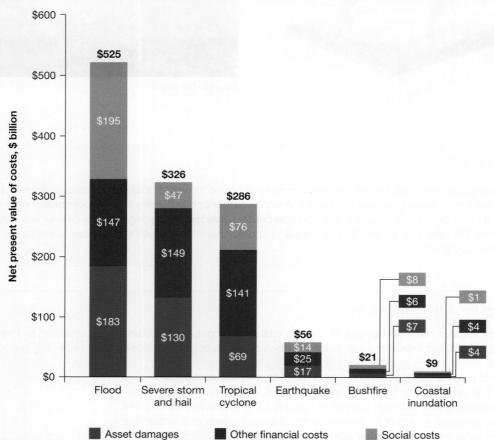

Source: Special Report: Update to the economic costs of natural disasters in Australia. Australian Business Roundtable for Disaster Resilience & Safer Communities. 2021. Deloitte Access Economics. Retrieved from https://www.deloitte.com/content/dam/assets-zone1/au/en/docs/services/economics/deloitte-au-economics-abr-natural-disasters-061021.pdf.

SKILLS ACTIVITY

Analyse figure 1.37 to answer the following questions.
1. Explain what might is meant by the terms 'asset damages', 'social costs' and 'other financial costs' by providing an example in each case.
2. Identify the amount in billions of dollars of social damage costs for each of (a) flood hazard, (b) severe storm and hail hazard and (c) tropical cyclone hazard.
3. Compare asset damage for each of the six hazards and arrange them in order from most to least.
4. Explain why floods cause more asset damage than the other hazards.

5. Identify a hazard where asset, social and other financial costs are quite similar.
6. Explain why asset damages due to coastal inundation are much lower than damage caused by flood inundation.
7. Consider why total damage due to earthquakes is more than double that caused by bushfires, particularly when Australia has a regular fire season.
8. Hypothesise why asset damage due to floods is more than ten times that caused by earthquakes.
9. Explain why social costs from flood and tropical cyclone hazards far exceed costs from severe storm damage.
10. Write a paragraph and discuss possible reasons the social costs due to flooding create the largest damage bill shown on the graph.

 Resources

🧩 **Interactivity** How a thunderstorm works (int-5615)

1.7 Exercise

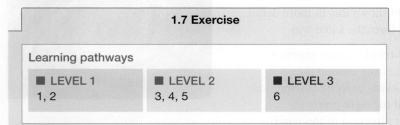

1.7 Exercise

Learning pathways

■ LEVEL 1	■ LEVEL 2	■ LEVEL 3
1, 2	3, 4, 5	6

These questions are even better in jacPLUS!
- Receive immediate feedback
- Access sample responses
- Track results and progress

Find all this and MORE in jacPLUS ▶

Explain and comprehend

1. **Identify** the three types of thunderstorms.
2. **Describe** the main hazards associated with thunderstorms.
3. **Explain** how economic factors or underlying health issues can affect an individual's vulnerability to thunderstorm events. Consider factors such as a susceptibility to thunderstorm asthma, housing security and physical mobility.
4. **Explain** how the formation of thunderstorms differs between warmer and cooler climates.

Analyse and apply

5. **Analyse** the impact of climate change on the frequency and cost of thunderstorm-related damage in Australia.
6. With reference to figure 1.37, answer the following.
   a. **Identify** which hazard has the highest economic costs.
   b. **Suggest** an explanation for why this might be.
   c. **Describe** how climate change may impact these costs in the future.

LESSON
1.8 Atmospheric hazards—tropical cyclones

LEARNING INTENTION

By the end of this lesson you should be able to explain, using a range of representations such as maps, conceptual models, block diagrams and cross sections, how natural hazards are the result of processes that occur in the atmosphere (atmospheric).

Source: Adapted from Geography General Senior Syllabus 2024 © State of Queensland (QCAA) 2024; licensed under CC BY 4.0.

1.8.1 Tropical cyclones

Tropical cyclones are very large storms (100–2000 kilometres in diameter) that bring heavy, driving rain and destructive winds to coastal and inland regions in tropical and subtropical parts of the world. Those that form over the Atlantic Ocean or eastern Pacific Ocean are called hurricanes, while those that form in the western Pacific Ocean and travel north are called typhoons. Figure 1.39 shows this in more detail. Regardless of what they are called, they all form the same way.

FIGURE 1.38 Cyclones form over the ocean, within the tropics.

Tropical cyclones usually form in the **inter-tropical convergence zone** (ITCZ), an area of low atmospheric pressure around the equator, because they require specific conditions to form: warmer sea temperatures, rising warm air, humidity and the right levels of **wind shear** (rapid change in the velocity or direction of the wind).

Once water temperatures exceed 26.5 °C and the surrounding air pressure falls below 990 hectopascals, low-pressure cells can develop into larger tropical storms, mostly between the 5° and 30° latitudes. These newly formed cells draw in more warm moist air from the ocean surface and increase significantly in size. As the huge storm clouds extend high into the troposphere, these systems take on their characteristic circular shape with an 'eye' in the middle. Winds around the cyclone become gale force and can reach speeds in excess of 280 km/h, but the eye remains calm. Depending on the category of cyclone, the eye may be anywhere between 40 and 100 kilometres wide. (See figure 1.40 for more detail on the formation of cyclones.)

The formation of cyclones is also affected by the Coriolis effect, the force that deflects winds clockwise in the southern hemisphere and anti-clockwise in the northern hemisphere. The effect of this force becomes more intense the further you move away from the equator, and is one of the key factors in shaping cyclones into their characteristic circular formation. Close to the equator, the Coriolis effect is weak. As a result, cyclones rarely form here.

inter-tropical convergence zone (ITCZ) the zone near the Equator where trade winds of the northern and southern hemispheres meet. The intense heat, warm water and high humidity create what is an almost permanent band of low pressure. The monsoon trough seen on weather charts is part of the ITCZ.

wind shear a sudden change in wind speed and/or direction over a relatively short distance in the atmosphere, generally due to altitude

Cyclones are well known for their erratic movement, particularly as they approach land, and may persist for several days offshore, but once they cross onto land they lose their energy source (the rising warm ocean air) and become a rain depression (low pressure area).

In previous years, the South Atlantic Ocean wasn't warm enough for hurricanes. Now, because of global warming, even deeper waters are getting warmer. This means unusual storms can happen in new places.

Hurricane Beryl hit the Caribbean, Mexico, and Texas in July 2024. It was the first Category 5 hurricane so early in the season and the strongest ever in early July. Hurricane Beryl caused a lot of damage, about 30 people died, and many areas flooded. Power was lost, and the US military had to help with rescue work. Scientists say global warming is causing these changes, and we'll see more of them in the future.

FIGURE 1.39 World distribution of tropical cyclones by names used in different regions

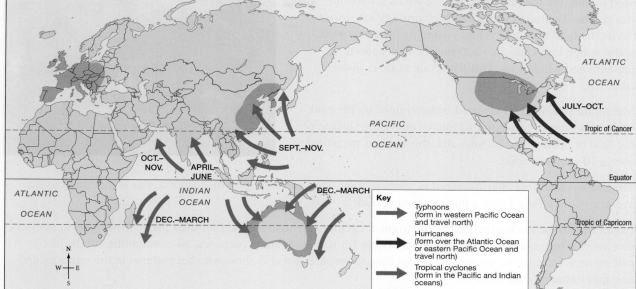

Key

→ Typhoons
(form in western Pacific Ocean and travel north)

→ Hurricanes
(form over the Atlantic Ocean or eastern Pacific Ocean and travel north)

→ Tropical cyclones
(form in the Pacific and Indian oceans)

Tornados/severe storms

Source: MAPgraphics

FIGURE 1.40 How a cyclone forms

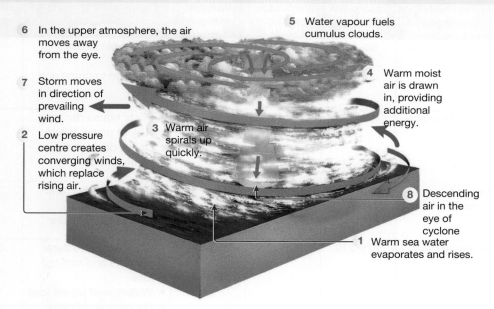

6 In the upper atmosphere, the air moves away from the eye.

5 Water vapour fuels cumulus clouds.

7 Storm moves in direction of prevailing wind.

4 Warm moist air is drawn in, providing additional energy.

2 Low pressure centre creates converging winds, which replace rising air.

3 Warm air spirals up quickly.

8 Descending air in the eye of cyclone

1 Warm sea water evaporates and rises.

on Resources

🔗 **Weblinks**   Tropical cyclone intensity
Coriolis effect
Rare south Atlantic tropical cyclone

Cyclone categories

When tropical cyclones make landfall, they can be very destructive, bringing gale force winds. Different scales are used to measure the intensity of these winds around the world. The Australian Bureau of Meteorology categorises the intensity of tropical cyclones using its own five-point scale, which also takes into account the atmospheric pressure and wind gusts. This five-point scale is shown in table 1.3.

Maximum sustained wind speed is determined by the peak mean wind speed, measured 10 metres above the surface of flat land or open water. In Australia, this mean is measured over a 10-minute period.

FIGURE 1.41 Satellite imagery showing tropical cyclone, depression and storm activity in the Pacific Ocean, August 2014

Tropical cyclones are also measured in terms of their wind gust strength. In Australia, this is measured as the average speed of wind over a three-second period.

In addition to measuring their intensity, tropical cyclones are also mapped for their paths. This helps to not only determine where cyclones most commonly occur for risk assessment purposes, so communities in the likely path of a cyclone can be warned and given time to prepare, but also show whether patterns in the intensity and frequency change over time.

TABLE 1.3 Australian Bureau of Meteorology tropical cyclone category system

Category	Australian category name	Strongest wind gust (km/h)	Average maximum wind speed (km/h)	Central pressure (hPa)	Effects	Saffir–Simpson Scale comparison (km/h)
1	Tropical cyclone	90–124	63–90	>985	• Negligible house damage • Damage to crops and trees	119–153
2	Tropical cyclone	125–164	90–125	985–970	• Minor house damage • Risk of power failure • Heavy damage to some crops	154–177
3	Severe tropical cyclone	165–224	125–165	970–955	• Some structural and roof damage • Likely power failure	178–208
4	Severe tropical cyclone	225–279	165–225	955–930	• Significant structural damage and roofing loss • Widespread power loss • Dangerous airborne debris	209–251
5	Severe tropical cyclone	>280	>225	<930	• Extreme danger • Widespread destruction	>252

Note: Storms with average maximum wind speeds between 52 and 54 km/h are referred to as tropical depressions, and tropical lows are storms with average maximum wind speeds between 56 and 61 km/h.

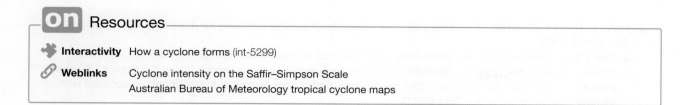

On Resources

Interactivity How a cyclone forms (int-5299)

Weblinks Cyclone intensity on the Saffir–Simpson Scale
Australian Bureau of Meteorology tropical cyclone maps

SKILLS ACTIVITY

1. Research the ten most powerful tropical cyclones on record and create a list.
2. Annotate a blank world map to show the locations of these tropical cyclones.

Climate change

Meteorologists have determined that climate change has contributed to the way tropical cyclones develop and behave. Firstly, tropical cyclones need warm ocean water and quite cool upper atmospheric conditions to form. If the air continues to warm, the difference (gradient) between surface temperatures and upper level temperatures will be reduced, so fewer cyclones may form. Secondly, increased surface temperatures over the ocean combined with higher levels of CO_2 (more CO_2 allows air to hold more moisture), provide cyclones with a much larger energy source, making them larger and more destructive.

Recent trends observed with tropical cyclones are:
- much higher volumes of rainfall near the centre when they make landfall
- an increase in high-category destructive storms
- slower movement due to weakening of the circulation forces that drive movement. (In northern Australia, cyclones now move almost 20 per cent slower than they did 70 years ago, making them more destructive to settlements and causing more flooding.)

DID YOU KNOW?

Due to continued warming of ocean waters, several hurricanes have travelled further into the north Atlantic than in previous years. Over the past decade, hurricanes reaching Portugal has become more common. However, it was extraordinary to see the remains of Hurricane Humberto reach the British Isles in 2019, and Hurricane Zeta the following year.

1.8 Exercise

1.8 Exercise

Learning pathways

■ LEVEL 1	■ LEVEL 2	■ LEVEL 3
1, 2	3	N/A

These questions are even better in jacPLUS!
- Receive immediate feedback
- Access sample responses
- Track results and progress

Find all this and MORE in jacPLUS ▶

Explain and comprehend

1. **Identify** between which lines of latitude most tropical cyclones occur.
2. **Explain** the difference between a tropical depression and a tropical storm.

3. Refer to figure 1.42, which shows the paths and intensity of tropical cyclones for more than 150 years until September 2006.
   a. **Identify** the categories on the Saffir–Simpson Scale that are most common across northern Australia.
   b. Which area of ocean does not seem to develop tropical cyclones, even though it is in the tropics? **Suggest** reasons cyclones do not develop in this region, and predict how rising sea temperatures associated with climate change might affect this pattern.
   c. **Identify** which ocean develops the most Category 5 cyclones on the Saffir–Simpson Scale. Based on your understanding of how cyclones form, **explain** the atmospheric patterns or features you would expect to find in this area.

FIGURE 1.42 Paths and intensity of tropical cyclones, tropical depressions and tropical storms

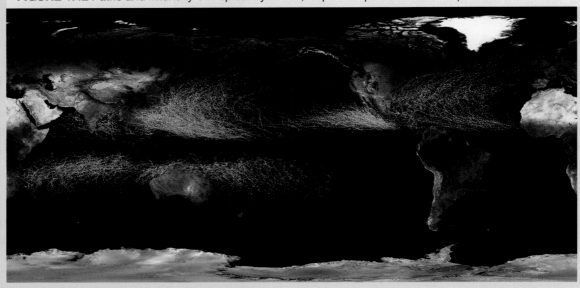

Source: Mogil, H. Michael. Extreme Weather: Understanding the Science of Hurricanes, Tornadoes, Floods, Heat Waves, Snow Storms, Global Warming and Other Atmospheric Disturbances, pp. 61. (Hardcover: Nov 13, 2007)

Sample responses are available in your digital formats.

LESSON
1.9 Atmospheric hazards — responding to tropical cyclones

LEARNING INTENTION

By the end of this lesson you should be able to understand why people in particular places have a greater or lesser capacity to respond to natural hazards and natural disasters.

Source: Adapted from Geography General Senior Syllabus 2024 © State of Queensland (QCAA) 2024; licensed under CC BY 4.0.

Improvements in public awareness and communication have enabled communities to be better prepared for storms, cyclones and related flooding than in the past. This is evident from the decreasing number of fatalities and injuries, despite such hazards becoming more powerful. However, an increasing population and more widespread settlements along the Australian coast have exposed community and government facilities to potentially greater economic damage.

Local authorities and emergency services play an important role in hazard response. They release information to help people prepare for and cope with the impact of atmospheric hazards in their area. In Australia, this information about cyclones comes as a Tropical Cyclone Warning Advice — either a tropical cyclone watch (24–48 hours before the onset) or tropical cyclone warning (onset within 24 hours). This advice relates important information including the area at risk (including a map), the intensity of the cyclone (using the Bureau of Meteorology five-level scale), the movement of the cyclone, the range and maximum strength of wind gusts expected, and advice about what action people should take to mitigate the effects of the cyclone.

FIGURE 1.43 Tropical Cyclone Larry hits Innisfail, March 2006

  Resources

    Weblink Bureau of Meteorology tropical cyclones warning services

CASE STUDY: Tropical Cyclones in northern Australia

Quick facts: Cyclones in Australia
- **Cyclone season** November to April
- **Strongest cyclone** Cyclone Mahina (1899)
- **Wettest cyclone** Cyclone Jasper (2023)
- **Average number per year** 11 (four or five of which reach land)

Typically, the cyclone season in northern Australia runs from November to April. This may vary by a few weeks in Pacific regions during a La Niña. According to the Australian Bureau of Meteorology, on average, between 10 and 13 tropical cyclones develop each year and at least one will cross the coast. Most come in from the Pacific Ocean, while a smaller number form in the monsoon trough north of Western Australia, the Northern Territory and the Gulf of Carpentaria. They all tend to move in a westerly direction. Coming out of a La Niña, the 2023–2024 season was notably intense, with the formation of five severe tropical cyclones and several other tropical low cells. The result was one death and over A$700 million in damage.

Historically, the far north of Australia has experienced many destructive cyclones. In 1974, Tropical Cyclone Tracy (category 3) struck Darwin with such intensity that 71 people were killed and the residential population forced to be evacuated. Tropical Cyclone Yasi (category 5) crossed the coast near Innisfail, north Queensland, in 2011 and became the most intense storm to hit the mainland. In 2017, Tropical Cyclone Debbie became the state's most expensive cyclone, causing damage over A$2.4 billion and the deaths of 14 people.

In December 2023, Tropical Cyclone Jasper crossed the coast near Port Douglas and became the wettest on record. Moving very slowly across the north, it dumped between 600 mm and 1500 mm of rainfall in many places during its five-day journey. Ironically, the resultant floods did far more damage than the winds and tidal surges. Just as residents were getting over some of the clean-up, Tropical Cyclone Kirrily formed in the Coral Sea about a month later. It crossed the coast just north of Townsville as a category 3 and headed west. Although Tropical Cyclone Kirrily moved quickly and weakened into a low rain depression, it inundated the region south of the gulf and towards Mt Isa, before tracking south through central Australia.

1.9.1 Preparedness strategies

Housing engineering standards have improved and can now better withstand the wind gusts of cyclones, but most places are not able to sustain the rapid flow of huge volumes of water from torrential rainfall, storm surges and flooding. Strategies to prepare for cyclones and severe storms include the following.

- Installing underground powerlines, which generally reduces the extent of outages and power failure in the event of a cyclone.
- Creating sea walls and shoreline sand buffers of at least 150 metres in coastal developments to help reduce the impact of tidal surges during cyclones.
- Restoring natural waterways. Many urban areas have reclaimed or changed their natural creek, river and wetland systems through development — for example, turning their natural creeks into cement drains. Removing these and creating a wetlands network can capture and slow the water flow during a flood.
- Capturing water for future use. Drinking water sources can be polluted by floodwaters after a cyclone. Very large buildings, such as shopping centres and industrial sheds, can be designed to capture water and have storage tanks for future use. This interception may seem small, but it can reduce stormwater flow in gutters and street drains.
- Increasing available green space in urban areas to allow for greater infiltration of rainfall. Cyclones and storms produce extremely high rainfall in a short period of time. Large ovals and outside sports areas can be designed to have run-off flow into suburban wetlands and ponds rather than all water ending up in stormwater drains.
- Designing multi-use public spaces. The destruction caused by cyclones often forces people to seek shelter or live away from home for long periods of time until the damage can be cleared and their homes made safe again. Multi-level car parks and large sports venues can be designed to also act as cyclone shelters and used as temporary storage facilities during emergency periods.

DID YOU KNOW?

Building codes and regulations for new buildings in Queensland aim to ensure they can withstand cyclones and other severe weather events. These regulations include requirements for structural integrity, materials and design to mitigate the risks posed by cyclones.

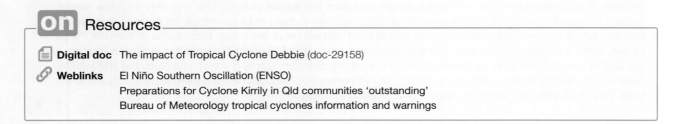

Resources

Digital doc The impact of Tropical Cyclone Debbie (doc-29158)

Weblinks El Niño Southern Oscillation (ENSO)
Preparations for Cyclone Kirrily in Qld communities 'outstanding'
Bureau of Meteorology tropical cyclones information and warnings

CASE STUDY: Typhoons in the western Pacific

Quick facts: Typhoons
- **Typhoon season** July to December
- **Strongest typhoon** Typhoon Tip (1979)
- **Average number per year** 25 to 30

Typhoons form the same way as tropical cyclones and have the same damaging effects. The term 'typhoon' is a regional name used to describe severe tropical storms that occur in the western Pacific and Asia. In central America (around the Caribbean Sea) and the eastern Pacific, these same storms are called hurricanes. Even though they have similar features, the way they affect people is often significantly different.

Because of the very large expanse of warm, tropical ocean in the western Pacific and the prevailing weather patterns, typhoons tend to be more frequent and more intense than tropical cyclones and hurricanes.

FIGURE 1.44 Debris from Super Typhoon Haiyan, Tacloban, Philippines

In November 2013, one of the world's largest ever storms, Super Typhoon Haiyan, hit the island nation of the Philippines. It then proceeded westward across the South China Sea, devastating coastal regions of northern Vietnam. Also known in the Philippines as Super Typhoon Yolanda, it originated in the western Pacific, generating highly destructive wind gusts of more than 300 km/h. Haiyan made landfall near Tacloban City — south of the capital, Manila — destroying almost everything in its path (see figures 1.44 and 1.45).

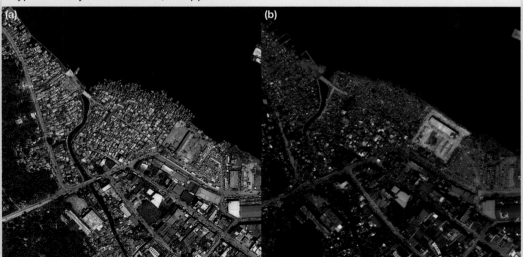

FIGURE 1.45 (a) Before and (b) after aerial images show the damage created by Super Typhoon Haiyan in Tacloban, Philippines.

Warm air and ocean temperatures ensured Typhoon Haiyan remained powerful as it moved towards Vietnam. Here, coastal and delta regions were exposed to storm surge waves higher than 5 metres. Despite weakening as it crossed the coast, the typhoon still had the power to kill or injure many people. Haiyan smashed houses and buildings, uprooted trees, knocked out water, electricity and transport infrastructure, flooded farms and fishing ports, and created havoc for residents over an area almost as large as Australia. Its devastation was extraordinary, destroying over 70 per cent of houses and infrastructure in its path and killing an estimated 6300 people. In total, about 11 million people were affected. Many people were forced to evacuate, having lost family members, homes, possessions and crops. The economic damage was estimated to be more than US$4.5 billion.

While countries around the world rallied to support survivors in the Philippines and Vietnam with aid, the level of despair and suffering after these events was incredible. Widespread panic and looting from desperate, hungry survivors made the distribution of essential aid difficult.

on Resources

Video eLessons	SkillBuilder: Understanding satellite images (eles-1643) SkillBuilder: Interpreting an aerial photo (eles-1654) SkillBuilder: Comparing aerial photographs to investigate spatial change over time (eles-1750)
Interactivities	SkillBuilder: Interpreting an aerial photo (int-3150) SkillBuilder: Comparing aerial photographs to investigate spatial change over time (int-3368)
Weblinks	Typhoon Hato Typhoon Haiyan Typhoon intensity and ocean warming Before and after Typhoon Haiyan

SKILLS ACTIVITY

TABLE 1.4 Number of tropical storms and cyclones, by region

Basin	Tropical storm or stronger (greater than 17 m/s sustained winds)			Hurricane/typhoon/severe tropical cyclone (greater than 33 m/s sustained winds)		
	Most	**Least**	**Average**	**Most**	**Least**	**Average**
Atlantic	28	4	12.1	15	2	6.4
NE/Central Pacific	28	8	16.6	16	3	8.9
NW Pacific	39	14	26	26	5	16.5
N Indian	10	2	4.8	5	0	1.5
SW Indian	14	4	9.3	8	1	5
Aus SE Indian	16	3	7.5	8	1	3.6
Aus SW Pacific	20	4	9.9	12	1	5.2
Globally	102	69	86	59	34	46.9

Note: 1981–82 to 2015–16 cyclone season for the southern hemisphere

Source: AOML/NOAA

1. Using the data in table 1.4, design a bar graph to display the average number of tropical cyclones that occur in each region per year.
2. Research and, as precisely as possible, on a map mark in the paths taken by both Typhoon Hato and Super Typhoon Haiyan. Insert two or three information boxes where extensive damage was done. For example, in Tacloban, Philippines. (A print-friendly version of a map to mark up has been included in the Resources tab.)

on Resources

Digital doc	South-East Asia map (doc-29161)
Video eLessons	SkillBuilder: Using multiple data formats (eles-1761) SkillBuilder: Comparing population profiles (eles-1704)
Interactivities	SkillBuilder: Using multiple data formats (int-3379) SkillBuilder: Comparing population profiles (int-3284)

1.9 Exercise

Explain and comprehend

1. Most parts of Australia experience severe thunderstorms. **Explain** how the risk to people and property from thunderstorms can be lessened by proposing three potential methods.
2. **Outline** which strategies authorities could put in place to mitigate the risk for vulnerable members of their community in Australia.

Analyse and apply

3. Which is more destructive: tropical cyclones, typhoons or hurricanes? Justify your answer with evidence or examples.

Propose and communicate

4. **Hypothesise** which factors apart from the strong wind gusts might have contributed to the scale of the destruction caused by Super Typhoon Haiyan in Tacloban in the Philippines.
5. Using figure 1.46, consider the age distribution of people in the Philippines in the years prior to Super Typhoon Haiyan. (A graph of Australia's age pyramid is provided for comparison.) **Suggest** how the age distribution of the population might affect the way a city or region can recover from a natural disaster.

FIGURE 1.46 (a) Australia's population pyramid 2023 and (b) Philippines' population pyramid 2023

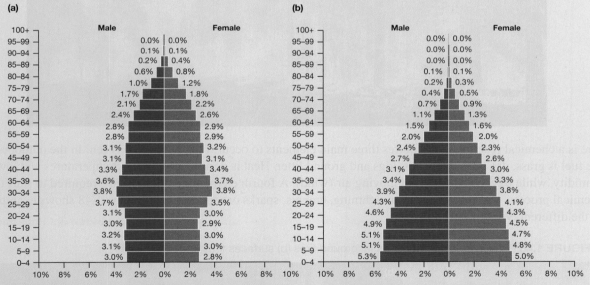

Source: © 2023 by PopulationPyramid.net, made available under a Creative Commons license CC BY 3.0 IGO: http://creativecommons.org/licenses/by/3.0/igo/

6. The Philippines is classified as a developing country by the United Nations. World Bank data from 2015 suggests that 21.6 per cent of Filipino people are living in poverty. **Evaluate** the impact the economic strength of a country would have on the ability to recover from a significant natural disaster like Super Typhoon Haiyan.

Sample responses are available in your digital formats.

LESSON
1.10 Atmospheric hazards — bushfires

1.10.1 When fire becomes a hazard

Fire has been essential for people to keep warm, hunt and cook since the middle Paleolithic Ages. First Nations Peoples of Australia also commonly used fire for regenerating certain bushland and grassy areas. However, if not in a controlled environment, fire is both a hostile and deadly adversary that kills people and animals as well as destroying property and the environment. The term 'bushfire' often refers to unplanned vegetation fires, so this section uses the term to also mean grass fires, forest fires and scrub fires. The term 'wildfire' refers to a bushfire that is out of control and cannot be contained.

FIGURE 1.47 Bushfires, or wildfires, can spread quickly and devastate huge areas of natural and human environments.

Fire is a chemical reaction that requires three main elements to occur: fuel, heat and oxygen. In the outdoors, the fuel is grass, dry trees, branches, logs and ground litter. Heat is derived from warm temperatures and low humidity, while oxygen comes from moving air (wind). A fourth element, ignition, is also required to start the chemical process, and may occur from lightning, embers, sparks or reflected heat. Figure 1.48 shows examples of the different types of bushfires.

FIGURE 1.48 Bushfires usually occur as (a) ground fires (b) surfaces fires and (c) crown fires.

Bushfires are natural hazards that create a large hazard zone as they advance. The most common forms of harm (death or injury) from bushfires include:

- burning due to direct contact with flames
- exposure to radiant heat
- exposure to smoke and air-borne particles
- burning caused by embers
- effects of falling trees or branches.

 Resources

 Weblink Where are fires burning?

DID YOU KNOW?

A rainforest catching on fire was once highly unlikely. The ground and layers of vegetation were too humid and too wet for fire to occur. However, prolonged periods of dry weather due to climate change are rearranging parameters.

CASE STUDY: Wildfires in Canada

Quick facts: Canada
- **Capital city** Ottawa
- **Population** 39 million
- **Forest coverage** 362 million hectares (40% of the country's land area)
- **Number of wildfires annually** 8000+

The boreal (coniferous pine) forest of North America is one of the world's most unique biomes. It is the largest remaining intact forest in the modern world. Spanning the far north of the continent from Alaska to Newfoundland, it covers an area of about 6 million km^2 (> 400 million ha). About 85 per cent of the total forest grows in Canada, much of which is still undisturbed. Consisting mostly of pine, fir, spruce, poplar, larch and balsam tree species, the boreal is also home to wolves, bears, fox, lynx and wood bison. As well as storing huge volumes of carbon, the boreal forests are important economically (providing timber, hydropower, minerals and energy) and culturally, with most indigenous peoples still residing in these regions. Because around 94 per cent of Canada's forests grow on publicly owned land, the government has been able to protect and manage their use.

FIGURE 1.49 Smoke from a wildfire fills the air in British Columbia, Canada.

However, 2023 saw major changes to this spectacular ecosystem. Large sections of forest became catastrophic infernos, with around 15 million hectares scorched by more than 6000 wildfires. Beginning in March 2023, large wildfires ravaged many remote areas of forest, affecting all 13 provinces of Canada. It was thought most fires started from lightning strikes, but some were caused by careless people and campers. A small number were due to arson.

Some fires were so large they created their own pyrocumulonimbus clouds, resulting in even more lightning strikes. The largest individual fire was in the Donnie Creek area of British Columbia, where around 5745 km^2 of forest was razed in June. A month later, 20 000 residents of Yellowknife (in the Northwest Territories) were evacuated as large fires converged on the town. Smoke and polluted air spread to many parts of North America, affecting human health and adding to the existing greenhouse gases.

Despite the efforts of 5821 domestic and 4990 international firefighters, large areas of boreal forest were destroyed. Some forest experts believe dominant fir and spruce trees may regrow, but many species have been lost. Scientists agree the wildfires can be attributed to climate change. After decades of disruption to climatic cycles, warmer and drier weather is more common, raising the risk of fire, with vast areas of vegetation becoming more flammable than in the past.

1.10.2 Bushfires in Australia

Australia is a very dry continent and has always been prone to bush and grass fires. In recent years, extreme weather events brought on by climate change and urban expansion have increased the risk of uncontrollable fires, particularly in Victoria, Tasmania, South Australia and the south-west of Western Australia. Bushfires or wildfires (fires that are out of control) are a climatological hazard because of their connection to hot, dry days, low humidity and gale force winds that make firefighting near impossible. Common causes of fire are careless campers, smokers, arson, vehicle accidents and powerline damage. Lightning strikes also start fires in some places.

Responding to the threat of bushfires is one of the most difficult challenges confronting people in fire-prone areas. The Australian bush consists of many combustible trees and includes ground fuel such as dry grass and fallen branches. Therefore, if a bushfire is threatening an area where people live, they must make a critical decision whether to leave early while escape routes are still open or stay. Leaving when a fire is close presents a significant risk because of low visibility, smoke, breathing problems, the danger of ambient heat and embers, the unpredictability of changing winds, and the likelihood of fallen trees and other road obstructions. Such conditions are extremely frightening and confusing, leading people to make potentially disastrous decisions due to fear or lack of experience. When high-intensity crown fires are assisted by powerful winds, ground fighting and even aerial bombing are almost impossible in such conditions.

FIGURE 1.50 Roadside fire warnings are common across Australia, alerting people to risk and advising appropriate behaviour.

Fire authorities recommend that residents in fire-prone areas have a safety evacuation plan in the event of a fire. They also recommend that homes adjacent to bushland are kept free of debris and suitable pumps and hoses be installed with a plentiful supply of water.

EXAMPLE: 2019–20 Black Summer

One of the most catastrophic fire seasons in Australian history was the summer of 2019 and 2020, which saw significant fires in all states and territories. Nearly 250 000 square kilometres of area was burnt, while close to 10 000 structures were destroyed, including 3500 homes.

The number of direct deaths was relatively low, at 34, although experts estimate that close to 450 people died as the result of smoke inhalation from the fires.

Economic cost estimates range, but some suggest these may have been as high as AU$3.65 billion.

FIGURE 1.51 Vast areas of forest were burnt in the 2020 bushfires.

SKILLS ACTIVITY

1. Using information from online research on bushfires and from your text, assign the following factors to the correct side of the provided Risk Management Triangle.
   - Death and harm to people, livestock, and crops
   - Depletion of wildlife and habitats
   - Damage to the environment (pollution)
   - Damage to critical infrastructure
   - Loss of work, income, and schooling
   - Loss of infrastructure
   - Access to external assistance
   - Location of houses, roads in fire-prone areas and forests
   - Monitoring fire-prone areas
   - Knowledge of fire behaviour and history
   - Trained fire-fighters, vehicles, and aircraft

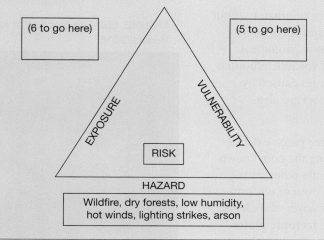

(6 to go here)

(5 to go here)

EXPOSURE

VULNERABILITY

RISK

HAZARD

Wildfire, dry forests, low humidity, hot winds, lighting strikes, arson

1.10 Exercise

1.10 Exercise

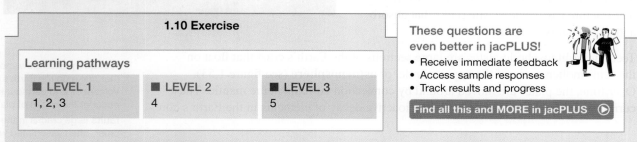

Learning pathways

■ LEVEL 1	■ LEVEL 2	■ LEVEL 3
1, 2, 3	4	5

These questions are even better in jacPLUS!
- Receive immediate feedback
- Access sample responses
- Track results and progress

Find all this and MORE in jacPLUS ▶

Explain and comprehend

1. **State** the elements that make fire.
2. **Identify** some important steps residents in rural areas should take to mitigate the risk of bushfire.
3. **Explain** why the 2023 Canadian wildfires were so difficult to fight or control.

Analyse and apply

4. **Discuss** why bushfires are now much more expensive to fight than in previous years.

Propose and communicate

5. **Evaluate** some of the effects of these fires (loss of forest, deaths and injuries to people, loss of homes and buildings, loss of wildlife, effects of carbon pollution, economic costs of fighting the fires as well as insurance and compensation) and write an extended paragraph on their impact.

Sample responses are available in your digital formats.

LESSON
1.11 Geological hazards

LEARNING INTENTION

By the end of this lesson you should be able to explain, using a range of representations such as maps, conceptual models, block diagrams and cross sections, how natural hazards are the result of processes that occur within the earth (geological).

Source: Adapted from Geography General Senior Syllabus 2024 © State of Queensland (QCAA) 2024; licensed under CC BY 4.0.

1.11.1 Understanding geological hazards

A geological hazard occurs or originates within the Earth, rather than on its surface or in its atmosphere. The most common geological hazards are earthquakes and volcanic eruptions, but tsunamis (large wave surges caused by earthquakes on the ocean floor) are also geological hazards.

FIGURE 1.52 The eruption of Klyuchevskaya Sopka in Russia, 2016

The Earth is divided into three parts: a dense core, a thick shell surrounding the core known as the mantle, and a thin, brittle outer crust. However, the crust is not an even or uniform surface. It is a very irregular cover of rock and soil, and is made up of large **tectonic plates**, uneven landforms and **faults** (huge cracks or weak points) that penetrate down into the upper mantle. The crust contains both hard, brittle rocks that fracture easily and elastic rocks that absorb and store energy.

Tectonic or lithospheric plates are large sections of the Earth's crust that float on the semi-molten rocks of the upper mantle, or **asthenosphere** (see figure 1.53). Over time, the plates are moved around by convection currents in the mantle. These currents are created by heat from the radioactive decay of elements in the Earth's core.

 Resources

📘 **Video eLesson** SkillBuilder: Interpreting a complex block diagram (eles-1746)

🧩 **Interactivity** SkillBuilder: Interpreting a complex block diagram (int-3364)

tectonic plates slow-moving plates that make up the Earth's crust. Volcanoes and earthquakes often occur at the edges of plates.

faults large cracks in the Earth's crust, often associated with the boundaries of the Earth's tectonic plates

asthenosphere the upper layer of the mantle, below the lithosphere, usually more than 100 km below the surface. It is where rock becomes molten and allows the solid tectonic plates to move over it.

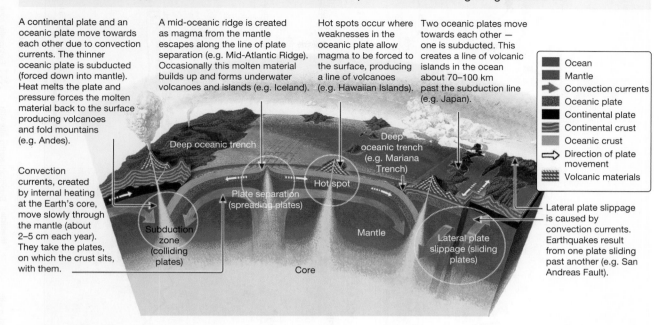

A continental plate and an oceanic plate move towards each other due to convection currents. The thinner oceanic plate is subducted (forced down into mantle). Heat melts the plate and pressure forces the molten material back to the surface producing volcanoes and fold mountains (e.g. Andes).

A mid-oceanic ridge is created as magma from the mantle escapes along the line of plate separation (e.g. Mid-Atlantic Ridge). Occasionally this molten material builds up and forms underwater volcanoes and islands (e.g. Iceland).

Hot spots occur where weaknesses in the oceanic plate allow magma to be forced to the surface, producing a line of volcanoes (e.g. Hawaiian Islands).

Two oceanic plates move towards each other — one is subducted. This creates a line of volcanic islands in the ocean about 70–100 km past the subduction line (e.g. Japan).

Convection currents, created by internal heating at the Earth's core, move slowly through the mantle (about 2–5 cm each year). They take the plates, on which the crust sits, with them.

- Ocean
- Mantle
- Convection currents
- Oceanic plate
- Continental plate
- Continental crust
- Oceanic crust
- Direction of plate movement
- Volcanic materials

Deep oceanic trench

Deep oceanic trench (e.g. Mariana Trench)

Hot spot

Plate separation (spreading plates)

Subduction zone (colliding plates)

Mantle

Lateral plate slippage (sliding plates)

Core

Lateral plate slippage is caused by convection currents. Earthquakes result from one plate sliding past another (e.g. San Andreas Fault).

1.11.2 Processes that create geological hazards

Earthquakes occur and volcanoes are formed when forces in the Earth move sections of tectonic plates, usually at faults. The Earth's lithosphere is divided into seven major tectonic plates and about 20 smaller ones (see figure 1.55). Most crustal movement happens when these plates push together, pull apart or pass each other in different directions. Faults are named according to the way plates or blocks of crust move along the surface. As shown in figure 1.54, there are generally three types.

- A diverging or normal fault creates a gap when two plates pull away from each other. These occur at diverging plate boundaries and a block either goes upward or downward, or **sea floor spreading** occurs on the ocean floor. An example of this on land is the East African Rift Zone.

- A converging or reverse fault, sometimes also called a thrust fault, occurs where two plates collide or are forced together at converging plate boundaries. On land, these collisions build fold mountains such as the Himalayas, whereas on the ocean floor the process is called **subduction**. Here, one plate may override another forming deep marine trenches, such as the Mindanao Trench near the Philippines.

- A transform or strike-slip fault occurs where two plates pass each other laterally with very little vertical movement. The San Andreas Fault is this type.

sea floor spreading the divergence of two oceanic crust plates

subduction a geological process where two tectonic plates collide at convergent boundaries

FIGURE 1.54 (a) Converging plate boundary (b) Diverging plate boundary (c) Transform fault boundary

(a)

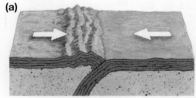

Converging plate boundary creates a reverse fault – two plates pushed together

(b)

Diverging plate boundaries creates a normal fault – two plates pulled part

(c)

Transform fault boundary creates a strike-slip fault – two plates slide past each other

FIGURE 1.55 World map of tectonic plates

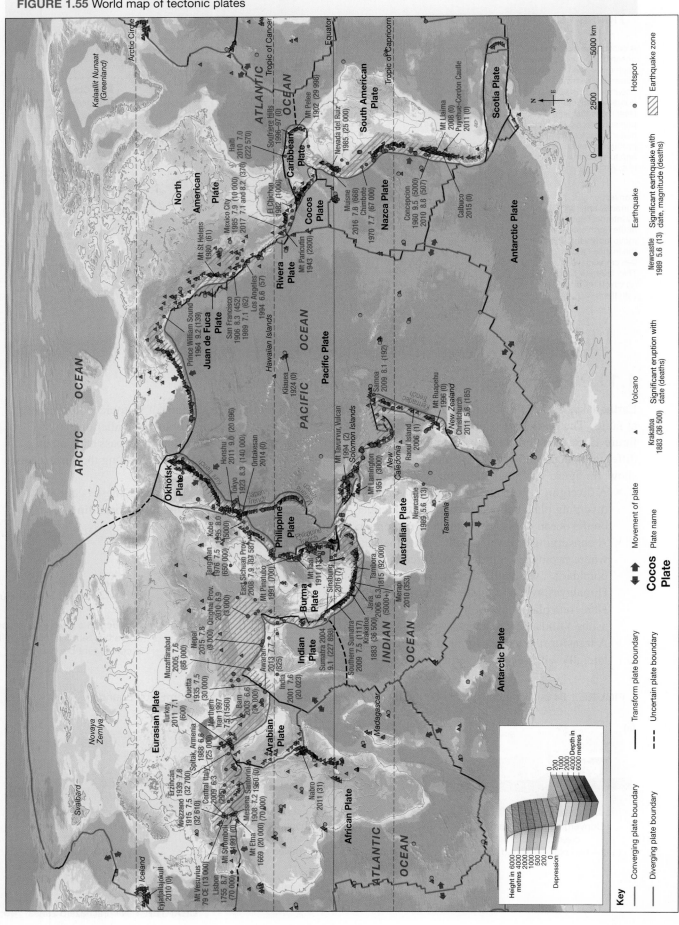

Source: Redrawn by Spatial Vision based on information from the Smithsonian National Museum of Natural History.

This process of plate tectonic movement involves two different types of crust: oceanic and continental. Oceanic crust is heavier, thinner and younger than continental crust. Most commonly, two diverging plates are both composed of oceanic crust. As the oceanic plates drift apart, molten rock or **magma** rises from the mantle below, cooling and forming a new oceanic crust. Such magma is basaltic in composition and creates Icelandic-type volcanoes and mid-ocean ridges. Earthquakes in areas near diverging plates tend to have relatively small magnitudes.

Over time, this tectonic movement puts enormous strain on crustal rocks, well beyond their level of strength. Eventually, these sections of plate slip or break, releasing an enormous amount of stored energy. This sudden and powerful slipping of sections of crust is called an earthquake. Because of the sheer size and force of the Earth's tectonic plates, surface movement is extremely powerful and people can't do anything except take preventative action to survive. Even though most earthquakes happen in the crust, they can also occur in the upper mantle, mostly near **subduction zones** where plates collide. The site where rock begins to break is called the **focus**. Most earthquake tremors occur directly on the surface at the **epicentre** (the location on the ground immediately above the focus).

The strength of the earthquake depends on how much rock 'breaks' and how far it is moved. Following a major quake, several 'adjustments' called aftershocks or tremors occur, adding to the fear and uncertainty of people affected. Earthquakes below the ocean can displace large areas of sea bed and water, sometimes causing a tsunami — for example, the tsunami that hit the coast of Sulawesi, Indonesia, in September 2018 after a magnitude 7.5 earthquake.

At the surface, seismic vibrations cause damage to infrastructure, particularly in populated areas. People can be killed or injured when buildings collapse or fires break out from the rupturing of fuel tanks and gas lines. They can also cause other types of hazards such as landslides and **liquefaction** (when the ground acts as though it is liquid, rather than solid).

Many active volcanoes are also associated with pressures at convergent and divergent plate boundaries. Subduction occurs at convergent boundaries, when one plate is pushed below another. When this occurs, surface material is forced down into the mantle where it softens. At some other weak point in the crust, this molten material can be carried upwards to the surface through a fissure (crack) as a volcano. Earthquake clusters can also occur in association with magma movements. At divergent boundaries, where crustal plates pull apart, large underwater ridges (mid-ocean ridges) form as hot molten lava oozes out across the ocean floor and cools.

magma hot molten rock formed below or within the Earth's crust. It reaches the surface through volcanic or plate tectonic activity and becomes lava and eventually igneous rock.

subduction zones the areas of the mantle in which convergent plates collide. Under the ocean, these areas are called trenches.

focus where an earthquake rupture occurs in the crust or mantle. Seismic waves radiate away from the focus.

epicentre the point on the Earth's surface directly above the focus when an earthquake has occurred

liquefaction when saturated or partially saturated soil loses its firmness and displays the properties of a liquid, such as when an earthquake shakes and loosens wet soil in low-lying areas, and the soil loses rigidity and moves like fluid, covering things in its path

1.11 Exercise

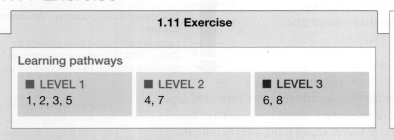

1.11 Exercise

Learning pathways

■ LEVEL 1	■ LEVEL 2	■ LEVEL 3
1, 2, 3, 5	4, 7	6, 8

These questions are even better in jacPLUS!
- Receive immediate feedback
- Access sample responses
- Track results and progress

Find all this and MORE in jacPLUS

Explain and comprehend

1. **Identify** which continents are largely unaffected by convergent faults and divergent faults.
2. **Identify** the tectonic plate on which Australia is located.
3. **State** which of the seven large tectonic plates covers the largest area.
4. **Identify** two continents that are severely affected by faults. From their position and the plate movement arrows shown on figure 1.55, **suggest** what type of faults affect these continents.

5. **Identify** which continent has a major subduction zone offshore and is located close to the deepest ocean trenches.

Analyse and apply

6. **Explain** the geological processes that result in earthquakes and volcanic activity in New Zealand.
7. If Australia continues to move north-east at about 6 centimetres per year, **describe** the approximate location of where Cape York will be in 50 million years' time.

Propose and communicate

8. Scientists have established the Pacific Ocean is getting smaller, the Atlantic Ocean is getting larger and the Himalayas are getting taller. **Hypothesise** what could be causing this to happen.

Sample responses are available in your digital formats.

LESSON
1.12 Geological hazards — earthquakes

LEARNING INTENTION

By the end of this lesson you should be able to explain, using a range of representations such as maps, conceptual models, block diagrams and cross sections, how natural hazards are the result of processes that occur within the earth (geological).

Source: Adapted from Geography General Senior Syllabus 2024 © State of Queensland (QCAA) 2024; licensed under CC BY 4.0.

1.12.1 Understanding earthquakes

When an earthquake occurs, **seismic waves** (energy vibrating through the Earth) travel outwards from the focus (also called the hypocentre). The fastest waves (known as **P-waves**, or primary waves) travel at about 6 km/sec through the Earth. These P-waves pass seismic recording stations, a network of sensors located in over 150 different places around the globe known as the Global Seismographic Network (GSN). The arrival time of the P-wave is detected and noted in real time by the network's computers, which collate the information about the speed and direction of the P-waves' travel to calculate the location of the earthquake. The epicentre location is often available less than one minute after an earthquake occurs.

The magnitude of an earthquake is determined from the strength of the seismic waves detected at each station. Several different formulas are used to calculate the magnitude. Most formulas depend on a measure of the shear waves (S-waves, or secondary waves), which have the largest amplitude and carry the most energy. S-waves travel more slowly than P-waves so it may take a few minutes to calculate an earthquake's magnitude. S-waves travel at between about 1 and 8 km/second, and P-waves between about 1 and 14 km/second. The speed of any seismic wave depends on variables such as the density and composition of the ground it is travelling through.

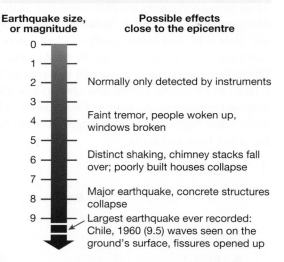

FIGURE 1.56 The Richter Scale

Earthquake size, or magnitude	Possible effects close to the epicentre
0	
1	
2	Normally only detected by instruments
3	
4	Faint tremor, people woken up, windows broken
5	
6	Distinct shaking, chimney stacks fall over; poorly built houses collapse
7	
8	Major earthquake, concrete structures collapse
9	Largest earthquake ever recorded: Chile, 1960 (9.5) waves seen on the ground's surface, fissures opened up

seismic waves waves of energy travelling away from an earthquake; are like huge vibrations and may travel through the Earth's mantle and crust or along the surface

P-waves also known as primary waves; high-frequency seismic waves that travel fastest and are measured first at a seismic station. P-waves can pass through solid rock and liquids.

One common method of measuring the strength of an earthquake is the Richter Scale. Invented by Charles Richter in 1935, it compares the amount of energy released from an earthquake using numbers on a logarithmic scale — a magnitude 8.0 is ten times more powerful than a 7.0. The largest earthquake ever recorded was in May 1960, when a magnitude 9.5 was measured near Valdivia, in southern Chile. Figure 1.57 shows some of the damage from this earthquake.

Another method of measuring earthquake intensity is the Modified Mercalli Intensity Scale, shown in table 1.5. This is also a numerical scale shown in Roman numerals with I at the low-intensity end and XII at the high-intensity end. On the Mercalli Scale, a value is assigned to a specific location based on how people were affected or how much structural damage occurred. Lower numbers match how or what people felt, while higher numbers are based on observed structural damage to infrastructure such as buildings or roads. This means that an earthquake might be low on the Richter Scale, but may rate high on the Mercalli Scale if it caused significant damage. This might occur, for example, in a developing country where communities do not have the resources to construct larger buildings that are strong enough to withstand weaker earthquakes, or in mountainous areas prone to landslips.

FIGURE 1.57 Valvidia after the 1960 earthquake.

TABLE 1.5 Modified Mercalli Intensity Scale, used to determine levels of earthquake damage (USGS)

Mercalli Scale	Shaking	Effects
I	Not felt	Not felt except by a very few persons under especially favourable conditions.
II	Weak	Felt only by a few persons at rest, especially on the upper floors of buildings.
III	Weak	Felt quite noticeably by persons indoors, especially on the upper floors of buildings. Many people do not recognise it as an earthquake. Standing motor cars may rock slightly. Vibrations similar to the passing of a truck. Duration estimated.
IV	Light	Felt indoors by many, outdoors by few during the day. At night, some awakened. Dishes, windows, doors disturbed; walls make cracking sound. Sensation like heavy truck striking building. Standing motor cars rocked noticeably.
V	Moderate	Felt by nearly everyone; many awakened during the night. Some dishes, windows broken. Unstable objects overturned. Pendulum clocks may stop.
VI	Strong	Felt by all, many frightened. Some heavy furniture moved; a few instances of fallen plaster. Damage slight.
VII	Very strong	Damage negligible in buildings of good design and construction; slight to moderate in well-built ordinary structures; considerable damage in poorly built or badly designed structures. Some chimneys broken.
VIII	Severe	Damage slight in specially designed structures; considerable damage in ordinary substantial buildings with partial collapse. Damage great in poorly built structures. Fall of chimneys, factory stacks, columns, monuments, walls. Heavy furniture overturned.
IX	Violent	Damage considerable in specially designed structures; well-designed frame structures thrown out of plumb. Damage great in substantial buildings, with partial collapse. Buildings shifted off foundations.
X	Extreme	Some well-built wooden structures destroyed; most masonry and frame structures destroyed with foundations. Rails bent.

Source: Abridged from The Severity of an Earthquake, USGS General Interest Publication 1989-288-913

CASE STUDY: Earthquakes in Mexico

Quick facts: Mexico
- **Capital city** Mexico City
- **Population** 128 million
- **Earthquakes annually** approximately 1500
- **GDP per capital** US$11 500

Mexico City is one of the world's most populated megacities, but due to its location, it is at high risk of earthquake hazards. On 19 September 1985, a powerful earthquake (8.1 on the Richter Scale and IX on Mercalli Scale) struck close to the city, causing many buildings in Mexico City to collapse and killing more than 10 000 people. Another 30 000 were injured and hundreds of thousands were left homeless. Because of this calamitous event, Mexico City improved its building and construction standards and established an emergency response team, *Brigada Internacional de Rescate Tlatelolco*, also known as *Los Topos* ('the moles'). In 2017, on the same date 32 years later, a 7.1 tremor struck Puebla near Mexico City killing 370 people. This occurred only 12 days after an 8.2 magnitude earthquake struck off the coast of Mexico near Chiapas, generating a tsunami alert. As buildings shook and collapsed, thousands fled onto the streets in fear. Because this quake struck during office and school hours, many casualties were workers and children. Rescue teams of Los Topos and volunteers sifted through the debris and damaged buildings looking for survivors feared buried.

The social and economic costs of earthquakes

A week after the 2017 quake, the death toll had risen to 400 as rescue teams worked to recover more bodies under the rubble. As well, the impact of injured people who would be unable to work was becoming clear in both social and economic terms. While the repair and construction of new buildings was already being planned, tens of thousands of Mexicans were confronted with the humanitarian disaster that followed. Some of these non-obvious events included the following.

- Initially, over half of the population (about 67 million people) were at risk of living in outright poverty.
- The emotional devastation of losing a loved one was even worse if that person was the family's main earner.
- Loss of access to food was likely to lead to illness and malnutrition, thus affecting the long-term health of family members.
- Many of the poor had paltry savings, or had assets tied up in the value of their house, their livestock or some other physical asset.
- The 2017 earthquakes came during the middle of a growing season for many farming households, thus affecting the agricultural capacity of southern Mexico.
- Lower agricultural output would have widespread consequences across the region, inevitably affecting food prices and reinforcing poverty.

FIGURE 1.58 Damage in the streets of Juchitan, Oaxaca, Mexico, a month after the September 2017 earthquake

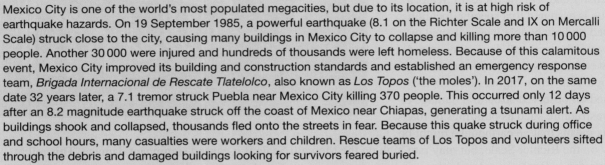

1.12.2 Responding to earthquakes

Earthquakes can happen at any time, with very little warning; many people are caught in the middle of the shaking. Most deaths and injuries occur from falling objects or buildings. People who live in earthquake zones practice safety drills, such as 'Drop (Duck), Cover, and Hold on!' Authorities suggest that when an earthquake strikes people should:

- keep calm
- if possible, turn off electricity, gas, and tap water
- protect yourself from falling objects such as signs, light fixtures and potted plants
- do not rush out of a building or use elevators.

In many communities at high risk of earthquakes, response and recovery efforts begin before an earthquake occurs — for example, by ensuring buildings are able to withstand predicted intensities of earthquakes in the area, and ensuring people know what to do when an earthquake strikes and that they have the appropriate equipment ready.

CASE STUDY: Earthquakes in Japan

Quick facts: Japan
- **Capital city** Tokyo
- **Population** 125 million
- **Earthquakes annually** approximately 1500
- **GDP per capital** US$34 000

Earthquakes are part of everyday life in Japan. In fact, the island nation is probably the most earthquake-proof country in the world. Japan's location makes it particularly vulnerable to tectonic plate movement and earthquakes. Situated at the intersection of four converging plates and surrounded by deep ocean trenches undergoing subduction, it is little wonder tremors occur daily.

In March 2011, Japan experienced one of its most severe natural disasters when a 9.0 quake lasting several minutes occurred in the Pacific Ocean 130 kilometres east of the island of Honshu. This intense quake triggered a tsunami that swamped hundreds of kilometres of coastline less than an hour later. It killed about 20 000 people, most of whom drowned in the surge of seawater. About 500 000 people were forced to evacuate their homes, creating a humanitarian crisis.

FIGURE 1.59 Earthquake and tsunami destruction in Japan

At the same time, seawater flooded the Fukushima nuclear power plant, forcing it into meltdown. This created an energy, economic and environmental disaster for the Fukushima region, which became contaminated with radioactive water and lay waste for years.

More recently, on New Year's Day 2024, another earthquake measuring 7.6 struck the Noto Peninsula on the west side of Japan, killing at least 200 people, injuring over 300, destroying buildings, knocking out power, and causing residents in coastal areas to flee to higher ground for fear of a tsunami. Many residents chose to remain outdoors in the cold, fearing further tremors, collapsing buildings and landslides. According to the US Geological Survey, it was the strongest quake on the west side of Japan for more than 40 years, almost levelling the historical city of Wajima, famous for its markets.

Some of the measures taken by people in Japan to manage the risks associated with frequent earthquakes include the following:
- designing buildings to cope with shock and movement
- implementing intense education programs for residents and practice drills for visitors
- maintaining immediate warning systems using phone alerts, sirens, and other signals
- being able to deploy rescue and recovery teams at short notice
- constructing a network of evacuation sites where people may need to shelter.

1.12 Exercise

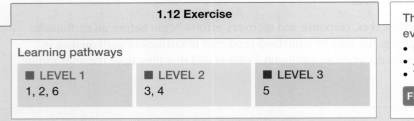

1.12 Exercise

Learning pathways		
■ LEVEL 1	■ LEVEL 2	■ LEVEL 3
1, 2, 6	3, 4	5

These questions are even better in jacPLUS!
- Receive immediate feedback
- Access sample responses
- Track results and progress

Find all this and MORE in jacPLUS ▶

Explain and comprehend

1. **Describe** how the 1985 earthquake in Mexico City led to improvements in the city's infrastructure and emergency response systems.
2. **Explain** the geological processes that result in earthquakes and volcanic activity in Japan.
3. **Describe** some of the social and economic challenges faced by people affected by earthquakes.

Analyse and apply

4. Consider both the 2011 and 2024 earthquakes in Japan. **Discuss** potential factors that contributed to the differing outcomes and responses in each event.

Propose and communicate

5. The San Andreas Fault is a continental transform fault running through California, the United States' most populous state (with over 37 million residents). The fault is about 1200 km long and forms part of the boundary between the Pacific Plate and the North American Plate. Both plates are moving to the north, but the Pacific Plate is moving slightly faster than the North American Plate. An enormous level of stress energy has been building up for over a hundred years, and seismic experts fear it has reached a critical state. With so many cities in the region, a large earthquake could be catastrophic.
   a. **Explain** how such a large population might affect the earthquake response.
   b. **Propose** two well-justified strategies to mitigate risk and to manage response if such an earthquake occurs.
6. **Design** a list of measures that Japan could take to reduce the risk of earthquake hazards.

Sample responses are available in your digital formats.

LESSON
1.13 Geological hazards — volcanic eruptions

1.13.1 Understanding volcanoes

Volcanoes exist on every continent. At present, several thousand are extinct or dormant and pose no threat to life. However, about 600 volcanoes above sea level are highly active. When they erupt, they may cause death and injury to people as well as destroy property and crops; however, volcanoes also produce mineral-rich soils, and so people settle around the base to make the most of the fertile farmland. In doing so, they put their lives and communities at risk.

Most volcanoes occur when magma can escape or be released at weak points of the Earth's crust. More than 60 per cent of all volcanoes occur adjacent to or near tectonic plate boundaries where either subduction or sea floor spreading is occurring. The most active volcanoes occur over subduction zones, many in the Pacific Ring of Fire. Volcanoes also form at isolated weak points in the crust known as 'hot spots', such as the Hawaiian Islands.

FIGURE 1.60 The soil near Mount Etna, in Sicily, Italy, is highly fertile.

Volcanoes differ according to the type of lava that is discharged from them. There are two main types.

- Basic lava cones are low with gentle slope like those in Hawaii and Iceland. This lava is low in silica and runs out quickly, and may travel a considerable distance from the volcano. These cones are commonly called shield volcanoes.
- Acidic lava volcanoes have tall, steep cones. This lava is high in silica, and is viscous (thick); it moves slowly and may even clog some of the volcano's vents, resulting in violent, explosive eruptions that can blast tonnes of rock and ash, and hot, poisonous gases into the air (see figure 1.63). Acidic volcanoes are an extremely dangerous hazard because of this explosive force. They are also referred to as **pyroclastic cones**.

FIGURE 1.61 The gentle slopes of Mauna Kahālāwai, a shield volcano in Hawaii

1.13.2 Impacts of volcanic eruptions

The most obvious effect of volcanic eruptions is the threat to life, but eruptions can also destroy forests and wildlife, houses, farmlands and rivers, as well as contaminate the atmosphere. They create lava flows that travel long distances and burn, bury or harm anything in their path, and are often associated with earthquakes, mudflows, lahars and avalanches. The gases they emit, such as carbon dioxide (CO_2) and sulfur dioxide (SO_2), are not only toxic and smelly, but also cause acid rain in regions some distance away. (Figure 1.64 illustrates these features of active volcanoes.) Large quantities of rock, ash and dust can bury things, cause roofs to collapse and make it difficult for living things to breathe. Volcanic ash also affects aircraft engines, sometimes even leading to the failure of key navigation equipment. If ash is sucked into aircraft engines and accumulates, it can potentially lead to engine failure. Because it is unsafe to fly through airborne volcanic ash, flight paths are often closed down near ash plumes, causing significant disruptions to travel and tourism. These hazards can cover significant distances, so volcanic exclusion zones (areas from which people are encouraged to evacuate in order to stay safe) can be quite large.

FIGURE 1.62 Pyroclastic cones in Guatemala

pyroclastic cones steep conical volcanic cones built by a combination of lava flows and cinder/ash from pyroclastic eruptions. They can form quickly and remain active for long periods.

FIGURE 1.63 Pyroclastic cloud, Mount Sinabung, Indonesia

FIGURE 1.64 Features of an active volcano

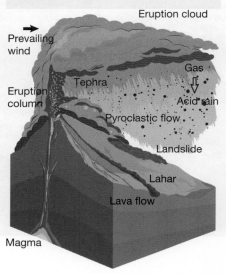

Eruption cloud

Prevailing wind

Gas

Tephra

Eruption column

Acid rain

Pyroclastic flow

Landslide

Lahar

Lava flow

Magma

CASE STUDY: Iceland — land of fire and ice

Quick facts: Iceland
- **Area:** 100 250 sq km
- **Capital city** Reykjavik
- **Population** 382 000 (2022)
- **GDP per capita** US$53 600
- **Number of active volcanoes** 32

The small island nation of Iceland is literally a land of fire and ice. Located in the far north Atlantic Ocean, it has more volcanic landscapes than any other country and is one of the top five countries in terms of active volcanoes. As well as furnace-like heat from magma, Iceland also has the contrasts of freezing air, snowfields and glaciers.

Volcanic landforms are ubiquitous, with over 30 active volcanoes and over 200 craters across the island. While some eruptions are dangerous, residents remain proud of these landscapes that harness geothermal energy as well as attracting millions of tourism and research dollars every year.

Iceland has a variety of volcanic types, such as red volcanoes with effusive eruptions, gray volcanoes with explosive eruptions, subglacial volcanoes and underwater volcanoes. As well as protruding volcanic craters, there are numerous geysers, hot springs, solfataras, maars, lava and moss fields. Some of the coloured rocky outcrops and landscapes are so unique they are not found anywhere elsewhere in the world.

Why are there so many volcanoes in Iceland?

Iceland is situated on a diverging fault (see figure 1.65) at the intersection of two tectonic plates at the northern end of the Mid-Atlantic Ridge. The two tectonic plates (North American and Eurasian plates) are pulling away from each other at a rate of 2 centimetres per year, which in seismic terms is quite fast. As expected, Iceland has not only regular eruptions but also some that last for days and others that go on for years.

Over the past 50 years, the most notable eruptions in Iceland were the following:
- Eldfell 1973
- Krafla 1980, 1981 and 1984
- Hekla 1980, 1991 and 2000
- Eyjafjallajökull 2010
- Grimsvötn 1996, 1998, 2004 and 2011
- Barðarbunga 2014
- Geldingadalur (Fagradalsfjall) 2021
- Meradalur (Fagradalsfjall) 2022
- Litli Hrútur (Fagradalsfjall) 2023
- Sundhnúkagígar (Grindavik) 2023

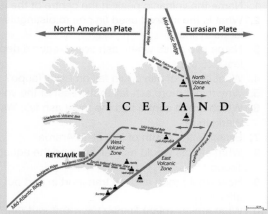

FIGURE 1.65 Iceland sits on a major divergent plate boundary

TABLE 1.6 Benefits and disadvantages of volcanic activity in Iceland

Benefits	Disadvantages
Geothermal energy is used to generate about a third of Iceland's low-cost electricity	Volcanoes are hazards that cause loss of life and damage; this is exacerbated by uncertainty of eruptions
Hot water is drawn directly and used for heating as well as in greenhouses where fruit and vegetables are grown all year round	People may have to evacuate towns (e.g. the fishing town of Grindavik was evacuated for months) or leave their homes
Research	Destruction of infrastructure

TABLE 1.6 Benefits and disadvantages of volcanic activity in Iceland *(continued)*	
Tourism	Ash clouds cause toxic pollution, reduce viability and make air-space dangerous for flight travel
	Toxic gases (CO_2 and SO_2) are dangerous to human health

on Resources

Video eLesson	SkillBuilder: Constructing and describing isoline maps (eles-1737)	
Interactivity	SkillBuilder: Constructing and describing isoline maps (int-3355)	
Weblinks	Unzen pyroclastic cloud	
	Smithsonian Institution Global Volcanism Program	

SKILLS ACTIVITY

An isopach map is one that shows lines connecting points of equal formation thickness. It can be used to show fallout from a volcano or the thickness of mining seams.

Refer to the isopach map in figure 1.66 to answer these questions.

1. Name the volcanic peak in the centre of the isopach lines.
2. What is the interval used (in cm) to distinguish between the isopach lines?
3. Name two cities where ash could reach a depth of at least 25 cm.
4. What depth of ash could fall on Lake Taupo?
5. Why might more ash fall to the east of the volcano?
6. Which urban centre had the most ash fall: Wellington or Wanganui?
7. Estimate the depth of ash at Te Araroa.
8. What effects would ash fall have on the aquatic life in Lake Taupo?
9. How would road and rail transport in and out of Te Araroa be affected?
10. What affects might Napier have suffered that Wanganui did not?

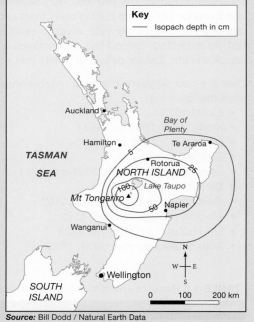

FIGURE 1.66 A hypothetical isopach map of the North Island of New Zealand showing potential depths of ash fallout

Source: Bill Dodd / Natural Earth Data

1.13.3 Responding to volcanic eruptions

When there is a volcanic eruption, and if residents or tourists have time, it is safest to move beyond the exclusion zone. However, if this is not possible or people fail to leave, certain steps need to be taken. People should protect themselves against ash and dust by wearing a mask or placing a wet cloth over their mouth and nose, wearing long-sleeved clothes and removing contact lenses. This helps to prevent acid-coated ash irritating their lungs and eyes. Keeping at least three days' supply of clean drinking water, a supply of food, a battery-operated radio, cash and a first aid kit is also important. People should also protect electronic equipment with plastic.

Responses to volcanic eruptions vary according to the severity and impact of the eruption, and the community's ability to respond. Consider the plan of action for the Tungurahua Volcano in Ecuador and the preparedness guide developed by the US Geological Survey found in the Resources tab.

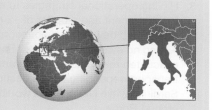

CASE STUDY: Mt Vesuvius, Italy

Quick facts: Italy
- **Capital City** Rome
- **Population** 59 million
- **GDP per capita** US$34 776
- **Number of active volcanoes** 3

Even though Italy's volcanoes have provided its farmers with rich, fertile volcanic soils, especially around Mt Vesuvius and Mt Etna, they are also a grim reminder of Earth's awesome authority.

In 79 CE, Vesuvius, a massive stratovolcano overlooking the Bay of Naples, erupted with overwhelming power, destroying the Roman cities of Pompeii and Herculaneum, killing many of their occupants. (Figure 1.67 shows the location of these two cities, as well as the areas of ash fall.) Vesuvius has erupted many times since 79 CE, most recently in 1944. It is still regarded as active and is monitored daily for gas and temperature changes. Today, Naples and the area adjacent to the base of Vesuvius is a heavily populated urban and tourist centre. More than 3 million people live within 12 kilometres of its cone and are exposed to the potential danger of Vesuvius, which continues to rumble and discharge smoke. Vesuvius's history of violent eruptions make it one of the world's most dangerous volcanoes.

Vesuvius evacuation plans today

Authorities in the areas surrounding Vesuvius have formulated plans to evacuate over 700 000 people if the volcano erupts. The Department of Civil Protection, which is charged with risk management and emergency response in Italy, has identified a 'red zone' for risk that includes 25 towns, all of which can be evacuated within 72 hours of a significant eruption. Another 63 towns, with a combined population of more than 1 000 000 people, lie in a 'yellow zone' that would be likely to suffer from falling ash and rock from the volcano.

The plan also includes four alert levels: basic, attention, pre-alert and alert. When the pre-alert level is activated, patients in care facilities and hospitals are relocated and heritage monuments are protected. Activation of the alert level means the plans to evacuate come into effect.

Risk management plans contain key strategies that need to be developed while the mountain is dormant. These include:
1. Delineating the three emergency zones based on proximity and potential fallout.
   a. The red zone is the area around the base of Vesuvius, where the risk of pyroclastic flow, heat and toxic gases is greatest.
   b. The yellow zone is further away and is the area most likely to be affected by fallout of ash and lapilli (rock fragments being launched from the volcano). Homes would be at risk of collapsing and residents would suffer from respiratory problems.

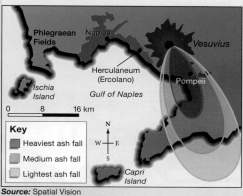

FIGURE 1.67 Mt Vesuvius and surrounding areas, 79 CE

Key
- Heaviest ash fall
- Medium ash fall
- Lightest ash fall

Source: Spatial Vision

FIGURE 1.68 Plate boundary, southern Italy

Key
- Plate boundaries

Source: MAPgraphics

c. The blue zone falls inside the yellow zone, but is considered at higher risk because of its topography. Here there is more chance of flooding or mudflows.

2. Educating and informing people about an orderly evacuation and removing complacency, without becoming alarmist. Money has been set aside for school education programs.
3. Having operational strategies to evacuate the entire red zone in 72 hours (12 hours for organisation, 48 hours for movement and a buffer of 12 hours if needed) using a fleet of 375 000 registered cars, 500 buses and 220 trains each day of the evacuation period.
4. Arranging evacuation centres. Each of the 25 local regions has been 'twinned' with another area of Italy where evacuated residents would be accommodated. Each township has been allocated a specific mode of transport depending on the destination of evacuees. For example, Pompeii residents would leave by boat to Sardinia, while Neapolitans would board trains that would take them to Lazio, the administrative region around Rome.

FIGURE 1.69 Looking across Naples towards Mt Vesuvius

FIGURE 1.70 Satellite image of the area around Mt Vesuvius

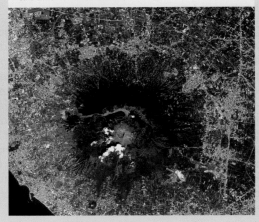

Source: © NASA Visible Earth

 Resources

📄 **Digital docs** Volcano warnings (doc-29160)
SWOT analysis grid (doc-29162)

🔗 **Weblinks** Mt Vesuvius Emergency Plan
What if Vesuvius erupted today?

1.13 Exercise

1.13 Exercise

Learning pathways		
■ LEVEL 1 1, 2, 3	■ LEVEL 2 4	■ LEVEL 3 5

These questions are even better in jacPLUS!
• Receive immediate feedback
• Access sample responses
• Track results and progress

Find all this and MORE in jacPLUS ▶

Explain and comprehend

1. **Explain** why a separate 'blue zone' is allocated inside the 'yellow zone' in the Naples response plan.
2. **Explain** why the volcanic activity in Italy (Vesuvius) is quite different from that occurring in Iceland.
3. **Explain** why the types of volcanic activity in Iceland are regarded as a tourist attraction rather than a severe threat.

4. **Discuss** whether education is the most successful strategy for emergency preparedness.

Propose and communicate

5. Read through the following description of Vesuvius erupting in 79 CE, as written by Pliny the Younger, and answer the questions that follow.

Extract from Pliny the Younger's account of the eruption of Vesuvius (24 August 79 CE):

> One night the earth shocks became so violent that it seemed the world was turned upside down . . . We decided to escape, and a panic-stricken crowd followed us . . . When we were clear of the houses we stood still . . . The sea seemed to roll back on itself . . . the shore had widened, and many sea creatures were beached on the sand. In the other direction appeared a horrible black cloud ripped by bursts of fire . . . Soon the cloud began to descend upon the earth and cover the sea . . . We were enveloped in night like the darkness of a sealed room without lights. We could hear the cries of women, the screams of children and the shouting of men. Some were calling to their parents, to their children, to their wives. Some lifted their hands to the gods, but a great number believed there were no gods now, and that this night was to be the world's last eternal one . . . We were immersed in darkness and ashes fell thickly upon us. From time to time we had to get up and shake them off for fear of being buried and crushed under the weight . . . Finally, a real daylight came. Before our terror-stricken sight everything was covered by a thick layer of ashes like a heavy snowfall. We returned to Misenum and had no thought of leaving until we received news about my uncle.

a. **Identify** what Pliny was describing in his account.
b. **Hypothesise** which hazardous events Pliny could have been describing when he said, 'The sea seemed to roll back on itself … the shore had widened, and many sea creatures were beached on the sand.'

Sample responses are available in your digital formats.

LESSON
1.14 Geomorphic hazards

LEARNING INTENTION

By the end of this lesson you should be able to explain, using a range of representations such as maps, conceptual models, block diagrams and cross sections, how natural hazards are the result of processes that occur on the surface of the earth (geomorphic).

Source: Geography General Senior Syllabus 2024 © State of Queensland (QCAA) 2024; licensed under CC BY 4.0.

1.14.1 Understanding geomorphic hazards

Geomorphic hazards are the result of processes that occur on the Earth's crustal surface or lithosphere. Often called 'mass wasting', they refer to **slope failure** (which occurs when the pull of gravity causes a hill or mountain slope to collapse), **subsidence** (when part of the land sinks or collapses), or **flow movement** (when rock, soil or sand mix with water and air and move downhill in a flow). The most common geomorphic hazards are landslides, mudslides and avalanches (snow). These usually happen when the force of gravity cannot be withstood any longer because of one of two changes:

- geological movement, particularly in an already unstable area
- human activity that has caused changes to the topography, such as vegetation clearing (figure 1.71), changes to drainage patterns and below-ground seepage or wild fires.

slope failure when the pull of gravity causes a hill or mountain slope to collapse

subsidence when part of the land sinks or collapses

flow movement when rock, soil or sand mix with water and air and move downhill in a flow

Even though mass wasting of rock, soil, mud or snow is less common than earthquakes and volcanic eruptions, it happens without warning and, in many cases, people are completely unprepared. The consequences can be catastrophic, causing widespread death and injury as well as burying towns and villages, destroying buildings and infrastructure (roads, tunnels, bridges and property), and even reshaping the surface topography. Across the world, landslides, mudflows and avalanches are some of the biggest killers of people in mountain regions, particularly where people live on or below hillsides.

FIGURE 1.71 Clearing of vegetation on steep land can increase the vulnerability to mass wasting.

The speed at which material travels downslope ranges from very slow (creep) to very rapid (rockfall). The momentum and force of a flow is usually determined by
- the mass of material (size and weight)
- the mix of material (clay, rocks, soil, debris)
- the amount of water or other fluid to lubricate and reduce friction
- the slope angle.

TABLE 1.7 Classification of flow movement

Average speed	Saturated flows	Non-saturated flows
Very slow (e.g. 1 cm/yr)	Solifluction	Soil creep
Slow (e.g. 1 m/day)		Earth flow
Moderate (e.g. 1–2 m/hr)	Debris flow	
Rapid (e.g. 50 km/hr)	Mudflow	
Very rapid (over 100 km/hr)		Debris avalanche

1.14.2 Processes that create geomorphic hazards

A **landslide** is a geomorphic event where rock, soil, mud or artificial fill move down a slope under the force of gravity (slope failure). Landslides may be caused by any number of natural processes, including:

- long-term weathering of rocks and scree, which produces regolith, loose weathered rock fragments and soil that overlay the bedrock
- soil erosion
- vegetation removal
- earthquakes
- heavy rainfall
- volcanic eruptions.

Human activities can also trigger landslides. The most common human causes are:

- road building
- construction on very steep slopes
- poorly planned changes to drainage
- mining
- disturbance of old landslide sites.

landslide the large-scale movement of rock, debris and soil down a slope due to unstable conditions, which may be caused by heavy rainfall, an earthquake or a volcanic eruption. A landslide under the ocean can cause a tsunami.

mudflow when large amounts of suspended silt and soil move quickly down a slope. They tend to occur mostly on steep slopes but can happen anywhere ground is unstable due to loss of vegetation.

EXAMPLE: Thredbo landslide, 1997

The most lethal landslide in Australia occurred in July 1997, when a large section of steep mountainside collapsed in Thredbo, New South Wales, carrying the Carinya ski lodge at high speed onto the Bimbadeen ski lodge below. Both buildings were destroyed and 18 people died beneath the rubble. One injured man survived for three days within the landslide debris before being rescued.

In comparison, a huge debris avalanche in Peru in 1970, which moved at speeds of up to 400 km/h, killed more than 20 000 people and destroyed the towns of Yungay and Ranrahirca.

FIGURE 1.72 rescue workers at Thredbo after the deadly landslide

A **mudflow** is a type of mass wasting where soil and debris become saturated (liquefied) and move rapidly down a slope (flow movement). Often, there is insufficient vegetation to hold the soil together due to land clearing or fires. In recent times, mudflows have happened due to very heavy rain falling in areas where slopes have been cleared for farming (for example, in the Philippines and Nepal) or razed by wildfires (for example, in California in the United States).

Heavy rainfall, such as that associated with tropical cyclones and severe thunderstorms,

FIGURE 1.73 mudflows in southern California, 2018

is a common trigger for mass wasting events. Large volumes of water add weight to soil and weathered rock particles, making slopes more unstable and susceptible to the influence of gravity. Water may also lubricate rock surfaces, thus reducing friction and allowing easier movement of overlying soil and rock. In countries that have cleared slopes from either farming or bushfire, heavy rain can trigger mudslides, often with tragic consequences.

Figure 1.74 illustrates the types of slope failures that can occur and the movement patterns of earth, soil and debris.

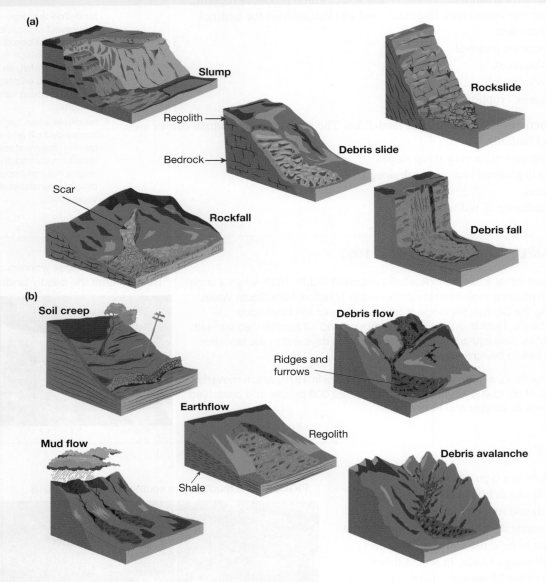

FIGURE 1.74 (a) Types of slope failure and (b) types of flow movement

1.14.3 Responding to geomorphic hazards

As with all other hazards, the best response to geomorphic hazards is understanding how to reduce harm, and preparedness for and after the event. Most people have limited choices about where they live, and humans have little or no control over natural events, so strong mitigation, prevention and adaptation strategies are required to minimise risk.

While the type, scale and scope of geomorphic hazards vary according to the location and proximity of people to the threat, they all have a common thread — they are potentially dangerous for everything in their path. The most significant variable in humans' ability to survive such a hazard is a country's level of development and the amount of money available for relief efforts when an event occurs. Figure 1.75 provides an example of the information and advice on landslides provided by the Philppines' National Disaster Risk Reduction & Management Council.

FIGURE 1.75 Information and advice on landslides for the Philippines

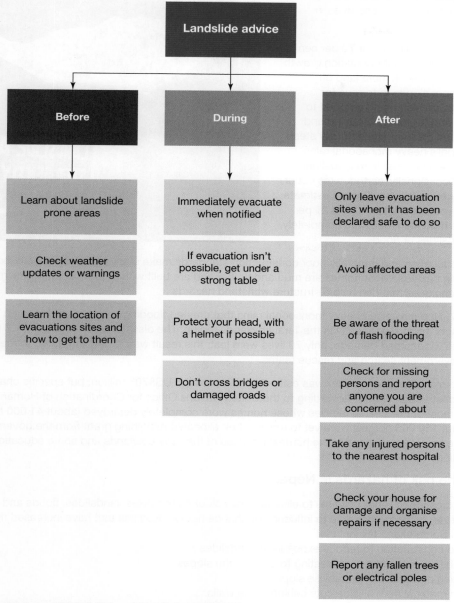

Landslide advice

Before
- Learn about landslide prone areas
- Check weather updates or warnings
- Learn the location of evacuations sites and how to get to them

During
- Immediately evacuate when notified
- If evacuation isn't possible, get under a strong table
- Protect your head, with a helmet if possible
- Don't cross bridges or damaged roads

After
- Only leave evacuation sites when it has been declared safe to do so
- Avoid affected areas
- Be aware of the threat of flash flooding
- Check for missing persons and report anyone you are concerned about
- Take any injured persons to the nearest hospital
- Check your house for damage and organise repairs if necessary
- Report any fallen trees or electrical poles

Source: NDRRMC Office of Civil Defense, Philippines

CASE STUDY: Landslides in Nepal

Quick facts: Nepal
- **Capital City** Kathmandu
- **Population** 30 million
- **GDP per capita** US$1336
- **Highest point** Mount Everest (8848 metres above sea level)

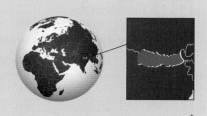

Nepal is a mountainous country wedged between India and China. It ranks among the poorest and least developed countries in the world.

Farming provides a livelihood for 70 per cent of the Nepalese people, but population growth is causing farmers to seek more land on higher and steeper slopes to terrace for farming. Forests are also being stripped to feed livestock, and for fuel for cooking and warmth. This land clearing, the naturally steep slopes and Nepal's heavy monsoonal rains combine to make it vulnerable to landslides and flooding. Occasionally, earthquakes occur, which also trigger landslides. Experts estimate that every five years, between 10 and 25 per cent of Nepal's mountain roads are completely lost due to landslides or floods.

FIGURE 1.76 Landslide warning sign, Nepal

Low levels of education and generally poor construction standards make many Nepalese people more vulnerable to mass wasting hazards. Most homes are mud and timber and are built without the benefit of the modern engineering techniques that can help a structure withstand hazards.

In August 2017, Nepal received heavy monsoonal rains that created flooding and landslide hazards in half the country's districts, most significantly in the Terai Lowlands region. The disaster affected around 1.7 million people. Despite widespread damage, only 70 lives were lost; this result was largely attributed to the quick response of government search and rescue teams.

The cost of recovery from the disaster was estimated to be around US$705 million, but specific challenges hampered reconstruction efforts. According to the United Nations Office for Coordination of Humanitarian Affairs, by May 2018, most of the eligible families whose homes were completely destroyed (about 41 000 homes) or damaged (about 150 000 homes) were yet to receive their allocated rebuilding grant from the government. Food shortages were being experienced in the hardest hit areas of the Terai Lowlands and some education, health and sanitation facilities were still inoperable.

Reducing the risk of hazards in Nepal

Given Nepal's location, it is not possible to eliminate the risk of earthquakes, landslides, floods and other natural hazards. However, it may be possible to influence or change human activities that have increased the risk of these hazards. Some examples include:
- better road construction to reduce exposure to landslides
- not using heavy machinery and blasting to cut through slopes
- using netting and grasses to stabilise slopes
- containing material from hillside cuts behind stone walls.

At present, the Annapurna Conservation Area Project is aiming to meet broader goals in land rehabilitation and tree planting, but could assist with the reduction in landslides too.

on Resources

Digital doc Landslides in Nepal (doc-29159)

Weblink Nepal floods and landslides

1.14 Exercise

Explain and comprehend

1. **Identify** a place in the flow chart shown in figure 1.77 where the first early warning signs of a landslide might occur.
2. Vegetation assists the process of interception in the water cycle. **Explain** what happens to rainwater and where it goes if vegetation is removed.
3. **Define** the phrase 'potential for disaster', as used in figure 1.77.
4. **Describe** how an increase in the frequency of summer storms might affect the stability of hill slopes and terraces.
5. **Outline** one intervention people could make to reduce the risk of a landslide.

Propose and communicate

6. **Suggest** which of the physical variables shown in figure 1.77 might be magnified due to climate change.

FIGURE 1.77 Flowchart showing events leading up to landslide

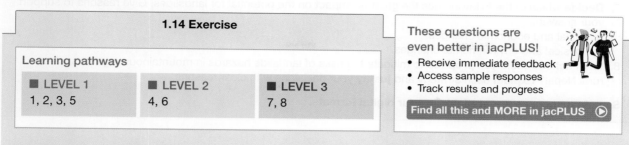

7. **Decide** which of the following has the greatest impact on the potential for landslides. Give reasons to support your answers.
   a. Social and economic changes
   b. Physical and environmental changes
8. What actions could you **propose** to mitigate the risks of landslide hazards in mountainous regions such as rural Nepal? **Explain** each strategy and **justify** your decisions with clear reasons.

Sample responses are available in your digital formats.

LESSON
1.15 APPLY YOUR SKILLS — Construction, interpretation and analysis of maps

LEARNING INTENTION

By the end of this lesson you should be able to use cartographic principles and data to create a map.

Source: Geography General Senior Syllabus 2024 © State of Queensland (QCAA) 2024; licensed under CC BY 4.0.

1.15.1 Analysing spatial and statistical data

TABLE 1.8 Deaths caused by major natural disasters (Australia's GDP per capita US$65 000 for comparison purposes)

Country	Deaths 1996–2015	Major natural disasters	GDP per capita ($US)
Haiti	229 699	Earthquake, hurricane	740
Indonesia	182 136	Tsunami, tropical storms	3570
Myanmar	139 515	Tropical storms	1275
China	123 937	Earthquake, floods	8123
India	97 691	Earthquake, tsunami, floods	1709
Pakistan	85 400	Earthquake, floods	1468
Russian Federation	58 545	Extreme weather, earthquake	8748
Sri Lanka	36 433	Floods, landslides	3835
Iran	32 181	Earthquake	4957
Venezuela	30 319	Floods	14 300

Source: EM-DAT, CRED / UCLouvain, 2024, Brussels, Belgium – www.emdat.be, accessed 2024-05-16; GDP per capita (current US$) – World Bank national accounts data, and OECD National Accounts data files. © 2024 The World Bank Group.

The data reveals that about 1.35 million people have died from natural hazards over the past 20 years. During the 7056 disasters recorded in that period, more than 50 per cent of fatalities occurred due to earthquakes and tsunamis. By far, the greatest number of deaths were from low- and middle-income nations (see table 1.8) According to the United Nations, the least developed countries had the most fatalities in terms of numbers killed per disaster and per 100 000 of population. However, several trends have emerged from figures gathered over the past two decades.

- The frequency of geological events such as earthquakes, tsunamis and volcanic eruptions remained constant and similar to previous patterns.
- Climate/weather-related events such as storms, cyclones, blizzards and heatwaves increased significantly. The number of these hazards was more than twice that of the previous 20 years.

- More extreme weather was experienced, and new temperature and rainfall records were set in many parts of the world.
- The number of fatalities doubled against the previous 20 years. It is uncertain whether this is due to more people living in disaster-prone areas or a spike from mega-disasters such as the tsunami in the Indian Ocean (2004) or the Haiti earthquake (2010).

SKILLS ACTIVITY

A choropleth map is a type of thematic map used to represent statistical data through various shading patterns or colours in predefined areas, such as countries, states or regions. The idea is to use different shades or colours to indicate the range of values in the data set, helping to visualise how a particular variable is distributed geographically. This makes it easy to see patterns and trends across different areas.

Key aspects of choropleth maps include the following.
- Areas with similar data values are shaded with the same colour or pattern intensity.
- Darker shades typically represent higher values, while lighter shades represent lower values.
- It is important to use a consistent scale to ensure accurate representation of the data.
- The map legend explains the colour or shading scheme used, allowing users to interpret the data correctly.
- Choropleth maps are particularly useful for comparing data across different regions and identifying geographical trends.
- As with all maps you create, you need to remember cartographic conventions. BOLTSS can be a useful acronym for remembering to include your map's border, orientation, legend, title, source and scale.

By using a choropleth map, you can quickly and effectively communicate complex data in a visually appealing and easily understandable format.
1. Use the blank map of the world in the resources tab and identify the countries that appear in table 1.8.
2. Using an appropriate legend, colour code the map to represent the deaths caused by major natural disasters.
3. Ensure that your map meets cartographic conventions.

Analyse your choropleth map

1. Describe the distribution of the countries with the highest number of deaths caused by major natural disasters.
2. Use the GDP per capita weblink to examine the choropleth map.
3. Describe the spatial association between high deaths from natural disasters and GDP per capita.
4. Explain possible reasons for this spatial association.
5. Research the locations of the Earth's tectonic plates. You can use the USGS weblink to help you.
6. Add the tectonic plate boundaries to your map.
7. Evaluate whether proximity to a plate boundary or GDP per capita have a higher impact on the likelihood of a country experiencing a high number of deaths from natural disasters.

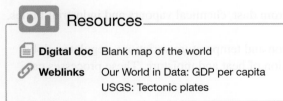

Resources

Digital doc Blank map of the world

Weblinks Our World in Data: GDP per capita
USGS: Tectonic plates

LESSON
1.16 Review

1.16.1 Summary

1.2 Natural hazards

- Natural hazards are extreme geophysical events with potential harm to people, living things, property and the environment, caused by powerful forces generating destructive energy.
- Hazards can cause death, injury and damage, transforming into disasters when resulting in significant fatalities, property loss and/or requiring long-term recovery plans.
- Hazard zones identify areas at risk and guide planning and insurance decisions.

1.3 Natural hazards in Australia

- Significant differences in hazard types occur between and within states in Australia, impacting how individuals perceive and assess risk.
- Cyclones are a common risk in the north of Australia, while bushfires are more prevalent in the south.

1.4 The impact of natural hazards

- Impacts of natural hazards are determined by more than the severity of the hazard itself. A range of factors have an influence, including cause, predictability, damage potential, prevention and preparedness, response, speed of onset, duration and frequency.
- Economic factors play a significant role in determining the severity of impacts of a natural hazard. Countries with strong economies can prepare for and respond to hazard events much more effectively than less wealthy countries.
- Climate change appears to be making many hazards more frequent and severe.

1.5 Assessing and responding to natural hazards

- Risk assessment involves understanding hazard effects, examining physical and human features, and evaluating community preparedness.
- Crichton's Risk Triangle highlights control factors in risk reduction are exposure and vulnerability.
- Historical data aids in forecasting hazards, while risk models guide emergency services and insurance considerations.

1.6 Atmospheric hazards

- Common atmospheric hazards include severe storms, blizzards, tropical cyclones, floods and droughts, occurring over short or long periods.
- Human-caused atmospheric hazards include air pollution from dust, chemical vapours and industrial fumes, posing toxicological risks to living organisms.
- Atmospheric circulation, driven by solar radiation absorption and temperature variations, regulates global climate patterns, influencing weather systems and distribution of heat and moisture. These processes can also lead to atmospheric hazards.

1.7 Atmospheric hazards — thunderstorms

- Thunderstorms are a common atmospheric hazard, forming when warm, moist air rises due to unstable atmospheric conditions.
- The main hazards associated with severe thunderstorms include torrential rain, flooding, hail, destructive winds and lightning strikes.
- Three main types of thunderstorms are single-cell storms, multicell thunderstorms, and supercells, each with distinctive features and varying severity of weather phenomena.

1.8 Atmospheric hazards — tropical cyclones

- Tropical cyclones, spanning 100 to 2000 kilometres in diameter, bring heavy rain and destructive winds to tropical and sub-tropical regions. They are known as hurricanes in the Atlantic and eastern Pacific, and typhoons in the western Pacific.
- Formed in the inter-tropical convergence zone (ITCZ), tropical cyclones require warm sea temperatures, rising warm air, humidity and specific wind shear conditions to develop, typically between 5° and 30° latitudes.
- Cyclones exhibit gale force winds, reaching speeds over 280 km/h, with a characteristic circular shape and calm 'eye' in the centre, influenced by the Coriolis effect, which shapes their circular formation and intensifies away from the equator.

1.9 Atmospheric hazards — responding to tropical cyclones

- Improved public awareness and communication have reduced fatalities and injuries from storms and cyclones despite their increasing power, though increasing population and coastal settlements pose greater economic risk.
- Strategies to prepare for cyclones and severe storms include installing underground powerlines, creating sea walls, restoring natural waterways, capturing water for future use, increasing green space for rainfall infiltration, and designing multi-use public spaces as cyclone shelters.

1.10 Atmospheric hazards — bushfires

- Fire, while historically essential for warmth, cooking and land management, can be a deadly adversary when not controlled, causing harm to people, animals, property and the environment.
- Bushfires, including grass fires, forest fires, and scrub fires, pose significant risks in many countries, including Australia and Canada, particularly during extreme weather events fuelled by climate change and urban expansion.
- Responding to bushfire threats is challenging due to the combustible nature of Australian bushland, requiring critical decisions about evacuation and preparation, including maintaining clear evacuation plans, clearing debris around homes and installing firefighting equipment such as pumps and hoses.

1.11 Geological hazards

- Geological hazards originate within the Earth, and include earthquakes, volcanic eruptions, and tsunamis. These pose significant risks to life, property and the environment.
- Tectonic plates, driven by convection currents in the mantle, cause geological phenomena such as earthquakes and volcanic activity, with fault types such as diverging, converging and transform faults shaping Earth's surface.
- Oceanic and continental crusts play distinct roles in plate tectonics, influencing the formation of features such as mid-ocean ridges and volcanic eruptions, while earthquakes result from the sudden release of stored energy in the crust, leading to seismic vibrations and potential tsunamis.

1.12 Geological hazards — earthquakes

- Seismic waves travel outward from the earthquake's focus, with P-waves being the fastest, detected by the Global Seismographic Network to calculate earthquake location within minutes.
- Earthquake magnitude, measured using formulas based on shear waves (S-waves), determines earthquake strength, commonly assessed on the Richter Scale, which compares energy release logarithmically.
- In earthquake-prone regions such as Japan and Mexico City, safety measures include building design resilience, education programs, warning systems and immediate response teams to mitigate casualties and infrastructure damage.

1.13 Geological hazards — volcanic eruptions

- Volcanoes, often found along tectonic plate boundaries and hot spots, differ in lava type, with shield volcanoes producing low-silica lava and pyroclastic cones emitting high-silica, viscous lava prone to explosive eruptions.

- Eruptions pose various hazards, including threats to life, destruction of infrastructure and farmlands, contamination of the atmosphere with toxic gases and ash, and disruptions to air travel and tourism.
- Responses to volcanic eruptions involve evacuation from exclusion zones, wearing protective gear against ash and dust, and ensuring access to essential supplies such as water, food, and first aid.

1.14 Geomorphic hazards

- Geomorphic hazards, such as landslides, mudslides and avalanches, result from processes occurring on the Earth's surface due to factors such as geological movement and human activities altering topography.
- These hazards pose catastrophic consequences, including loss of life, injury, destruction of infrastructure and property, and reshaping of surface topography, especially in mountainous regions.
- Triggered by factors such as heavy rainfall, human activities such as deforestation and construction on steep slopes exacerbate the risks associated with geomorphic hazards, highlighting the importance of mitigation, prevention and adaptation strategies to minimise harm.

1.16.2 Key terms

asthenosphere the upper layer of the mantle, below the lithosphere, usually more than 100 km below the surface. It is where rock becomes molten and allows the solid tectonic plates to move over it.

atmospheric hazards a potentially damaging natural event generated in the troposphere, such as a severe storm, tropical cyclone (typhoons and hurricanes), tornado, blizzard and wind storm

biophysical environment both living (biotic) and non-living (abiotic) surroundings of an organism or population, made up of the elements of the atmosphere, hydrosphere, lithosphere and biosphere

climatological hazard a hazard that occurs due to the climatic conditions of an area, such as bushfires, droughts and heatwaves

demographic profile a detailed description of a group of people, including information such as their ages, genders, ethnicities, incomes, education levels, jobs, family sizes and where they live

epicentre the point on the Earth's surface directly above the focus when an earthquake has occurred

exposure the degree or likelihood of a place, person or thing being affected by a hazard, in terms of risk assessment

extreme weather event a weather event that is rare at a particular place and/or time of year, with unusual characteristics in terms of magnitude, location, timing or extent

faults large cracks in the Earth's crust, often associated with the boundaries of the Earth's tectonic plates

flow movement when rock, soil or sand mix with water and air and move downhill in a flow

focus where an earthquake rupture occurs in the crust or mantle. Seismic waves radiate away from the focus.

geological hazards a potentially damaging natural event occurring in the Earth's crust, such as a volcanic eruption, earthquake or tsunami

geomorphic related to the formation of the Earth's surface and its changes

geomorphic hazards a potentially damaging event on the Earth's surface — such as an avalanche, landslide or mudslide — that is often caused by a combination of natural and human processes

hazard zone an area that may be affected by a natural hazard; for example, areas vulnerable to flooding based on past events or areas likely to be affected by pyroclastic flows from a volcano

hydrological hazard an extreme event with a high-water component, such as flash flooding, cyclones, ice melt, storm surges and tsunamis

inter-tropical convergence zone (ITCZ) the zone near the Equator where trade winds of the northern and southern hemispheres meet. The intense heat, warm water and high humidity create what is an almost permanent band of low pressure. The monsoon trough seen on weather charts is part of the ITCZ.

landslide the large-scale movement of rock, debris and soil down a slope due to unstable conditions, which may be caused by heavy rainfall, an earthquake or a volcanic eruption. A landslide under the ocean can cause a tsunami.

liquefaction when saturated or partially saturated soil loses its firmness and displays the properties of a liquid, such as when an earthquake shakes and loosens wet soil in low-lying areas, and the soil loses rigidity and moves like fluid, covering things in its path

magma hot molten rock formed below or within the Earth's crust. It reaches the surface through volcanic or plate tectonic activity and becomes lava and eventually igneous rock.

magnitude a measure of size; for example, earthquakes are measured according to magnitude on the Richter Scale

mudflow when large amounts of suspended silt and soil move quickly down a slope. They tend to occur mostly on steep slopes but can happen anywhere ground is unstable due to loss of vegetation.

natural disaster a large natural event, such as a cyclone, flood, earthquake or landslide, that causes considerable loss of life, damage to property and infrastructure, and/or destroys sections of the environment

natural hazard an extreme event occurring either in the lithosphere or in the atmosphere. It can be highly destructive and cause considerable harm to living things and property. Examples include tropical cyclones, tornadoes, earthquakes and volcanoes.

P-waves also known as primary waves; high-frequency seismic waves that travel fastest and are measured first at a seismic station. P-waves can pass through solid rock and liquids.

permafrost soil, rock or sediment that remains frozen for two or more years, commonly found in polar regions and high mountain areas

polar vortex a large area of low pressure and cold air that typically resides over the polar regions during the winter months, but can occasionally shift southward, bringing frigid temperatures and winter weather to lower latitudes

pyroclastic clouds rapidly moving currents of hot air, gases and ash that run from the crater down the sides of a volcano. They are extremely lethal due to their high speed and lack of sound.

pyroclastic cones steep conical volcanic cones built by a combination of lava flows and cinder/ash from pyroclastic eruptions. They can form quickly and remain active for long periods.

risk the potential for something to go wrong. This is a subjective assessment about actions that may be predictable or unforeseen.

risk management strategies and actions to reduce or mitigate risk based on the known consequences of encountering a hazard

sea floor spreading the divergence of two oceanic crust plates

seismic waves waves of energy travelling away from an earthquake; are like huge vibrations and may travel through the Earth's mantle and crust or along the surface

slope failure when the pull of gravity causes a hill or mountain slope to collapse

subduction a geological process where two tectonic plates collide at convergent boundaries

subduction zones the areas of the mantle in which convergent plates collide. Under the ocean, these areas are called trenches.

subsidence when part of the land sinks or collapses

tectonic plates slow-moving plates that make up the Earth's crust. Volcanoes and earthquakes often occur at the edges of plates.

toxicological related to the negative impacts of chemical substances

vulnerability the degree of risk faced by a place, person or thing, based on an approaching hazard's potential impact, the place's degree of preparedness, and the resources available to respond

wildfire an uncontrolled fire that spreads through vegetation

wind shear a sudden change in wind speed and/or direction over a relatively short distance in the atmosphere, generally due to altitude

KEY QUESTIONS

1. What is a natural hazard?
2. Where do natural hazard zones occur and why?
3. What are atmospheric, geomorphic and geological hazards?
4. What factors affect the severity of impact of a natural hazard?
5. What factors affect a community's response to a natural hazard?
6. What factors affect a community's vulnerability to the risks of a natural hazard?
7. How are people in developed and developing communities affected differently by natural hazards?
8. How do people in developed and developing communities respond to natural hazards?

1.16.3 Exam questions

▶ Question 1 (4 marks)

Source: QCE 2019, Senior Geography Unit 1 Sample Assessment, Q.1; © QCAA

Explain the geographical processes that result in earthquakes and the secondary hazard of tsunami that contribute to the hazard zones shown in figure 1.78

FIGURE 1.78 Quake epicentres: 358 214 events, 1963 – 1998

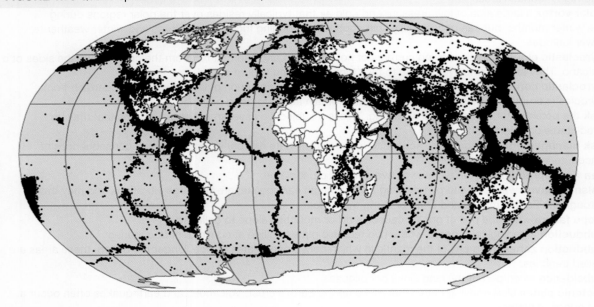

Source: By NASA, DTAM project team — http://denali.gsfc.nasa.gov/dtam/seismic/, Public Domain, https://commons.wikimedia.org/w/index.php?curid=35429.

In your response you should:
- describe the general pattern of hazard zones
- identify specific hazard zones for earthquakes and tsunamis and explain the processes for these locations.

Answer in no more than 150 words.

Question 2 (4 marks)

Explain how residents might respond to the common impacts of the natural hazard shown in figure 1.79.

FIGURE 1.79 Severe tropical cyclone Olga, off the coast of Western Australia.

1.16 Section II — Extended response question

Question 3 (16 marks)

In a written response of approximately 450–600 words, respond to the following:
- Analyse and interpret the data presented in the stimulus material (figures 1.80 to 1.82) and explain why many parts of Australia are now vulnerable to the effects of wildfire hazard.
- Comment on the impacts of the hazard on residents, emergency crews, property, crops and wildlife, and assess whether the effects of climate change are increasing exposure and expanding the hazard zone for these events.

FIGURE 1.80 Estimated changes in days of high to catastrophic fire danger

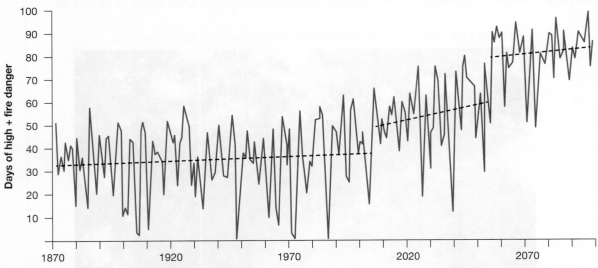

Source: JJones, RN, Young, CK, Handmer, J, Keating, A, Mekala, GD & Sheehan, P 2013, Valuing adaptation under rapid change, National Climate Change Adaptation Research Facility, Gold Coast, 184 pp.

FIGURE 1.81 Forest Fire Danger Index

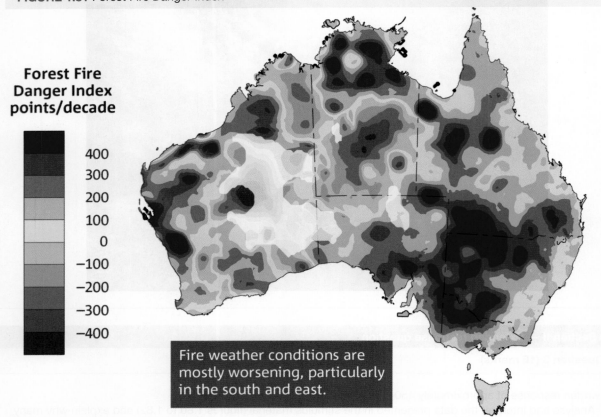

Source: © Commonwealth of Australia 2024, Bureau of Meteorology. State of the Climate 2018. Retrieved from http://www.bom.gov.au/state-of-the-climate/2018/australias-changing-climate.shtml.

FIGURE 1.82 The frequency of extreme heat events is increasing

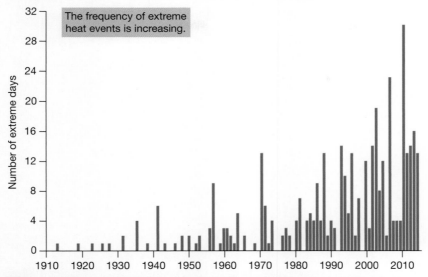

The frequency of extreme heat events is increasing.

Source: © Commonwealth of Australia 2024, Bureau of Meteorology. State of the Climate 2018. Retrieved from http://www.bom.gov.au/state-of-the-climate/2018/australias-changing-climate.shtml.

Sample responses are available in your digital formats.

2 Ecological hazard zones

SUBJECT MATTER

In chapter 2, students:

- **explain** the geographical processes that result in ecological hazards and hazard zones
- **recognise** the spatial patterns of hazard zones and the implications for people and environments
- **investigate** natural and anthropogenic factors to **identify** why some places are more vulnerable to ecological hazards than other places.

Students conduct a case study of a selected ecological hazard to:

- **understand** the factors that contribute to the spread of the hazard and the resulting primary, secondary and tertiary impacts on communities
- **propose** action for managing the impacts of a selected ecological hazard
- **understand** that ecological hazards evident in the world today pose significant management challenges.

LESSON SEQUENCE

Fully worked solutions for this topic are available in the Resources section at www.jacplus.com.au.

LESSON
2.1 Overview

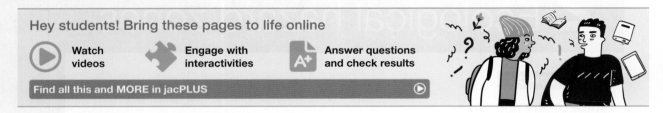
2.1.1 Introduction

The World Health Organization defines an **ecological hazard** as an interaction between living organisms or between living organisms and their environment that could have a negative effect. Ecological hazards are substances or activities that place people, habitats or an environment at risk of illness, injury, damage or disruption. This includes negative social and economic impact, as well as physical harm.

In this topic, you will examine different types of ecological hazards and the zones in which they are most likely to occur. This includes examining the different factors that affect the onset and severity of a specific ecological hazard, and the potential ways of managing the different challenges it presents.

You will also explore both biological and anthropogenic (human-generated) causes for hazards, and consider why some hazards pose more of a threat than others. This involves analysing vulnerability data and explaining strategies for hazard preparedness, mitigation and response.

As part of this unit of work, you will examine case studies of ecological hazards, make evidence-based observations about the impacts of specific hazards, and suggest effective strategies to reduce risk.

ecological hazard an interaction between living organisms or between living organisms and their environment that could have a negative effect

FIGURE 2.1 Oil spill clean-up operations in Thailand

2.1.2 Syllabus links

KEY QUESTIONS

1. What is an ecological hazard?
2. Where do ecological hazards occur and why?
3. What biological and anthropogenic factors influence a community's vulnerability to specific ecological hazards?
4. What factors affect the severity of impacts of an ecological hazard?
5. What actions can be taken to reduce risk?
6. What factors affect a community's response to an ecological hazard?
7. How are people in developed and developing communities affected differently by ecological hazards?
8. Why might different communities seek different solutions to the challenges of ecological hazards?
9. How does climate change affect the impact of some ecological hazards?

LESSON
2.2 Ecological hazards

LEARNING INTENTION

By the end of this lesson you should be able to explain how ecological hazards are the result of biological and anthropogenic processes.

Source: Adapted from Geography General Senior Syllabus 2024 © State of Queensland (QCAA) 2024; licensed under CC BY 4.0.

2.2.1 Types of ecological hazards

Ecology is a field of study that deals with the relationships of organisms to one another and to their non-living surroundings.

In our everyday lives, at times we are exposed to ecological hazards, either intentionally or by accident. For example, you may have a conversation with a sick person who coughs and sneezes frequently, or a veterinary surgeon may be attending to a sick fruit bat that has lyssavirus. These circumstances have an element of risk of the spread of pathogens — that is, organisms such as bacteria or viruses that can cause disease. Because these are living things that can be risky for people to have contact with, they are considered to be **hazards**.

This means that ecological hazards are sources of danger or harm to people and environments. Ecological hazards are potentially dangerous phenomena, substances, human activities or conditions that may cause the following:

- loss of life
- injury or other health impacts
- property damage
- loss of livelihoods and services
- social and economic disruption
- environmental damage.

hazard something that has the potential to cause harm. It may be obvious (for example, a flooded section of road) or not obvious (for example, a damaged hidden electrical wire).

biological processes that are vital for organisms to live, for example, plants require the process of photosynthesis to survive

anthropogenic processes processes that involve human activity; for example, the burning of fossil fuels to produce electricity

Ecological hazards are the result of **biological** and **anthropogenic processes**. Biological processes are processes that are vital for organisms to live; for example, the process of photosynthesis. Anthropogenic processes involve human activity; for example, the burning of fossil fuels to produce electricity.

Ecological hazards involve human interactions with living organisms and the environment, or interactions between living organisms that have the potential to adversely affect the social and economic wellbeing of people. Ecological hazards might also include substances or activities that pose a threat to a habitat or an environment, such as acid rain caused by air pollution, toxic chemicals released by industry into waterways or high levels of radiation caused by nuclear accidents.

FIGURE 2.2 A journalist checks radiation levels in the Fukushima region in Japan, where a tsunami led to a nuclear power plant meltdown in 2011.

Diseases

Diseases, both **infectious diseases** such as influenza and **vector-borne diseases** such as malaria (see section 2.10), are one type of ecological hazard. In this case, the hazard is a result of human exposure to living organisms, such as bacteria, viruses and parasites, and affects human health and wellbeing. While infectious diseases are an outcome of biological processes, human activities may contribute to their outbreak, spread and severity. Some examples of hazardous diseases are shown in table 2.1.

TABLE 2.1 Examples of hazardous diseases

Infectious diseases	Vector-borne diseases
Influenza	Malaria
Measles	Dengue fever
Cholera	Ebola
Hepatitis	Yellow fever
HIV/AIDS	Lyme disease

Plant and animal invasions

Other ecological hazards may have an impact on the physical environment as well as on people. Such hazards include environmental **plant** and **animal invasions** (see table 2.2 and lesson 2.4), and **pollutants** of the lithosphere, atmosphere and hydrosphere (see table 2.3 and lesson 2.6). Often, these ecological hazards are a result of people's activities; that is, of anthropogenic processes. People are responsible for many cases of plant and animal invasions, for example, where the invasive species has been deliberately introduced, as were rabbits and the cane toad in Australia.

TABLE 2.2 Examples of invasive plants and animals in Australia

Invasive plants	Invasive animals
Lantana	Cane toad
Groundsel	Fox
Rubber vine	Rabbit
Prickly pear	Carp
Paterson's curse	Feral pig

infectious diseases contagious or communicable diseases that are spread by being passed from one person to another

vector-borne diseases diseases that are carried by organisms, such as mosquitos or fleas, that are capable of transmitting disease-producing bacteria, viruses and parasites from one person to another

plant invasions the introduction of a non-indigenous plant to an area that negatively impacts the environment

animal invasions the introduction of a non-indigenous animal to an area that negatively impacts the environment

pollutants substances introduced into the environment that are potentially harmful to human health and the natural environment

Pollution

Pollution of the biophysical environment is frequently a result of human activities. For example, motor vehicles are a common source of air pollution in large cities, rising carbon dioxide levels in the atmosphere are contributing to climate change, and concern about the amount of plastic waste washing into our waterways and into the world's oceans is increasing. As the world's human population has grown and spread, and as economies have expanded, so too has the pollution associated with people's activities.

FIGURE 2.3 Pollution near the Panama Canal in South America

TABLE 2.3 Examples of environmental pollutants

Lithosphere	Atmosphere	Hydrosphere	Biosphere
Salt	Carbon dioxide	Salt	Herbicides
Radiation spills	Sulphur dioxide	Plastics	Pesticides
Oil spills	Nitrogen oxides	Heavy metals	Plastics
Farm chemicals	Particulates	Fertilisers	Oil spills

2.2.2 Ecological hazard zones

The specific areas, spaces or places at risk of experiencing an ecological hazard are known as ecological hazard zones. At a local scale, even schools can be at risk of becoming an ecological hazard zone. A school may be severely affected by influenza outbreaks during the 'flu season', for example, and be forced to close when the risk of the spread of disease becomes too great.

EXAMPLE: Cholera in Yemen (2016–2021)

Yemen, a country already devastated by years of civil war, became an ecological hazard zone due to the outbreak of cholera between 2016 and 2021. The conflict had severely damaged infrastructure, including water and sanitation systems. This created the perfect conditions for cholera, an acute diarrhoeal disease caused by the ingestion of food or water contaminated with the bacterium vibrio cholerae. By the end of 2020, over 2.5 million cases of the disease had been reported, making it one of the largest cholera outbreaks in recent history.

The collapse of healthcare services, combined with widespread malnutrition and displacement of millions of people, exacerbated the situation. The rapid spread of cholera was fuelled by the lack of access to clean drinking water, inadequate sanitation facilities, and the breakdown of health services due to ongoing conflict.

However, an extensive vaccination program that began in 2021 helped to significantly reduce the outbreak.

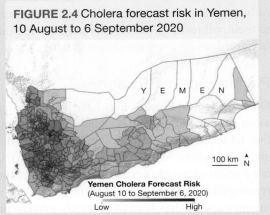

FIGURE 2.4 Cholera forecast risk in Yemen, 10 August to 6 September 2020

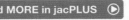

Source: NASA Earth Observatory map by Lauren Dauphin using data from Antar Jutla and Moiz Usmani, University of Florida.

2.2 Exercise

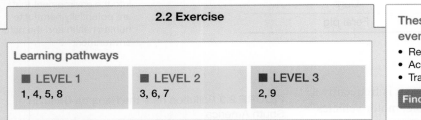

2.2 Exercise

Learning pathways

■ LEVEL 1	■ LEVEL 2	■ LEVEL 3
1, 4, 5, 8	3, 6, 7	2, 9

These questions are even better in jacPLUS!
- Receive immediate feedback
- Access sample responses
- Track results and progress

Find all this and MORE in jacPLUS ▶

Explain and comprehend

1. **List** two examples of ecological hazards mentioned in the lesson.
2. **Discuss** which ecological hazards present a risk where you live. Can you **identify** any patterns for when and where these hazards occur? (For example, is there a typical flu season? Is a specific weed more likely to affect crops at a certain time of year?)
3. **Explain** why living organisms that can cause disease are considered ecological hazards.
4. **Identify** the impact of ecological hazards on human health and the environment.
5. **Suggest** an ecological hazard that might affect people living in developing communities that might not be as significant in developed communities.

Analyse and apply

6. **Describe** the difference between biological and anthropogenic processes, providing examples of each.
7. **Compare** and contrast the roles of biological and anthropogenic processes in the emergence of ecological hazards.
8. **Identify** the factors that might affect the way different communities manage the following ecological hazards.
   a. The introduction of a plant disease that damages important crops
   b. The accumulation of plastics on beaches
   c. The spread of a dangerous new strain of influenza (the flu)

LESSON
2.3 The impacts of ecological hazards

LEARNING INTENTION

By the end of this lesson you should be able to:
- explain how the severity of the impacts of ecological hazards is influenced by different factors
- explain how climate change may affect the severity and incidence of some ecological hazards, and increase risk.

Source: Adapted from Geography General Senior Syllabus 2024 © State of Queensland (QCAA) 2024; licensed under CC BY 4.0.

2.3.1 Factors affecting the severity of impact

The impacts of ecological hazards aren't always the same. A range of different factors influence the severity of an ecological hazard's impact.

Speed of onset

Hazards, both natural and ecological, can occur rapidly or develop over a longer period — that is, they may be rapid-onset or slow-onset hazards. The collapse of the Samarco iron ore mine tailings dam in Brazil in November 2015 is an example of a rapid-onset hazard. Heavy-metal-laden mud from the dam inundated the surrounding area, and contaminated community water supplies and the Doce River basin.

A close link often exists between the speed of onset of a hazard and its predictability. Rapid-onset hazards may be difficult to predict. Communities are often more vulnerable to the impacts of hazards when they have little or no time to prepare. Because slow-onset hazards tend to be much more predictable, people generally have more time to prepare for and manage the impact of the hazard; for example, by evacuating if at risk.

Magnitude

Magnitude is a measure of the strength or extent of a hazard. A similar type of ecologically hazardous event may be more or less severe, depending on its magnitude. An oil spill from a recreational boat, for example, is of much smaller magnitude and, therefore, less severe than an oil spill from a large tanker.

FIGURE 2.5 When a dam broke in the Samarco iron ore mine, 19 people were killed and nearby villages were suddenly flooded.

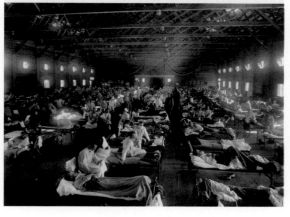

FIGURE 2.6 The 1918–20 influenza pandemic is believed to have killed up to 50 million people.

Frequency

Frequency refers to how often a hazardous event occurs; it is the return interval of hazards of certain sizes. In most cases, smaller hazardous events tend to be more frequent than large-scale events and are usually less severe. For example, small outbreaks of influenza tend to occur frequently. However, an outbreak of a particularly virulent type of the flu is much rarer and much more severe in its impact. For example, the 1918–20 influenza pandemic killed more people than World War I because experts weren't used to treating an outbreak of this type.

Duration

Duration is the length of time that a hazard lasts. Usually, the longer the hazard lasts the greater the impact is likely to be, although this is not always the case. Some infectious diseases may cause their most serious effects early in an outbreak when the most vulnerable people are at greatest risk and strategies to prevent spread have not been implemented yet. This was the case during the early period of the COVID-19 pandemic, before any vaccines had been developed.

Temporal spacing

Temporal spacing refers to the sequencing and seasonality of events — that is, whether the event is random or regular. Many diseases, such as influenza, tend to be seasonal, while others, such as measles, are much more irregular.

Mobility and location

The mobility and areal extent (the area affected) of a hazard can also have an impact on its severity. For example, a disease outbreak affecting just a single, isolated village is much less severe than an outbreak of disease in a large city or over several cities or countries.

Some invasive animal species, such as rabbits, are very mobile and adaptable, and can cover large distances in short periods of time. Others may have limited ability to move to different locations because they require very specific conditions to survive.

FIGURE 2.7 The European rabbit is an invasive species in Australia, capable of spreading rapidly.

Climate change

Climate change may affect the severity and incidence of some ecological hazards and increase risk. For example, climate change may increase the range of habitats available to disease-carrying insects such as mosquitoes and, therefore, increase the areal extent and the risk of outbreaks of mosquito-borne diseases.

Climate change might also increase the range of invasive plants and animals and their associated risks. It might also be possible for the incidence of some types of pollutants to increase with a changing climate.

FIGURE 2.8 Climate change can change the distribution of mosquitoes.

2.3.2 Risk management

Risk is the chance that any hazard will cause harm to people or the environment. It is the probability of harmful consequences or expected losses (such as deaths, injuries, damage to property or damage to the environment) from ecological (and natural) hazards. Risk is closely associated with the **vulnerability** of particular people and places to hazards. Topic 1 explained how the risk posed by a hazard might be mitigated by reducing the level of exposure and/or vulnerability.

EXAMPLE: Vulnerability to cholera

Some places and people are more vulnerable to ecological hazards than others. For example, cholera, a water-borne infectious disease, rarely affects people in wealthier countries where treated water is the norm.

People in poorer countries, where sanitation and water treatment may be inadequate, are much more vulnerable to outbreaks of cholera.

FIGURE 2.9 Access to clean water is crucial in preventing cholera outbreaks.

Managing the impacts of ecological hazards involves actions that aim to prevent or reduce the damage or harm caused by the hazard.

These actions usually include the following:
- identifying the hazard
- if possible, putting in place measures to prevent the hazard
- where prevention is not possible, preparing for the hazard, including the use of hazard mitigation to control the risk.

Risk management of ecological hazards will depend on the nature of the hazard. Risk management for an invasive plant will be different from managing the risk of a disease outbreak. It may be possible to detect early outbreaks of potentially invasive weeds and to eradicate them before control mechanisms are needed. An outbreak of influenza may be impossible to prevent, although the use of vaccinations may mitigate the risk.

risk the potential for something to go wrong. This is a subjective assessment about actions that may be predictable or unforeseen.

vulnerability the degree of risk faced by a place, person or thing, based on an approaching hazard's potential impact, the place's degree of preparedness, and resources available to respond

CASE STUDY: Water extraction from the Murray–Darling Basin

Quick facts: Murray–Darling Basin
- **Location** Four Australian states and one territory (South Australia, Victoria, New South Wales, Queensland and the ACT)
- **Size** 1 million km²
- **Provides drinking water for** over 3 million people
- **Supplies water for** over 40% of Australia's agricultural produce

FIGURE 2.10 The Murray River provides water for a significant proportion of Australia's agricultural output.

Large amounts of water are extracted from rivers across the Murray–Darling Basin for irrigation of crops and water supplies for the basin's communities. Water is also lost through evaporation, seepage and leakage. During periods of drought, water levels are further reduced. Climate change will also likely cause the basin to be drier overall and increase the likelihood of extremes in drought and flooding conditions.

Over time, water extraction has threatened the basin's wetland and river ecosystems, and increased the risk of salinity and toxic blue-green algal bloom hazards. In the case of salinity, communities and environments in the lower section of the Murray River are more vulnerable, because salinity levels increase downstream. Climatic conditions worsen the risk from water extraction of blue-green algae: the algae mostly bloom during droughts, when water levels are lower and temperatures higher. Increased concentrations of nitrogen fertilisers in the water, the result of run-off from farmland, also encourage algal growth. During 1991, drought conditions led to the basin's most extensive and severe blue-green algae hazard event, when the Darling River experienced a 1000-kilometre-long outbreak, leading to the declaration by the NSW government of a state of emergency.

Although the degradation of the natural environment and the hazard risks of salinity and blue-green algae in the Murray–Darling Basin are widely recognised, mitigating the risk has been difficult. This is because of the many competing interests in the basin: First Nations people are the traditional custodians of the basin's water and land; irrigation farmers rely on water extraction for their livelihood; urban communities rely on the rivers for their water supply; ecologists remind us that increased water flows are essential to sustain the basin's rivers and wetlands. Often, the difficulty of managing these competing interests is increased by differences in viewpoint among the five governments responsible for different parts of the basin.

Over time, mitigation measures such as improved farming and irrigation practices to reduce salt movement into rivers, and the use of interception schemes have greatly reduced the risks associated with salinity. Because blue-green algae occur naturally in rivers and lakes, and blooms are usually related to climate conditions, mitigation of this hazard is more difficult. Releasing water may help flush out the algae, but once blooms begin, they are difficult to stop, so managing risk involves monitoring of algal concentrations and awareness of the risk when water is extracted.

Limiting water extraction and increasing the amount of water returned to the basin's rivers is critical to the health and sustainability of the river system. The Murray–Darling Basin Plan requires that 3200 gigalitres of water be returned to basin rivers. To help meet this target, limits are in place on how much water can be taken from the Murray–Darling rivers (termed 'sustainable diversion limits'). Upgraded and more efficient irrigation systems on farms, and the reduction of water loss from irrigation channels help. The Australian government has also introduced a water buy-back scheme to purchase water and return it to rivers. By February 2024, 1247 gigalitres of water had been recovered through this scheme.

SKILLS ACTIVITY

Use the blank map of Australia included in the Resources tab to complete this activity.
1. Add the Murray River and the Darling River to the map, as well as both of their basins.
2. Identify the five largest population centres within the basin and mark them on your map.
3. Ensure that your map correctly uses cartographic principles (consider BOLTSS).

on Resources

Digital doc	Blank map of Australia (doc-42349)
Video eLesson	SkillBuilder: Reading and describing basic choropleth maps (eles-1706)
Interactivity	SkillBuilder: Reading and describing basic choropleth maps (int-3286)
Weblinks	The Murray–Darling Basin Plan
	The science behind the Murray–Darling Basin Plan
	Pumped: Who's benefitting from the billions spent on the Murray–Darling? (2017)
	Travel the length of the Murray–Darling system

2.3 Exercise

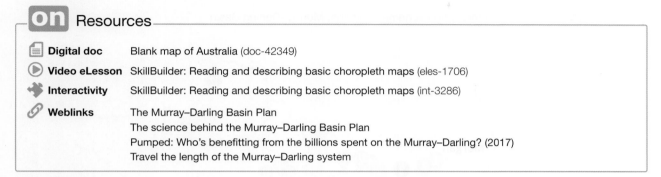

2.3 Exercise

Learning pathways

■ LEVEL 1	■ LEVEL 2	■ LEVEL 3
1, 2	3, 4, 5	6, 7

These questions are even better in jacPLUS!
- Receive immediate feedback
- Access sample responses
- Track results and progress

Find all this and MORE in jacPLUS ⊙

Explain and comprehend

1. **List** three human processes or actions that might cause an outbreak of influenza to spread.
2. **Identify** which states are at risk from the hazards of excess water extraction from the Murray–Darling Basin.
3. **Describe** a strategy that the Australian government has put in place to prevent unsustainable water extraction from the Murray–Darling Basin.

Analyse and apply

4. Complete a table similar to table 2.4 to **identify** the different factors that influence the severity of the impact of salinity and blue-green algae in the Murray–Darling Basin.

TABLE 2.4 Factors affecting the severity of salinity and blue-green algae in the Murray–Darling Basin

Factor	Salinity	Blue-green algae
Speed of onset		
Magnitude		
Frequency		
Duration		
Temporal spacing		
Mobility and location		
Climate change		

Propose and communicate

5. Salt interception schemes involve pumps that divert salty groundwater away from rivers and into evaporating basins. In the Murray–Darling Basin, 14 of these have been installed at locations where water in the soil (groundwater) is particularly high in salt. These can be seen in figure 2.11.
   a. **Describe** the pattern of salt interception scheme locations in the Murray–Darling Basin.
   b. **Suggest** possible reasons for the pattern identified.

FIGURE 2.11 Salt interception scheme locations, Murray–Darling Basin

Source: Murray–Darling Basin Authority. 2023. [Managing salinity, https://www.mdba.gov.au/climate-and-river-health/water-quality/salinity/managing-salinity]. © Australian Government (Murray–Darling Basin Authority)

6. a. **Explain** the potential impacts for Australia if the water in the Murray–Darling Basin dries up or the water quality becomes unsuitable for its current uses.
   b. **Suggest** what impacts this might have on Australia as a country socially, economically, politically and environmentally.

7. **Discuss** how the economy of a place could affect how vulnerable it is to an ecological hazard such as a disease outbreak.

Sample responses are available in your digital formats.

LESSON
2.4 Plant and animal invasions

2.4.1 Types of plant and animal hazards

An **invasive plant** or **animal** is defined by the Australian Department of Climate Change, Energy, the Environment and Water as 'a species occurring, as a result of human activities, beyond its accepted normal distribution and which threatens valued environmental, agricultural or other social resources by the damage it causes'.

Invasive plants and animals have three characteristics.
- They are not **indigenous** (native) to the area in which they are found. (They are exotic or alien species.)
- They are a consequence of anthropogenic activities. (They have been introduced, either accidentally or deliberately, through the actions of people.)
- They pose a threat to the environment and to people and their activities; that is, they are hazards.

Since British settlement, many exotic plants and animals have been introduced to Australia, either deliberately or accidentally. Some were brought for economic reasons; for example, as crops or farm and work animals. Others, such as cats, were brought in as pets. Many garden plants and animals were introduced by acclimatisation societies in the 1800s to make Australia more like Europe. Domesticated species, such as lantana and donkeys, also became wild. These are known as feral plants or animals.

Unfortunately, many of Australia's introduced species have caused extensive environmental and economic harm. Foxes, feral cats, rabbits and feral pigs are among the worst of Australia's invasive animal species, while around half of our introduced plants have invaded native vegetation and around a quarter are regarded as, or have the potential to become, serious environmental weeds. These include the rubber vine, lantana, groundsel and water hyacinth (see table 2.5 and figure 2.13). Scientists have identified 32 weeds of national significance, based on an assessment of their invasiveness, potential for spread and environmental, social and economic impact.

FIGURE 2.12 Cane toads were introduced into Australia in an effort to control insect pests that attacked sugar cane.

invasive plant a non-indigenous plant species that has been introduced to a specific area by people (either intentionally or accidentally) and has multiplied to an extent that it threatens to damage or is damaging the economic, environmental or social value of a place

invasive animal a non-indigenous animal species that has been introduced to a specific area by people (either intentionally or accidentally) and has multiplied to an extent that it threatens to damage or is damaging the economic, environmental or social value of a place

indigenous something that is native to or originates from a specific place; for example, kangaroos are indigenous to Australia. Note that the word often has a different meaning when capitalised. The term 'Indigenous' refers to First Nations Peoples of Australia of Aboriginal or Torres Strait Islander descent.

TABLE 2.5 Some of Australia's most noxious weeds

Alligator weed	European blackberry	Fireweed
Gamba grass	Mimosa	Athel pine
Prickly pear	African boxthorn	Prickly acacia
Bitou bush	Chilean needle grass	Mesquite

FIGURE 2.13 Invasive water hyacinth.

Invasive plants and animals will have many or all the following characteristics.
- They grow and mature rapidly, producing large numbers of seeds or offspring.
- They are highly successful at spreading to and colonising new areas.
- They can thrive in different types of habitats.
- They can outcompete native species and have few or no natural enemies.
- They are very costly or difficult to remove or control.

2.4.2 The impact of invasive plants and animals

Invasive plants and animals have environmental, economic and social impacts.

Environmental impacts

Perhaps the most significant impact of invasive species is on Australia's biodiversity. As one of 17 'megadiverse' countries, Australia is one of the most biodiverse places in the world. However, this biodiversity is under threat. Around 1700 species of plants and animals are listed by the Australian government as at risk of extinction. Already, 30 native mammals have become extinct since European settlement.

FIGURE 2.14 Major invasive species and number of protected terrestrial species threatened

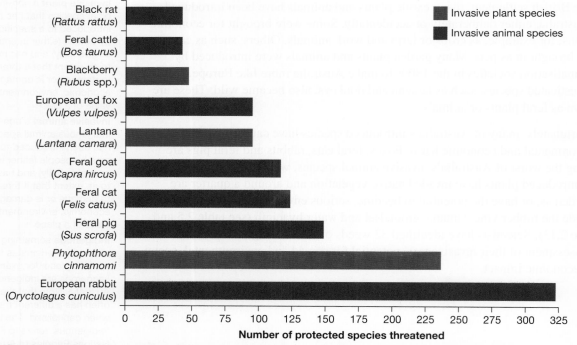

Source: Based on data from Kearney SG, Carwardine J, Reside AE, Fisher DO, Maron M, Doherty TS, Legge S, Silcock J, Woinarski JCZ, Garnett ST, Wintle BA & Watson JEM (2018). 'The threats to Australia's imperilled species and implications for a national conservation response'. *Pacific Conservation Biology* 25(3):231–244. Figure 3, pp. 235.

While land clearing and fire are in part responsible for the extinction threat, the most significant causes involve invasive species: predation by feral cats and foxes; habitat destruction by feral herbivores such as pigs, goats and rabbits; and the spread of invasive weeds. Invasive species threaten natural ecosystems and biodiversity. Frequently, they crowd out and replace natural species because they can outcompete, prey on or poison native species or may carry diseases. Invasive species also alter habitats and cause land and water degradation, making it difficult for native species to survive.

Economic impacts

Invasive plants and animals have both direct and indirect economic costs. The direct costs are due to monitoring and controlling the pest species, including the physical removal of pests, baiting, poisoning, fencing and research. Indirect costs occur when agricultural production is reduced through damage done to fences and crops, degradation of soils and water, herbivore competition with farm animals and reduction in fish stock in marine environments.

FIGURE 2.15 Red fire ants are a prolific invasive species that severely impact natural habitats, farms and livestock.

Social impacts

The social impact of invasive species can include the loss of amenity in national parks, impact on Australia's First Nations Peoples and their way of life and traditional practices, health impacts and reduction in recreational fishing. Ecological hazards such as mouse plagues can also have a considerable social impact. For farmers who have to manage the impact on their crops, mice can take a significant toll on their wellbeing by infesting and reducing their yield.

2.4.3 Invasive plant and animal hazard zones

As figure 2.16 shows, every continent other than Antarctica is currently threatened by plant and animal invasions. Australia and New Zealand are especially vulnerable to the risk of invasion. Table 2.6 shows that island countries are over-represented among countries with the highest numbers of invasive species — six out of the top ten countries are islands. A recent study of the world's alien species hotspots identified the Hawaiian Islands, the North Island of New Zealand and the Lesser Sunda Islands of Indonesia as the places with the highest numbers of established alien species. In fact, about half of New Zealand's plant life consists of invasive species.

FIGURE 2.16 The global spread of invasive species

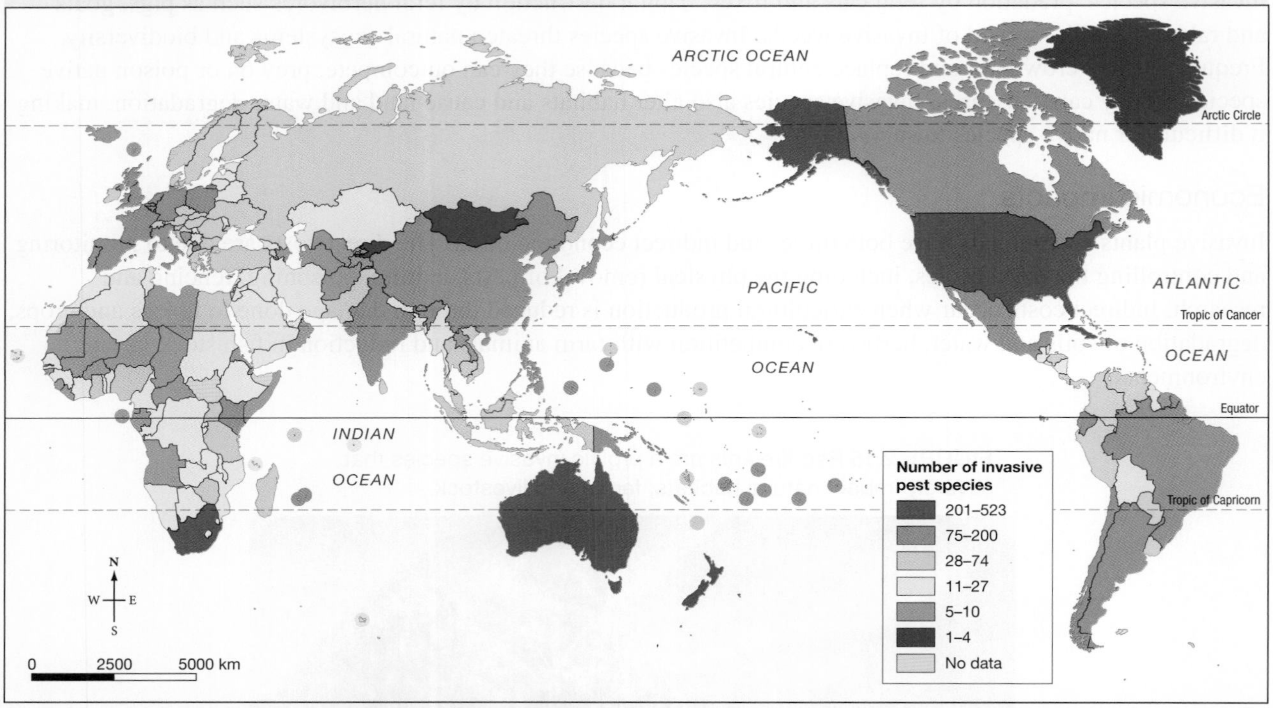

Number of invasive pest species
201–523
75–200
28–74
11–27
5–10
1–4
No data

Source: 'Mapping the global state of invasive alien species: patterns of invasion and policy responses' by Anna J. Turbelin, Bruce D. Malamud and Robert A. Francis, Global Ecology and Biogeography, (Global Ecol. Biogeogr.) (2017) 26, 78–92, Fig 1.

In addition, all ten countries in table 2.6 have been European colonies. European settlers — in these cases, the British, French, Spanish and Dutch — introduced alien invasive species to the countries they colonised, sometimes deliberately, such as the rabbit in Australia, and sometimes accidentally, such as the mouse.

TABLE 2.6 Countries with the highest number of invasive alien species

Country	Number of species	Density of species (species per 100 000 km^2)
1. United States of America	523	5.7
2. New Zealand	329	124.9
3. Australia	322	4.2
4. Cuba	318	298.8
5. South Africa	208	17.1
6. French Polynesia	190	5191.3
7. New Caledonia	183	1001.1
8. Reunion	173	6889.7
9. Fiji	167	914.1
10. Canada	166	1.8

Source: 'Mapping the global state of invasive alien species: patterns of invasion and policy responses' by Anna J. Turbelin, Bruce D. Malamud and Robert A. Francis, Global Ecology and Biogeography, (Global Ecol. Biogeogr.) (2017) 26, 78–92, Fig 1.

However, invasive plant and animal hazard zones are often not countrywide. Like most species, invasive species are usually adapted to particular physical environments in their place of origin. They pose the greatest threat to similar environments in the invaded country. Many of Australia's invasive plant species are tropical, so they mostly threaten the northern parts of Australia. Some invasive species are adapted to coastal environments, so predominantly threaten those environments. As the climate changes, the areas potentially susceptible to invasion also change.

SKILLS ACTIVITY

1. Use the map key in figure 2.16 to complete column 1 of the following table. Add the names of three or four countries at each interval to column 2 of the table.

Number of alien invasive species (range)	Country examples

2. Create a new table, based on table 2.6, which re-ranks countries according to the density of alien species.
   a. Describe how Australia's ranking changed compared with its ranking based on overall numbers.
   b. Make a list of factors that might influence the density of invasive species in these ten countries.
   c. Suggest which of these factors might apply specifically to Australia.
   d. Digitally construct a multiple column or bar graph to illustrate the economic costs and total costs of the invasive species listed in table 2.6.

CASE STUDY: Containing and eradicating red fire ants in South East Queensland

Quick facts: Red fire ants
- **Species** *Solenopsis invicta*
- **Origin** South America
- **First sighting in Australia** Port of Brisbane, 2001
- **Potential cost if not eradicated** $2 billion per year

Red fire ants are a highly invasive and aggressive species of ant. They possess a venomous sting, prey on vertebrates and invertebrates, eat plants, and attack en masse. Since they first arrived in Australia in the 1990s and were detected in 2001, red fire ants have spread across South East Queensland, including to Minjerribah (North Stradbroke Island) in 2023. In December 2023, fire ants were for the first time detected in northern NSW, in South Murwillumbah, and in January 2024 were discovered nearly 80 kilometres further south at Wardell.

FIGURE 2.17 A red fire ant

More worryingly, fire ants were discovered in April 2024 near Oakey on the Darling Downs. From here, they could possibly spread south along watercourses and into the Murray–Darling Basin, posing a risk to larger areas of Australia.

Fire ant invasions have environmental, economic and social impacts (see table 2.7).

To manage the impact of fire ants in Australia, a National Red Imported Fire Ant Eradication Program was launched in 2001. The first aim of this program was to contain fire ant populations to South East Queensland, and to prevent their spread to other parts of Australia. In the longer term, the aim is to reduce infestations and, finally, to eradicate red fire ants from South East Queensland. The eradication of fire ants involves surveillance and the detection of fire ant nests, destroying any fire ant colonies found through baiting and use of insecticides, and then ongoing monitoring of the sites. The containment of fire ants involves restrictions on the movement of materials such as soil, mulch, fodder and pot plants that may carry fire ants. If containment of fire ants fails, fire ants will likely continue to spread, as they have done to NSW in 2023–2024.

TABLE 2.7 Some examples of environmental, economic and social impacts of fire ants

Environmental impacts	Because of their aggressiveness and venomous sting, fire ants can have a devastating effect on lizards, frogs, mammals and birds, and may cause irreparable damage to ecosystems.

(continued)

TABLE 2.7 Some examples of environmental, economic and social impacts of fire ants *(continued)*

Economic impacts	As well as the enormous cost currently of detection and eradication programs, around $200–300 million a year, fire ants could potentially reduce agricultural output by 40% and cost Australia's economy $2 billion a year.
Social impacts	For people, the severe irritation and painful, burning stings may limit outdoor activities, such as picnics and sporting events if ovals, parks and playgrounds become infested.

on Resources

📄	**Digital doc**	Fire ant hazard risk assessment (doc-29164)
▶	**Video eLesson**	SkillBuilder: Creating and reading compound bar graphs (eles-1705)
🏹	**Interactivity**	SkillBuilder: Creating and reading compound bar graphs (int-3285)
🔗	**Weblinks**	Biosecurity Queensland Fire Ant Identification
		Queensland Government fire ant information
		Fire ants' aggression

2.4 Exercise

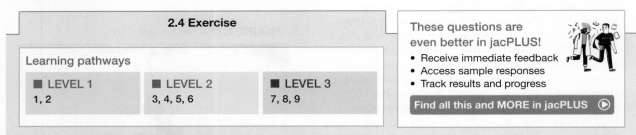

2.4 Exercise

Learning pathways

■ LEVEL 1	■ LEVEL 2	■ LEVEL 3
1, 2	3, 4, 5, 6	7, 8, 9

These questions are even better in jacPLUS!
- Receive immediate feedback
- Access sample responses
- Track results and progress

Find all this and MORE in jacPLUS ▶

Explain and comprehend

1. **Describe** the global pattern of alien invasive species numbers illustrated in figure 2.16. Refer to continents in your answer, and identify which continents have the most and least invasive species.
2. a. **Identify** the two countries that show the greatest difference to the overall pattern of their continent.
   b. **Suggest** features, trends or strategies that might have led to these countries having more or fewer invasive species.

Analyse and apply

3. **Suggest** reasons rabbits and feral pigs are responsible for the highest economic costs to Australia.
4. **Suggest** reasons for the high costs to Australia's natural environment of foxes and feral cats.
5. **Explain** why a particular invasive species might have a high social impact.
6. **Explain** why the social impact of invasive species might be so difficult to quantify.

Propose and communicate

7. **Propose** some advice that you could offer people and governments in areas most at risk of fire ants in South East Queensland.
8. **Suggest** what measures might be put in place to reduce the risk of further spread of fire ants in northern New South Wales.
9. In 2017, Australian state, territory and federal governments pledged $411.4 million over 10 years to fight fire ants in south East Queensland. **Evaluate** whether this cost is justified. Give reasons to support your decision.

Sample responses are available in your digital formats.

LESSON
2.5 Reducing the risk of plant and animal invasions

2.5.1 Prevention and control of invasive species

The best way to mitigate the risk of plant and animal invasions is prevention: keep the invaders out. In Australia, we now have strict biosecurity measures to reduce the risk of alien species being introduced. Anyone entering Australia from overseas is required to complete customs declaration forms. These declarations are part of our management of the risk of introduced species, and considerable penalties apply to people who illegally bring in banned products or do not declare them.

The Biosecurity Act (2015) requires people bringing goods into Australia to declare any prohibited or potentially prohibited items, such as food, animal products or plant materials. It is also prohibited to take some kinds of food into some areas of Australia, to prevent the spread of invasive species from one part of the country to another.

A second step in the management process involves the early control and eradication of any invasive species. It may be possible to completely eradicate invaders if they are detected early enough. If that proves impossible, early detection and control may be able to keep the numbers of the invaders at reasonably low levels. The ongoing attempt to keep invasive fire ants confined to small areas in South East Queensland is an example of this, although success is proving difficult and expensive (see the case study in lesson 2.4). Once invasive species are established, three control methods are possible: chemical, mechanical and biological.

FIGURE 2.18 Parts of Australia enforce quarantine regulations to prevent the spread of invasive species.

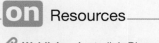

Resources

🔗 **Weblinks** Australia's Biosecurity Act
Interstate quarantine in Australia

Chemical control

Chemical control involves the use of herbicides and pesticides to kill invaders, as shown in figure 2.19. Chemicals can be an effective management technique, but may also have negative consequences.

These include:
- Chemicals can kill non-target species, including native species, as well as target species.
- Chemicals can be expensive.
- Pollution of soil and water can occur.
- Invasive species may develop resistance to the chemicals used.

Mechanical or physical control

Mechanical control involves the use of machinery and people to remove the invaders, as shown in figure 2.20. This is often effective in small areas, and can reduce local invasive populations. Physical control includes techniques such as trapping, shooting and fencing to control numbers of invasive animal species and weeding of invasive plant species.

Biological control

Biological control involves the use of an invasive species' natural enemy or a disease. Prickly pear cactus was brought under control in Australia with an introduced cactoblastis moth, whose caterpillar feeds on the cactus. (These caterpillar are shown in figure 2.21.) Rabbits have been controlled through the use of myxomatosis and calici viruses. Other biological control methods involve genetic engineering and breeding intervention programs.

Management processes

Figure 2.22 provides an overview of the costs associated with managing invasive species over time. This management process involves the following elements.
- Prevention (stopping the species from invading) is the most cost-effective way of dealing with invasive plants and animals.
- Eradication is the complete removal or killing of a species.
- Containment refers to the measures taken to prevent the further spread of a pest species.
- Asset-based protection is the control of a pest species based on the threat it poses to environmental and human assets in an area.

As any invasion progresses, management techniques become more expensive, and the benefits compared with the management costs decrease. It is important, then, that Australia continues with its program of prevention and early eradication of any new invasive species.

FIGURE 2.19 Chemical removal of invasive weeds

FIGURE 2.20 Mechanical removal of aquatic weeds

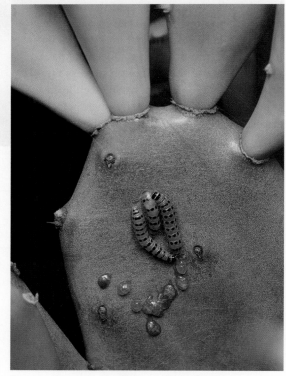

FIGURE 2.21 Caterpillar of the Argentinian moth *Cactoblastis cactorum* feeding on prickly pear

Source: Wayne Lawler/Science Photo Library

FIGURE 2.22 Cost curve for management of invasive species

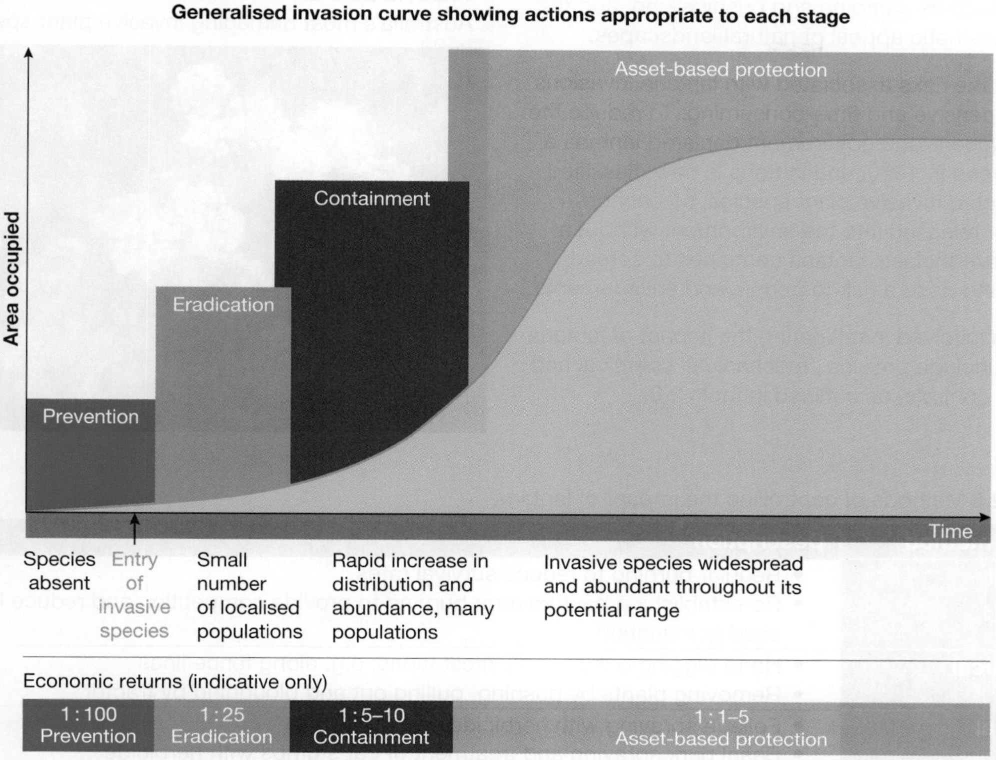

Generalised invasion curve showing actions appropriate to each stage

Area occupied →

Asset-based protection

Containment

Eradication

Prevention

Time →

| Species absent | Entry of invasive species | Small number of localised populations | Rapid increase in distribution and abundance, many populations | Invasive species widespread and abundant throughout its potential range |

Economic returns (indicative only)

| 1:100 Prevention | 1:25 Eradication | 1:5–10 Containment | 1:1–5 Asset-based protection |

Note: A ratio of 1:x indicates that for every $1 spent, the economic return is $x. For example, a ratio of 1:100 means that for every $1 spent in prevention, the return to the economy is $100.

Source: Harris, Stephen & Elliott, Craig & Woolnough, Andrew & Barclay, Candida. (2018). A heuristic framework for invasive species research planning and measurement. Developing an invasive species research strategy in Tasmania. No. 117. 13 pages.

CASE STUDY: Containing and eradicating the risk of lantana invasions in South East Queensland

Quick facts: Lantana
- **Species** *Lantana camara*
- **Origin** Central and South America
- **First grown in Australia** Adelaide Botanic Gardens, 1841
- **First reported as a weed species** 1879 in Queensland
- **Spatial distribution** 5 million hectares in most coastal and hinterland areas of Australia

Lantana is one of Australia's worst invasive plant species and most likely South East Queensland's most widespread weed. It is an aggressive invader of natural ecosystems and of grazing and farming land, so has environmental, economic and social impacts.

Lantana can be toxic to animals and other plants, and, when mature, forms into dense, impenetrable thickets 2–4 metres in height. It is frequently located in cleared areas near the margins of South East Queensland's conservation areas and national parks. It can shade out, smother and kill smaller native plant species and can climb trees and smother their canopies. In some places, lantana dominates the understory of invaded forests.

Lantana has been responsible for a reduction in Australia's biodiversity. At least 1400 native plant and animal species, many of which were already identified as threatened, are at risk because of lantana invasions.

Invasions lead to the loss of grazing and farming land, stock poisoning and reduction in land values. Experts estimate lantana invasions may cost Australia's grazing industry over $100 million per year in lost productivity. Controlling infestations also increases costs.

Social impacts include the reduction in recreational activities such as camping and bushwalking, and the reduced aesthetic appeal of natural landscapes.

Managing the risks associated with lantana invasions can be expensive and time-consuming. To reduce the risk, the Queensland government declared lantana a noxious weed in 1920, and lantana is now classified as a restricted invasive plant species, so cannot be sold, or released into the environment without a permit. Nevertheless, lantana continues to spread naturally and pose a risk to people and environments.

Strategies involved in mitigating the impact of lantana invasions include physical, mechanical, chemical and biological controls, as outlined in table 2.8.

FIGURE 2.23 Lantana (*Lantana camara*) is one of Australia's most damaging invasive plant species.

TABLE 2.8 Methods of controlling the impact of lantana

Control method	Description
Physical	• Regular burning to reduce survival rates • Re-establishing pasture after burning to provide competition and reduce lantana seed germination
Mechanical	• Hand digging out of small infestations; e.g. along fence lines • Removing plants by pushing, pulling out and ploughing by tractor
Chemical	• Foliage spraying with herbicide • Basal bark spraying and treatment of cut stumps with herbicide
Biological	• Introduced sap-sucking and leaf-mining insects that reduce the growth and vigour of lantana plants

 Resources

 Weblink Lantana distribution and control in Queensland

2.5 Exercise

2.5 Exercise

Learning pathways

■ LEVEL 1	■ LEVEL 2	■ LEVEL 3
1, 3	2, 4	5, 6

These questions are even better in jacPLUS!
- Receive immediate feedback
- Access sample responses
- Track results and progress

Find all this and MORE in jacPLUS ▶

Explain and comprehend

1. **Identify** the most cost-effective way to mitigate the risk of plant and animal invasions.
2. Refer to figure 2.24.
   a. **Explain** the impact that lantana is having on this environment.
   b. **Suggest** how the impact of lantana might be reduced in this location.

FIGURE 2.24 Lantana weed spreading in an Australian forest

3. Create a table like the one provided.
   a. **Identify** positives and negatives of each of the methods listed on the table.

Control method	Positives	Negatives
Physical		
Mechanical		
Chemical		
Biological		

   b. **Explain** why the use of one method of controlling the spread of lantana might not be effective.
4. **Explain** how the use of chemical control methods for managing invasive species can have both positive and negative impacts.

Analyse and apply

5. **Compare** and **contrast** the effectiveness and potential drawbacks of chemical, mechanical, physical and biological control methods for managing invasive species.
6. **Discuss** the importance of continued prevention and early eradication efforts in managing invasive species.

Sample responses are available in your digital formats.

LESSON
2.6 Pollutants

2.6.1 Types of pollutants

Pollutants are substances introduced into the environment that are potentially harmful to human health and to the natural environment. While pollution may occur naturally — for example, from the ash and gases of volcanic eruptions — most pollutants are a result of human activity. Virtually every human activity generates waste and, therefore, is a potential source of solid, liquid or gaseous pollution.

Anthropogenic pollution can come from a variety of sources, including:
- industrial sources, such as pollutants released from factories into the air or water, or leached into the soil
- transport, such as the exhaust emissions from various types of motor vehicles or spills from transport accidents
- agricultural sources, such as farm chemicals and animal wastes
- mines and quarries, such as dust and mining wastes
- domestic sources, such as smoke from cooking fires or household waste.

FIGURE 2.25 Motor vehicles are a major source of anthropogenic pollution in many parts of the world, including Australia.

Pollution sources are usually divided into two categories: **point sources** and **non-point sources**. Point sources are particular locations and include an industrial site, a mine or quarry, a sewerage treatment plant or an oil storage tank. Non-point sources involve broad areas and include run-off from agricultural or urban areas. Because motor vehicles release pollutants into the air, which can then spread over wide areas, transport is an example of a non-point source.

Pollution occurs in each of the Earth's four biophysical systems:
- atmosphere, through the release of chemicals and particulates into the air
- hydrosphere, through the release of wastes, chemicals and other contaminants into surface water and groundwater
- lithosphere, through the release of wastes of various types on or into soil
- biosphere, through the effects of waste of various types on living things.

point sources a particular location from which a pollution hazard originates; for example, an industrial site or an oil tank

non-point sources broad areas a pollution hazard originates from; for example, run-off from city streets

2.6.2 Reducing the risk of pollutants

The two broad approaches to reducing the risk of pollution are prevention and control. Options for the prevention and control of pollution will vary, depending on the type of pollutant and the source. Point sources of pollution are much easier to manage because they can usually be identified and monitored. Non-point source pollution often comes from a large number of small sources. These small sources can build up pollution over a large area, sometimes to unmanageable levels. Urban air pollution from motor vehicles is such a case.

The extent to which anthropogenic wastes constitute an ecological hazard depends on a range of factors. The amount of waste produced is important. Very small amounts of even potentially harmful chemical wastes may not be hazardous, while larger concentrations are hazardous. Some wastes quickly **biodegrade** (break down through natural processes), so may not pose a risk to people or the natural environment, but some pollutants are very persistent and remain hazardous for long periods of time.

A further issue for human health is that some pollutants, especially water pollutants, can enter the food chain in a process known as **bioaccumulation**. The pollutant may first be consumed by plankton, which are in turn eaten by fish, which are then consumed by humans. If there is a process of biomagnification, pollutants increase in concentration as they move along the food chain, and this may then increase the risk for humans at the end of the chain.

Some pollutants also create secondary pollutants. Acid rain, for example, is a secondary pollutant. It is produced when sulphur dioxide and nitrogen oxides (the primary pollutants) react in the atmosphere to produce sulphuric acid and nitric acid, which are absorbed into rain water. Sulphur dioxide and nitrogen oxides are produced when fossil fuels, such as coal and oil, are burnt in factories, power stations or motor vehicles.

FIGURE 2.26 (a) Los Angeles blanketed in pollution (b) Water pollution from copper mining in Romania (c) Oil pollution from pumps near Baku, Azerbaijan

biodegrade to break down through natural processes

bioaccumulation the process by which pollutants enter and concentrate through the food chain by passing from one food source to another by being eaten

FIGURE 2.27 World pollution map

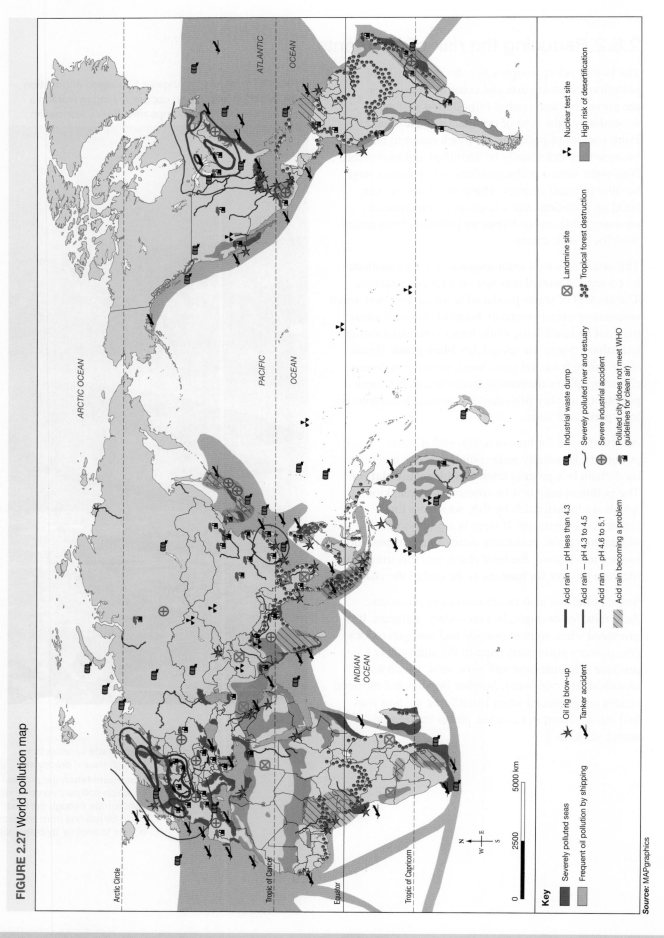

Key

Severely polluted seas

Frequent oil pollution by shipping

Oil rig blow-up

Tanker accident

Acid rain — pH less than 4.3

Acid rain — pH 4.3 to 4.5

Acid rain — pH 4.6 to 5.1

Acid rain becoming a problem

Industrial waste dump

Severely polluted river and estuary

Severe industrial accident

Polluted city (does not meet WHO guidelines for clean air)

Nuclear test site

High risk of desertification

Landmine site

Tropical forest destruction

ARCTIC OCEAN

ATLANTIC OCEAN

PACIFIC OCEAN

INDIAN OCEAN

Arctic Circle

Tropic of Cancer

Equator

Tropic of Capricorn

N
W — E
S

0   2500   5000 km

Source: MAPgraphics

2.6 Exercise

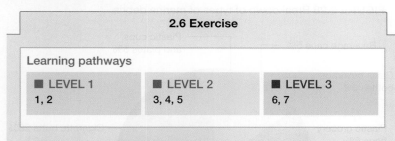

2.6 Exercise

Learning pathways

■ LEVEL 1	■ LEVEL 2	■ LEVEL 3
1, 2	3, 4, 5	6, 7

These questions are even better in jacPLUS!
- Receive immediate feedback
- Access sample responses
- Track results and progress

Find all this and MORE in jacPLUS ▶

Explain and comprehend

1. **Define** the term 'pollutant'.
2. **List** two examples of anthropogenic sources of pollution.
3. **Explain** the difference between point sources and non-point sources of pollution.
4. **Describe** the four broad types of pollution, providing examples of each.
5. **Describe** the process of creating secondary pollutants using acid rain as an example.

Analyse and apply

6. **Compare and contrast** prevention and control as approaches to reducing the risk of pollution, according to the text.
7. **Discuss** potential risks to human health associated with bioaccumulation of pollutants in the food chain.

Sample responses are available in your digital formats.

LESSON
2.7 Hydrospheric pollutants

2.7.1 Types of hydrospheric pollutants

Hydrospheric pollution can occur in both surface water bodies (rivers, lakes and reservoirs, seas and oceans), and groundwater and **aquifers**. Figure 2.28 outlines a breakdown of types of marine debris. However, unlike air and soil, water can also dissolve pollutants, as well as carry them in suspension. Many hydrospheric pollutants come directly from a polluting source, such as factories, farms, mines, sewage plants, power stations and urban settlements. Hydrospheric pollutants can also first be released by industry and vehicles into the atmosphere, and then fall back into rivers, lakes and the sea. For example, sulphur dioxide released when fossil fuels are burnt can dissolve in water to produce sulphuric acid. Carbon dioxide can also dissolve in water to produce carbonic acid, so a secondary effect of atmospheric CO_2 is ocean **acidification**.

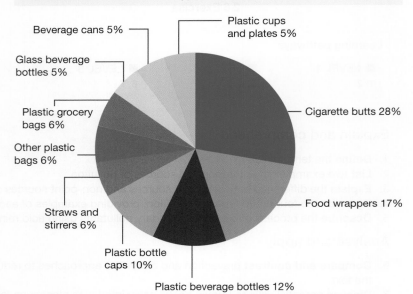

FIGURE 2.28 Breakdown of types of marine debris

- Beverage cans 5%
- Plastic cups and plates 5%
- Glass beverage bottles 5%
- Cigarette butts 28%
- Plastic grocery bags 6%
- Other plastic bags 6%
- Straws and stirrers 6%
- Food wrappers 17%
- Plastic bottle caps 10%
- Plastic beverage bottles 12%

Note: Data is the result of 25 years of surveying debris collected by volunteers in annual debris clean-ups in over 100 countries.

2.7.2 The impact of hydrospheric pollutants

Water pollution can be hazardous to human health. Around 1.4 million premature deaths around the world occur through polluted drinking water or unsafe sanitation. Most of these deaths are a result of diarrhoeal diseases, especially in children under five years of age.

People living in developing countries are most vulnerable to the hazard of polluted water. These countries are more likely to not only have unsafe drinking water and poor sanitation facilities, but also have poorly managed industrial sources of water pollution.

While human deaths and disease are the most severe effects of water pollution, a range of other environmental, social and economic impacts are also possible, some of which are outlined on table 2.9.

aquifers layers of rock that can hold large quantities of water in the pore spaces

acidification an increase in hydrogen ions, which lowers the pH of the water

TABLE 2.9 Sources and effects of water pollutants

Water pollutant	Main sources	Main effects
Sewage and wastewater	• Sewage treatment plants • Poorly managed sanitation	• Unsafe drinking water • Human illness and death • Algal growth and eutrophication
Fertilisers	• Agricultural runoff • Industry	• Algal blooms • Eutrophication
Heavy metal (e.g. arsenic, mercury)	• Industrial runoff • Mining	• Illness and death of people and wildlife
Salt	• Land clearing • Poor agricultural practices	• Salty drinking water • Reduced crop yields
Fuels and oil	• Fuel spills • Damaged oil pipelines and oil tankers	• Death of wildlife such as fish and seabirds
Plastics and other rubbish	• Litter from urban areas and ships	• Clogging of watercourses • Accumulation on beaches and in the ocean • Potential ingestion by fish, turtles and other animals • Biomagnification of microplastics
Chemicals (e.g. pesticides, herbicides, fire retardants)	• Industry • Agricultural run-off • Run-off from fire control	• Illness and death of native plants and animals • Human illness
Carbon dioxide	• Burning of fossil fuels in power stations, industry and motor vehicles	• Ocean acidification
Sediment	• Land clearing and deforestation • Poor farming practices	• Reduction in photosynthesis by plants • Clogging of fish gills • Accumulation on sea and lake floors, covering plant and animal life

Source: The Ocean Conservancy

EXAMPLE: Iowa fish kill, March 2024

On 11 March 2024, the accidental release of over one million litres of liquid nitrogen fertiliser into a drainage ditch by an agricultural company in Iowa, United States, polluted nearly 100 kilometres of the East Nishnabotna and Nishnabotna rivers downstream of the spill.

Almost all aquatic life was killed because of the spill, including around 750 000 fish, as well as frogs, snakes, mussels and earthworms.

FIGURE 2.29 The Nishnabotna River in Iowa, United States

SKILLS ACTIVITY

Refer to table 2.10 to complete the task.

TABLE 2.10 Sources of marine debris by region

Source	Africa (%)	North America (%)	Central America (%)	South America (%)	Caribbean (%)	South-East Asia (%)	Western Asia (%)	Europe (%)	Oceania (%)
Land-based litter	76.1	55.2	84.9	69.7	82.6	72.4	60.5	60.4	75.2
Ocean-based litter	12.7	5.0	4.8	12.0	6.6	12.7	9.7	24.9	5.2
Smoking-related litter	8.4	37.2	8.0	15.4	7.7	11.2	27.7	11.1	19.7
Legal and illegal dumping of garbage and waste	1.8	1.9	1.1	2.0	1.8	1.6	1.5	2.8	1.8
Medical and personal hygiene litter	1.0	0.7	1.2	1.0	1.3	2.0	0.6	0.8	1.1

1. Create a ternary graph using data from table 2.10 to represent three types of marine debris in the ocean for three regions. To complete this task you will need to transform the data, creating a new table displaying the total (as a per cent) for land-based, ocean-based and other sources of litter for your three chosen regions. Refer to SkillBuilder: Constructing ternary graphs in the Resources tab to complete this task.

 Resources

▶ **Video eLesson**   SkillBuilder: Constructing ternary graphs (eles-1728)

✦ **Interactivity**   SkillBuilder: Constructing ternary graphs (int-3346)

CASE STUDY: Hydrospheric pollution in the Great Barrier Reef

Quick facts: Great Barrier Reef
- **Location** 10°S to 24°S, between Torres Strait and Gladstone, Queensland
- **Size** around 2300 km in length and 348 000 km² in area
- **Comprising** 3000 individual reefs, around 400 hard and soft coral species, and 900 islands
- **Provides habitats for** 1500 fish, 4000 molluscs, 240 bird species, and a great diversity of sponges, anemones, marine worms, crustaceans

While coral bleaching and climate change are the greatest threats to the Great Barrier Reef, land-based run-off has a major impact on water quality in inshore areas of the reef and threatens many inshore reefs and marine ecosystems. The main pollutants that threaten the reef include fine sediment, excess fertiliser nutrients, and chemicals such as pesticides and herbicides. Mostly these pollutants come from agricultural activities (grazing, sugar cane, grain crops, irrigated cotton and horticulture), although some come from the various urban settlements located in the 35 river catchments that drain into the reef's inshore waters.

Discharge from large rivers such as the Burdekin, Fitzroy and Herbert create large sediment plumes and can affect waters and reefs for hundreds of kilometres. For example, one study of the Burdekin River showed that during 2011, 275 reefs up to 450 kilometres north of the mouth were impacted by sediment and nutrient pollution.

Sediment in rivers is the result of the erosion of topsoil and riverbanks. Deforestation and poor land management practices can lead to accelerated erosion and increased sediment loads.

FIGURE 2.30 Aerial view of the Great Barrier Reef

The discharge of excessive sediment frequently occurs in wet seasons and is a major risk during floods. Sediment can smother coral, seagrass and benthic (bottom-dwelling) species, and reduce sunlight available for photosynthesis.

Excessive fertiliser applied to crops can wash into rivers and then into inshore reef waters. Nitrogen from the fertiliser contributes to the growth of algae, leading to algal blooms. These blooms can be harmful to animals, block sunlight and reduce coral diversity. Nitrogen fertilisers have also been linked to outbreaks of crown-of-thorns starfish, which prey on coral.

Pesticides and herbicides are toxic, so they pose a risk to marine plants and animals living in inshore reef waters. Herbicides washed into marine waters affect non-target plants, especially seagrass, and the animals such as dugong and turtles that depend on seagrass.

CASE STUDY: Pollution of the hydrosphere in Senegal

Quick facts: Senegal
- **Location** Northwest Africa, on the Atlantic coast
- **Population** 18.2 million
- **Capital city** Dakar
- **Human Development Index (HDI)** 0.51 (ranked 169 out of 193)

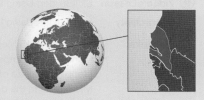

Like many developing countries in Africa, Senegal experiences a number of water pollution hazards. Agricultural run-off, industrial effluents and mining wastes contaminate surface water in many places. Human wastewater is also a major issue; fewer than 60 per cent of the population use at least basic sanitation services, so waterborne diseases are common in Senegal, and diarrhoea is the leading cause of under-five child mortality.

Water pollution is particularly an issue for Senegal's coastal areas and larger cities, especially around Dakar, where 80 per cent of the country's industries are located. More than 70 factories, including chemical plants, an oil refinery and an abattoir, discharge effluent into Hann Bay, and raw sewage often goes directly into the ocean. Until 2011, the city was serviced by one single sewage treatment plant, and local services were sporadic, with some garbage collections occurring weeks apart. Water pollution is at such a high level, it is considered dangerous to swim in any of the coastal waters near Dakar; consequently, the tourism and fishing industries in the area have suffered significant losses.

To prevent any increase in pollution hazards in Hann Bay and to reduce existing hazards, the government has developed an action plan to construct infrastructure for the collection, treatment and discharge of effluents. Because government funding can be a challenge for Senegal, overseas development assistance is often necessary. For example, the French Development Agency and the European Investment Bank are providing around $55 million for a project to clean up the bay. One project, the construction by a French company of a wastewater plant on the coast to treat 26 000 cubic metres a day for 500 000 people was due to start in 2025, but some of the work has been suspended.

FIGURE 2.31 Pollution in Saint-Louis, Senegal, on the Senegal River, where coastal tourism makes up a significant part of the local economy

On the other hand, the Water Supply and Sanitation Sector Project, due to end in December 2024, has helped to improve access to drinking water in three rural areas, and the cities of Dakar and Ziguinchor. Nevertheless, challenges to reduce pollution of the hydrosphere remain for Senegal.

on Resources

📄 **Digital doc**	Marine pollution in Bali and Senegal (doc-29165)
▶ **Video eLesson**	SkillBuilder: Using latitude and longitude (eles-1652)
✜ **Interactivity**	SkillBuilder: Using latitude and longitude (int-3148)
🔗 **Weblinks**	Milestone 2015 Mega Expedition
	Marine debris: Impacts and solutions
	Mapping the size of the Great Pacific Garbage Patch
	Microfibres in the ocean
	Sunscreens and coral health
	Bali waters polluted with plastic
	Pollution levels in Senegal
	Microplastics a threat to wildlife
	Bacteria evolve to eat plastic
	The Great Green Wall

2.7 Exercise

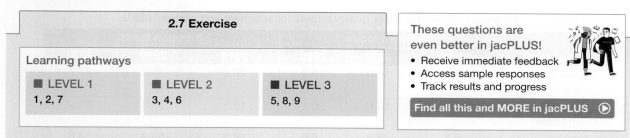

2.7 Exercise

Learning pathways

■ LEVEL 1	■ LEVEL 2	■ LEVEL 3
1, 2, 7	3, 4, 6	5, 8, 9

These questions are even better in jacPLUS!
- Receive immediate feedback
- Access sample responses
- Track results and progress

Find all this and MORE in jacPLUS ▶

Explain and comprehend

1. **Identify** three main pollutants that threaten the Great Barrier Reef.
2. **List** two types of land-based activities that contribute to the pollution of inshore reef waters.
3. Refer to figure 2.32.
   a. **Describe** the trends in the condition of coral and seagrass shown in figure 2.32.
   b. **Explain** possible reasons for the trends.
   c. Refer to the scores in the note to figure 2.32. **Explain**, with reasons, what these scores indicate about the overall condition of the inshore reef area.

FIGURE 2.32 Inshore reef indicators over time

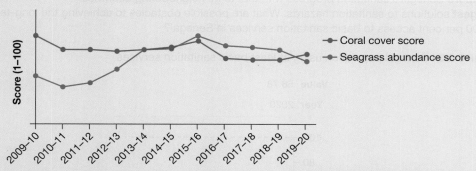

Note: 2009–10 scores are as follows: coral cover, 53; seagrass abundance, 24. 2019–20 scores are as follows: coral cover, 40; seagrass abundance, 35.

Analyse and apply

4. Refer to figure 2.33 and table 2.11.
   a. **Analyse** and interpret the data provided in these sources to **explain** the trends in sediment and nitrogen loads for the three catchments and Great Barrier Reef as a whole.
   b. **Explain** how these trends represent a risk for people and environments in the Great Barrier Reef.
   c. **Generalise** about the potential impacts of sediment and nutrient loads for people and environments in two specific locations.

FIGURE 2.33 Sediment loads report card, 2020

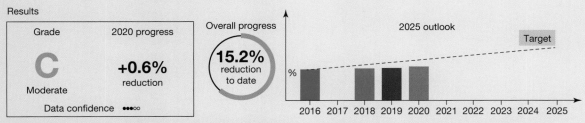

Source: Reef Water Quality Report Card, © The State of Queensland (Department of Environment and Science).

TABLE 2.11 Progress in achieving water quality targets for sediment and nitrogen loads

Location	Sediment loads			Nitrogen loads		
	2016	2019	2020	2016	2019	2020
Great Barrier Reef region	13.9	14.6	15.2	20.9	25.5	27.7
Burdekin catchment	19.0	19.8	20.1	15.1	21.6	25.2
Herbert catchment	8.8	8.9	9.0	44.1	53.8	54.6
Tully catchment	9.8	10.1	10.4	7.1	13.7	19.2

Data source: Reef Water Quality Report Card, © The State of Queensland (Department of Environment and Science).

Targets: sediment, 25% reduction by 2025; nitrogen, 60% reduction by 2025.

5. Refer to figures 2.34 and 2.35.
   a. **Describe** the potential hazards associated with the absence of basic sanitation services.
   b. **Analyse** the graphs and describe the progress that Senegal has made in providing at least basic sanitation services to its people.
   c. **Compare** Senegal's sanitation progress with that of its neighbouring countries.
   d. **Suggest** solutions to sanitation hazards. What are possible obstacles to achieving the long-term objective of 100 per cent access to basic sanitation services in Senegal?

FIGURE 2.34 Senegal population using at least basic sanitation services

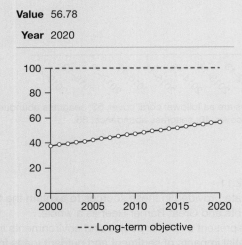

Value 56.78

Year 2020

- - - Long-term objective

Source: Based on the data from 'People using at least basic sanitation services (% of population)'. WHO/UNICEF Joint Monitoring Programme (JMP) for Water Supply, Sanitation and Hygiene (washdata.org). © 2024 The World Bank Group, All Rights Reserved. Licensed under CC-BY 4.0.

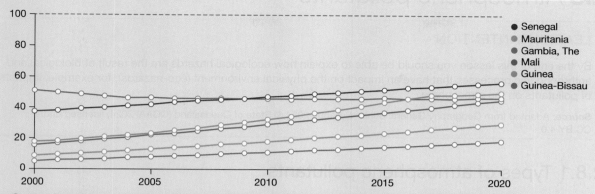

FIGURE 2.35 Percentage of population using at least basic sanitation services, various countries

- Senegal
- Mauritania
- Gambia, The
- Mali
- Guinea
- Guinea-Bissau

Source: Based on the data from 'People using at least basic sanitation services (% of population)' - Senegal, Mauritania, Gambia, The, Mali, Guinea, Guinea-Bissau. WHO/UNICEF Joint Monitoring Programme (JMP) for Water Supply, Sanitation and Hygiene (washdata.org). © 2024 The World Bank Group, All Rights Reserved. Licensed under CC-BY 4.0.

6. Refer to figure 2.36.
   a. **Analyse** the graph and describe the progress that Senegal has made in achieving clean ocean waters.
   b. **Explain** possible factors that might account for Senegal's lack of progress in achieving clean ocean waters.

FIGURE 2.36 Ocean Health Index: Clean Waters score, Senegal

Value 43.81

Year 2022

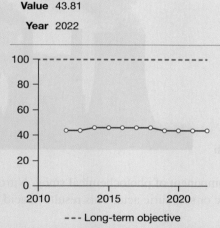

--- Long-term objective

Source: Halpern BS, Longo C, Hardy D, McLeod KL, Samhouri JF, Katona SK, et al. (2012) 'An index to assess the health and benefits of the global ocean'. *Nature.* 2012;488: 615–620. doi:10.1038/nature11397.

7. **Describe** the potential hazards associated with the absence of clean ocean waters.

Propose and communicate

8. **Suggest** solutions to Senegal's ocean water hazards. What are possible obstacles to achieving its long-term objective of 100 per cent clean ocean water?

9. **Suggest** why people living in developing countries are particularly vulnerable to the hazards of polluted water, considering factors such as access to safe drinking water, sanitation facilities and industrial pollution management.

Sample responses are available in your digital formats.

LESSON
2.8 Atmospheric pollutants

LEARNING INTENTION

By the end of this lesson you should be able to explain how ecological hazards are the result of biological and anthropogenic processes that have an impact on the physical environment (eco-hazards); for example, impacts of pollutants on the atmosphere.

Source: Adapted from Geography General Senior Syllabus 2024 © State of Queensland (QCAA) 2024; licensed under CC BY 4.0.

2.8.1 Types of atmospheric pollutants

Atmospheric pollution involves substances in the air that exceed naturally occurring levels and may be hazardous to human health and wellbeing, and the biophysical environment. Primary air pollutants come directly from a polluting source, such as motor vehicles, factories or power stations. Secondary pollutants are formed when primary pollutants combine, or chemical reactions occur involving primary pollutants. Table 2.12 provides a list of the main types of atmospheric pollutants, as well as their main sources and effects.

Ozone is an example of a secondary pollutant. It forms in sunlight through chemical reactions between nitrous

FIGURE 2.37 Coal power stations, like this one in Germany, are a source of primary air pollutants.

oxides and hydrocarbons, and is a component of photochemical smog. Nitrous oxides and sulphur dioxide can combine with rainwater to form nitric or sulphuric acid. This results in acid rain, another example of a secondary pollutant.

Solid particles or particulate matter can also pollute the air, so air pollution can be a gas, a liquid or a solid. Particulate matter is usually measured according to the size of the particles. PM2.5 is 2.5 micrometres or less in diameter; PM10 ranges between 2.5 and 10 micrometres. (In comparison, a human hair is around 100 micrometres.) Because very fine particles can be inhaled and travel deep into the lungs, they are a significant cause of lung and heart diseases.

TABLE 2.12 Sources and effects of major air pollutants

Air pollutant	Main sources	Main effects
Nitrogen oxides (NO_x)	<ul><li>Motor vehicles burning diesel or petrol</li><li>Power stations burning fossil fuels</li><li>Industry</li></ul>	<ul><li>Respiratory problems</li><li>Throat and lung infections</li><li>A major contributor to photochemical smog and acid rain</li><li>Can affect growth/damage plants</li></ul>

Sulphur dioxide (SO$_2$)	• Power stations burning fossil fuels • Industry	• Respiratory problems and severe coughing • Eye irritation • Circulatory and heart problems • Major contributor to acid rain
Carbon monoxide (CO)	• Motor vehicles burning petrol	• Reduces oxygen in the blood • Headaches and vomiting • Large amounts are lethal • Forms carbon dioxide (greenhouse gas)
Hydrocarbons	• Incomplete burning of petrol in motor vehicles • Industry • Petrol stations and oil refineries	• Contribute to photochemical smog (haze)
Ozone (O$_3$)	• Motor vehicle exhaust fumes • Other pollutants in the presence of sunlight	• Component of photochemical smog (haze) • Eye, throat and lung irritation • Breathing difficulties • Large amounts can be fatal • Affects growth of plants
Particulate matter (PM2.5, PM10)	• Combustion of fossil fuels in industry and motor vehicles • Building and road construction	• Eye irritation, breathing difficulties and lung damage • Discolours paint and fabrics

Air quality indexes differ around the world and measure a wide range of air pollution hazards, including specific pollutants such as carbon dioxide levels and general levels of particulate matter in the air. One commonly used measure of air quality is the US-EPA 2016 standard, shown in table 2.13, which measures fine particulate matter (PM2.5). These particles are typically heavy metals and compounds released into the air from car exhaust, burning landfill and industrial processes.

TABLE 2.13 Air Quality Index (AQI) scale using PM2.5 levels (US-EPA 2016 standard)*

AQI	Air pollution level	Health implications	Cautionary statement
0–50	Good	Air quality considered satisfactory; air pollution poses little or no risk	None
51–100	Moderate	Air quality acceptable; some pollutants may present moderate health concern for very small number of people (e.g. those with asthma or respiratory disease)	Active children and adults, and people with respiratory disease, should limit prolonged outdoor exertion
101–150	Unhealthy for sensitive groups	Members of sensitive groups may experience health effects; general public unlikely to be affected	Active children and adults, and people with respiratory disease, should limit prolonged outdoor exertion

(continued)

TABLE 2.13 Air Quality Index (AQI) scale using PM2.5 levels (US-EPA 2016 standard)* *(continued)*

AQI	Air pollution level	Health implications	Cautionary statement
151–200	Unhealthy	Everyone may begin to experience health effects; members of sensitive groups may experience more serious health effects	Active children and adults, and people with respiratory disease, should avoid prolonged outdoor exertion; everyone, especially children, should limit prolonged outdoor exertion
201–300	Very unhealthy	Health warnings of emergency conditions; entire population is more likely to be affected	Active children and adults, and people with respiratory disease, should avoid all outdoor exertion; everyone, especially children, should limit outdoor exertion
300+	Hazardous	Health alert: everyone may experience more serious health effects	Everyone should avoid all outdoor exertion

Note: *Air Quality Index shows levels of PM2.5 as µg/m3 (micrograms per square metre)

Source: © 2008–2016 World Air Quality / United States Environmental Protection Agency

2.8.2 The impact of air pollution

Air pollution can be hazardous to human health. According to the World Health Organization, air pollution is the cause of over one-third of deaths from stroke, lung cancer, and chronic respiratory disease, and one-quarter of deaths from ischaemic heart disease worldwide. The WHO estimates that over 90 per cent of the world's population lives in places where air quality does not meet recommended levels. People living in low- and middle-income countries are most vulnerable to the hazard of air pollution. China and India, with the world's largest populations, also have the largest number of deaths associated with air pollution. (Figure 2.38 shows an example of air pollution in India.)

Air pollution also has a negative impact on the natural environment. Animals can be affected in similar ways to people and plants are vulnerable to changes in ozone, nitrogen oxides and sulphur dioxide, which can produce acid rain (see figure 2.39). Acid rain has a negative impact on soils and water courses, so may contribute to lithospheric and hydrospheric pollution, as shown in figure 2.40.

FIGURE 2.38 Air pollution in Delhi, capital of India

FIGURE 2.39 The impact of acid rain on coniferous forests

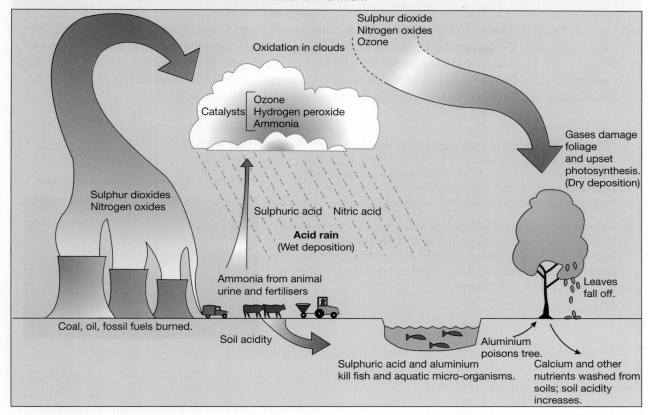

FIGURE 2.40 The effects of acid rain on the natural environment

Sulphur dioxide
Nitrogen oxides
Ozone

Oxidation in clouds

Catalysts

Ozone
Hydrogen peroxide
Ammonia

Sulphur dioxides
Nitrogen oxides

Gases damage foliage and upset photosynthesis. (Dry deposition)

Sulphuric acid   Nitric acid

Acid rain
(Wet deposition)

Ammonia from animal urine and fertilisers

Leaves fall off.

Coal, oil, fossil fuels burned.

Soil acidity

Aluminium poisons tree.

Sulphuric acid and aluminium kill fish and aquatic micro-organisms.

Calcium and other nutrients washed from soils; soil acidity increases.

Air pollution is most hazardous in large cities. Air polluting industries are frequently located in cities. More importantly, they have the greatest concentration of motor vehicles, which are the leading source of many air pollutants. Most cities around the world monitor their air quality and, if necessary, provide health warnings based on air quality index levels.

2.8.3 Factors affecting air pollution

Air pollution is affected by a range of economic, environmental, social and political factors.

Economic factors: level of economic development

The most hazardous levels of air pollution occur in the megacities of middle- and low-income countries. Delhi, Beijing, Cairo and Dhaka are among the most polluted of these cities. This is a consequence of the number of people and vehicles, as well as of the difficulty of controlling emissions in poorer countries, especially with large populations. Often, these very large cities are also the location of the country's major industries and commerce, and these also contribute to air pollution.

Economic factors: industrial activities

The manufacturing industry can significantly contribute to air pollution in urban areas. Factories, power plants and industrial facilities release a variety of pollutants into the atmosphere, including particulate matter, sulphur dioxide, nitrogen oxides and volatile organic compounds. These pollutants can have detrimental effects on human health and the environment. In urban centres, where manufacturing and industrial activities are often concentrated, the impact of these emissions can be particularly pronounced.

Environmental factors: topography and weather

Air pollution can also be affected by topography and by climate and weather conditions. For example, cities such as Kathmandu and Mexico City are located in mountain basins, and this increases the risk of hazardous levels of air pollution. Cold mountain air sinks into the basins, creating temperature inversions which then trap air pollutants. The clear, sunny weather associated with high pressure systems and subtropical climates provides the conditions needed for the formation of photochemical smog and ozone. This means that cities such as Los Angeles, Karachi and Delhi are very susceptible to smog conditions and ozone pollution.

EXAMPLE: Brisbane's topography, weather patterns and air pollution

There are mountain ranges to the north-west, west and south of Brisbane, which form a basin-like structure, so Brisbane experiences temperature inversions, especially in winter.

In summer, air pollution is affected by the recirculation of polluted air. Evening sea breezes move polluted air inland along the Brisbane Valley towards Ipswich, where it stagnates. Early the next morning, breezes move the polluted air back towards the south-western parts of Brisbane.

FIGURE 2.41 Air pollution over Brisbane

Environmental factors: climate

In many cities, air pollution is also affected by climatic conditions. Cities such as Brisbane and Mexico City experience subtropical climates with intense sunshine all year round. This provides the necessary conditions to produce photochemical smog. In summer, heavy rain and storms do wash many pollutants from the atmosphere. In winter, however, there are lower rainfall totals. High pressure systems dominate both cities, producing clear, sunny and relatively calm weather conditions. The colder weather also contributes to the formation of temperature inversions. This means that air pollutants can build up over many days, so the incidence of ozone (a component of photochemical smog) and other pollutants tends to peak in winter months in both cities.

Social factors: motor vehicles

Motor vehicles are also the main source of air pollution in wealthy countries such as Australia. Figure 2.42 shows that all of Australia's capital cities are car dependent. Brisbane had around 2.2 million motor vehicles in 2017, with 70 per cent of people's journeys to work being by car and 93 per cent of all travel by car. Approximately 70 per cent of Brisbane's air pollutants are produced by motor vehicles.

FIGURE 2.42 Annual travel by private and public transport in Australian cities

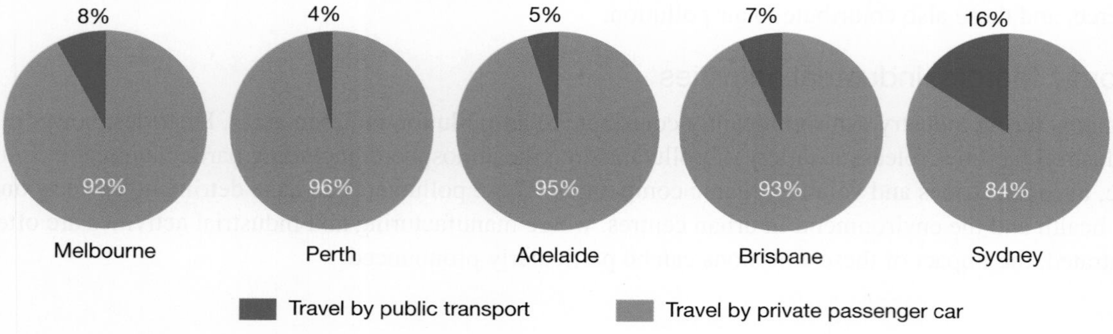

	Melbourne	Perth	Adelaide	Brisbane	Sydney
Public transport	8%	4%	5%	7%	16%
Private passenger car	92%	96%	95%	93%	84%

■ Travel by public transport   ■ Travel by private passenger car

Australia has relatively high vehicle emission standards and recently introduced measures to improve these standards. It also has high-quality fuels, and hybrid and electric vehicles are growing in number. The investment in road infrastructure, such as Brisbane's network of tunnels, does assist in reducing congestion and air pollution, but does not reduce car dependence.

Public transport networks play an important role in reducing car dependence and air pollution in all cities.

EXAMPLE: Public transport in Brisbane

Since the 1990s, Brisbane has developed a busway network with separate rapid transit bus-only corridors, adding to Brisbane's existing rail and bus networks. A Brisbane Metro subway system and a Cross River Rail project have also been planned to ease congestion in the inner parts of the city. The City Council introduced a CityCycle scheme, followed by e-bike and e-scooter hire, and invested in bikeways to encourage more people to cycle to work rather than use a private motor vehicle. Brisbane City planning now incorporates transit-oriented developments, mostly high-density residences close to train stations, in order to encourage people to use public transport rather than cars.

FIGURE 2.43 Brisbane city bus

SKILLS ACTIVITY

Refer to **SkillBuilder: Reading topographic maps at an advanced level** and **Creating a transect of a topographical map** in the Resources tab. (A printer-friendly version of figure 2.44 can also be found in the Resources tab.)

Explain and analyse the role of topography in pollution

1. Locate Cerro Madin (2470 metres above sea level) and Xaltepec Volcano (2690 metres above sea level) in figure 2.44. Draw a cross-section between these two locations. What is the compass direction from left to right on the cross-section?

▶

FIGURE 2.44 Mexico City topographic map

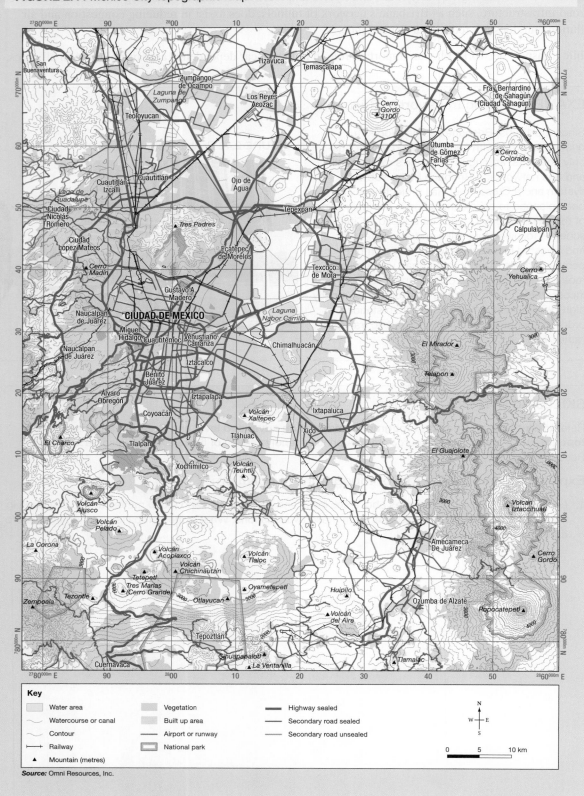

Key

Water area		Vegetation	▬▬▬	Highway sealed
Watercourse or canal		Built up area	▬▬▬	Secondary road sealed
Contour	▬▬▬	Airport or runway	▬▬▬	Secondary road unsealed
Railway		National park		
▲ Mountain (metres)				

0   5   10 km

Source: Omni Resources, Inc.

2. Using your cross-section, calculate the approximate altitude of the centre of Mexico City. How far below the summits of the two mountains does the city lie? (The interval between contour lines on the map is 100 metres.)
3. Identify five more volcanoes on the map. Determine their altitudes.
4. Use your cross-section and the contour patterns on the map to describe the topography of Mexico City and its surroundings.
5. Using a digital topographic map of Brisbane (for example, from the **QTopo** weblink in the Resources tab), draw a 100 kilometre north–south and east–west cross-section of Brisbane. Create a transect of each cross-section. (Use the SkillBuilder in the video eLessons of your online Resources to revise this skill.)
6. Using your transects and the topographic map, describe the topography of Brisbane.

Identify the impact of topography

7. Write a paragraph describing how Brisbane's topography might have an impact on the city's air pollution.

 Resources

▶ **Video eLessons**   SkillBuilder: Reading topographical maps at an advanced level (eles-1749)
SkillBuilder: Constructing and describing a transect on a topographical map (eles-1727)
SkillBuilder: Understanding satellite images (eles-1643)

✦ **Interactivities**   SkillBuilder: Reading topographical maps at an advanced level (int-3367)
SkillBuilder: Understanding satellite images (int-3139)
SkillBuilder: Constructing and describing a transect on a topographical map (int-3345)

🔗 **Weblink**   QTopo

CASE STUDY: Mitigating atmospheric pollution in Mexico City

Quick facts: Mexico City
- **Location** Valley of Mexico on central Mexican plateau
- **Altitude** 2240 metres
- **Population** 22.5 million
- **Vehicle numbers** 6.2 million registered vehicles in 2022
- **Main industries** food, chemicals, textiles, plastics, paper, machinery, employing 20% of the workforce

In 1992 Mexico City was identified by the United Nations as the world's most air-polluted city, a result of decades of industrialisation, population growth and the presence of millions of motor vehicles without pollution controls. Consequently, people suffered from serious pollution-related health conditions, including asthma, respiratory infections and cardiovascular disease.

Mexico City's elevated mountain basin topography and weather patterns contribute to the city's air pollution risk. Mountain basins and their associated temperature inversions trap air pollutants. The intense sunshine of the subtropical climate, and clear, calm weather associated with high pressure systems provide the conditions necessary for the formation of photochemical smog and ozone.

FIGURE 2.45 Mexico City has been able to reduce the risk of air pollution from high numbers of cars and the manufacturing industry.

By 2023, Mexico City's global air pollution ranking had significantly improved, such that over 900 cities were considered more polluted than Mexico City. Nonetheless, the city continues to experience air pollution problems, especially from particulate matter. In 2022, the city's average annual level of PM2.5 was more than four times the World Health Organization limit.

The reduction in Mexico City's hazardous air pollution over time occurred for a number of reasons. The most polluting industries relocated, or were forced to move or close by the government. In 1990, a Comprehensive Program against Air Pollution commenced, followed by a second program, ProAire, in 1995. By 2011, the city government and the Metropolitan Environmental Commission were using ProAire IV to help reduce the use of cars, promote public transport and environmental education, reduce energy consumption, and encourage the use of cleaner and more efficient energy.

Because transportation, especially cars, was the greatest source of air pollution, much of the hazard mitigation involved motor vehicles. This included vehicle restrictions, stricter emissions standards, the use of less polluting fuels and catalytic convertors, and regular emissions testing of cars. Mexico City's vehicle restriction program, known as Hoy No Circula (Cars Don't Circulate), or 'no drive days', was introduced in 1989, and restricts a fifth of vehicles on rotating days between Monday and Saturday, based on the last number of the car's registration plate.

Mexico City also invested heavily in public transport to provide cleaner alternatives to motor cars. Its Metro system was greatly expanded, and a new Metrobús network, with rapid bus transit lines, was established. Taxis and minibuses were renovated, and the city also has a large bike sharing scheme.

To evaluate the effectiveness of mitigation, and provide warnings when pollution becomes hazardous, 34 automatic monitoring stations, known as Red Automática de Monitoreo Atmosférico, or RAMA, are located across the city. These stations measure a range of pollutants, including ozone and particulate matter (PM2.5 and PM 10), and release hourly data for each of the pollutants.

 Resources

 Weblinks Impacts of air pollution
Mexico City air quality
Queensland air quality

2.8 Exercise

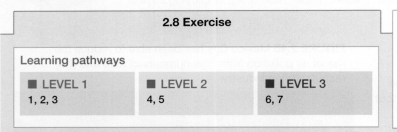

2.8 Exercise
Learning pathways

■ LEVEL 1	■ LEVEL 2	■ LEVEL 3
1, 2, 3	4, 5	6, 7

These questions are even better in jacPLUS!
- Receive immediate feedback
- Access sample responses
- Track results and progress

 Find all this and MORE in jacPLUS ⊙

Explain and comprehend

1. **Explain** the difference between primary and secondary pollutants.
2. **Describe** the impact of the manufacturing industry on air pollution in urban areas.
3. Refer to figure 2.46.
   a. **Describe** the spatial pattern of PM2.5 concentrations shown on the map.
   b. **Suggest** reasons for this pattern.
   c. **List** and **explain** implications of hazardous concentrations of PM2.5 pollution.

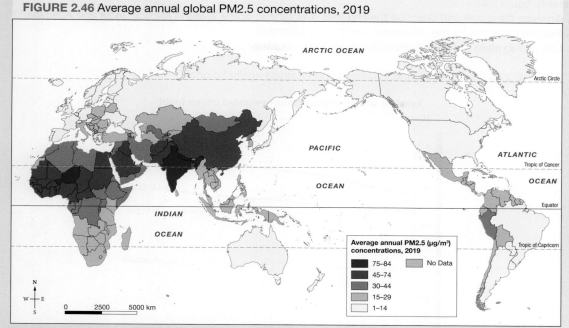

FIGURE 2.46 Average annual global PM2.5 concentrations, 2019

ARCTIC OCEAN

Arctic Circle

PACIFIC

ATLANTIC

Tropic of Cancer

OCEAN

OCEAN

INDIAN

Equator

OCEAN

Tropic of Capricorn

Average annual PM2.5 (µg/m³) concentrations, 2019

75–84	No Data
45–74	
30–44	
15–29	
1–14	

N
W—E
S

0 2500 5000 km

Source: Health Effects Institute, 2020. State of Global Air 2020. Data source: Global Burden of Disease Study 2019. IHME, 2020. Map redrawn by Spatial Vision.

Analyse and apply

4. Refer to figure 2.47.
   a. **Describe** the spatial pattern of the share of deaths attributed to air pollution as shown on the map.
   b. **Suggest** reasons for this pattern.
   c. **Compare** the patterns shown on the map with those shown on figure 2.46. **Identify** any relationship between PM2.5 concentration and share of deaths attributed to air pollution. If a relationship can be seen, use specific places as examples.

FIGURE 2.47 Share of global deaths attributed to air pollution, 2019

ARCTIC OCEAN

Arctic Circle

PACIFIC

ATLANTIC

Tropic of Cancer

OCEAN

OCEAN

Equator

INDIAN

OCEAN

Share of deaths attributed to air pollution, 2019

More than 18%	No Data
16–18%	
13–15%	
10–12%	
7–9%	
4–6%	
1–3%	

N
W—E
S

0 2500 5000 km

Tropic of Capricorn

Source: Based on data from IHME, Global Burden of Disease (2019). Map redrawn by Spatial Vision.

5. Refer to figure 2.48.
   a. **Analyse** the graph and describe the changes over time in ozone levels in Mexico, India and Australia.
   b. **Suggest** reasons for the differences among the countries.
   c. **List** and **explain** the hazard implications for each country.

FIGURE 2.48 Ozone levels, India, Mexico and Australia against global

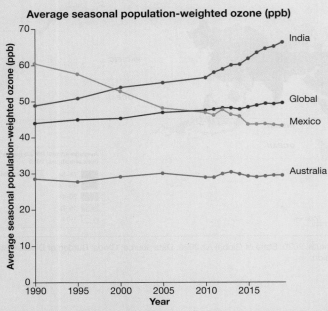

Source: Health Effects Institute. 2020. State of Global Air 2020. Data source: Global Burden of Disease Study 2019. IHME, 2020.

6. Refer to figure 2.49.
   a. **Analyse** the graph and describe the changes over time in air pollution levels in Mexico City.
   b. **Explain** the implications of these changes for people living in Mexico City.
   c. **Suggest** reasons for differences in levels of nitrogen oxides and sulphur dioxide over time.

FIGURE 2.49 Mexico air pollution, 1990 to 2015

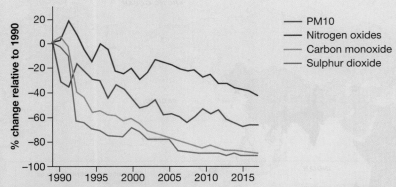

Source: SEDEMA (Secretaría del Medio Ambiente del Distrito Federal). 2017. Data from SIMAT (Sistema de Monitoreo Atmosférico de la Ciudad de México). Graph drawn by Climate Interactive. © 2018 Climate Interactive. https://www.climateinteractive.org/

Propose and communicate

7. **Evaluate** the effectiveness of current strategies for mitigating air pollution in cities with subtropical climates.

Sample responses are available in your digital formats.

LESSON
2.9 Lithospheric pollutants

LEARNING INTENTION

By the end of this lesson you should be able to explain how ecological hazards are the result of biological and anthropogenic processes that have an impact on the physical environment (eco-hazards); for example, impacts of pollutants on the lithosphere.

Source: Adapted from Geography General Senior Syllabus 2024 © State of Queensland (QCAA) 2024; licensed under CC BY 4.0.

2.9.1 Types of lithospheric pollutants

Lithospheric pollution involves substances in the soil that exceed naturally occurring levels and may be hazardous to human health and wellbeing, and the biophysical environment. The pollutants can be organic (for example, oil-based chemicals and some pesticides), or inorganic (for example, salt, and metals such as lead, mercury and arsenic).

Most lithospheric pollutants are a result of human activities such as manufacturing, agriculture, mining and urban waste disposal. Lithospheric pollutants often come directly from a polluting source, but can also be transported, sometimes over long distances, from their original source. For example, electronic waste products are often transported from wealthier countries to poorer countries, where they can then pollute the environment and damage people's health.

FIGURE 2.50 Landfills can be a source of lithospheric pollution.

Table 2.14 outlines the sources and effects of lithospheric pollutants.

TABLE 2.14 Sources and effects of lithospheric pollutants

Lithospheric pollutant	Main sources	Main effects
Heavy metals (e.g. lead, arsenic, mercury)	• Industrial waste disposal • Mining and smelting • Agricultural chemicals • Electronic waste • Lead in paint	• Toxic to people and wildlife, both plants and animals • Pose health risks to people and can affect children's development • Can leach into waterways
Asbestos	• insulation material in houses and factories • Disused mine sites	• Carcinogenic when fine particles are inhaled
Fuels and oil	• Industrial and motor vehicle spills • Oil wells and refineries • Damaged oil pipelines	• Toxic to soil organisms • Human illness and death — can be carcinogenic • Death of wildlife

(continued)

TABLE 2.14 Sources and effects of lithospheric pollutants *(continued)*

Lithospheric pollutant	Main sources	Main effects
Chemicals (e.g. pesticides, herbicides)	• Industry • Agricultural run-off • Landfill sites	• Toxic to non-target organisms • Health risks to people
Salt	• Land clearing • Poor agricultural practices	• Reduced crop yields • Leaching into water courses
Plastics	• Litter from urban areas • Waste disposal • Landfill sites	• Degrade into microplastics rather than decomposing • Microplastics absorbed by plants can bioaccumulate in the food chain

2.9.2 The impact of lithospheric pollutants

Lithospheric pollution can be hazardous to people, plants and animals, and the biophysical environment. Pollutants can enter people's bodies through direct contact with the skin, breathing in of contaminated dust particles, or ingesting food grown in contaminated soil. Possible effects include illness and disease, permanent damage to the nervous system and death.

Many soil contaminants affect the growth of plants, including crops, and can be toxic. Animals that consume plants grown in polluted soil can absorb hazardous amounts of toxins through a process of bioaccumulation. Some lithospheric pollutants degrade very slowly, if at all, and can persist and remain hazardous for long periods of time in the environment. Lithospheric pollutants can also leach into groundwater and rivers, or become airborne as dust particles, so contribute to hydrospheric and atmospheric pollution.

People living in developing countries are most vulnerable to the hazard of lithospheric pollution. Not only are people often directly exposed to hazardous pollutants, but poor or unsafe waste management also increases the risk of exposure to the pollutants.

EXAMPLE: Asbestos in New South Wales and Queensland, 2024

In early 2024, asbestos-contaminated mulch was found in more than 60 sites across Sydney, including schools, hospitals and parks. The resulting health emergency led to the closure of parks and schools. Contaminated mulch was also found at a school and a dog park in Brisbane in early 2024.

In both NSW and Queensland, asbestos-related health risks were minimised through early detection and remediation of affected locations.

FIGURE 2.51 Asbestos warning sign

CASE STUDY: E-waste pollution hazards at Agbogbloshie, Accra, Ghana

Quick facts: Ghana
- **Location** West Africa
- **Population** 34.4 million
- **Capital:** Accra
- **Human Development Index (HDI)** 0.62 (ranked 145 out of 193)
- **Sustainable Development Index (SDI)** 61.8/100 (ranked 122 out of 166)

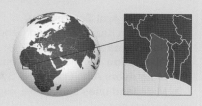

Electronic waste (discarded electrical and electronic equipment), is the fastest growing type of waste worldwide. In 2022, around 62 million tonnes of e-waste were produced, of which only around 22 per cent is known to have been properly collected and recycled. By 2030, e-waste produced is likely to total 82 million tonnes.

By the time the e-waste scrapyard in Agbogbloshie was demolished by the Ghanaian government in 2021, it was thought to be the largest e-waste dump, and among the top ten most polluted places, in the world. The site received discarded electronic goods and household appliances — such as computers, phones, microwaves, monitors and other electronic goods — that had not been repaired.

FIGURE 2.52 Rubbish being burnt in Agbogbloshie

By 2014, Agbogbloshie had already received 192 000 tonnes of waste. A significant industry has developed based on repairing and selling these products, but about 15 per cent of the electronic equipment is unusable or unrepairable, and is recycled for component parts. At Agbogbloshie, this work was done by the estimated 40 000 individuals who worked the dump site, including young children and teenagers from poor families who came to the capital seeking work. Most of these workers lived in the nearby informal settlement of Old Fadama.

The process of recycling e-waste involved workers dismantling some parts of the products, but extraction of the most precious metals from many of the components required burning off the casing and insulation — largely plastics — in fires set around the dump. The remaining melted metals were collected from the ashes and sold on to traders. The dump also 'disposed' of tyres, which were burnt for the metal reinforcements embedded in the rubber. While other more environmentally friendly ways of extracting the metals from these products are possible, burning the waste was the quickest and most cost-efficient method.

Many of the workers from the dump site continue to suffer from serious illnesses, such as cancers, from exposure to the toxic components of the electronics. Burnt e-waste releases a range of toxic substances and gasses, contaminating the soil and air. The dump was located on a former wetland, on the banks of the Odaw River, so some of the e-waste components also ended up in the river system.

The demolition of the scrapyard has not involved any decontamination of soils polluted with toxic heavy metals. Nor has it ended work on e-waste. Instead, this has been pushed closer to where people live and into people's homes. This means that many informal e-waste sites have begun to emerge across Accra and nearby cities, driven by the need for poor Ghanaians to earn a living. As a consequence, they continue to be exposed to the serious health risks associated with e-waste.

---

on Resources

 Weblinks Lead in Australia
The disposal of e-waste in Ghana
Australian e-waste found in Agbogbloshie
Asbestos information

CASE STUDY: Lithospheric pollution — Aral Sea region, Central Asia

Quick facts: Aral Sea region
- **Location** Kazakhstan and Uzbekistan in Central Asia
- **Basin area** 2.5 million km²
- **Population** 40 million in the basin, 4 million in Aral Sea region
- **Main rivers** Amu Darya and Syr Darya
- **Main land uses** agriculture, especially cotton, mining of gold, uranium and natural gas

The shrinking of the Aral Sea, once the world's fourth largest lake, is considered by some to be one of the world's worst ecological disasters.

From the 1960s, rivers that flowed into the sea began to be diverted for agriculture, especially the irrigation of cotton. By the 1980s, almost 90 per cent of water flowing to the Aral Sea had been diverted, and by 1997 the sea had shrunk to about 10 per cent of its original size.

The sea's desiccation destroyed its natural water-based and wetland ecosystems and led to the loss of a previously thriving commercial fishing industry. As the sea receded, it became increasingly saline, and was polluted by toxic agricultural chemicals and heavy metals. Once the sea floor was exposed, these pollutants were left in the lithosphere and became hazardous to people living near the area.

In addition, the exposed seabed may contain remnants of anthrax, smallpox and plague from a decommissioned bioweapons testing site on the former Aral Sea island of Vozrozhdeniya, exposing people to deadly disease hazards.

FIGURE 2.53 Changes to the Aral Sea, 1960–2017

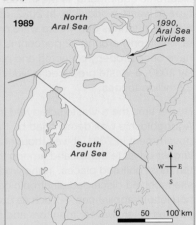
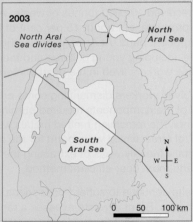
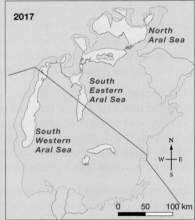

Source: MAPgraphics with data from the U.S. Geological Survey

Because of the region's arid climate, wind-blown dust is common. Contaminated soil from the dry sea floor is picked up by strong winds and spread across the region. Over time, people living near the Aral Sea were exposed to contaminants and began to suffer from higher rates of respiratory, heart and kidney disease and cancers. Children also suffer higher levels of birth defects and stunted growth. Agricultural soils have been polluted by wind-blown dust, so local food with high levels of contamination is consumed by people.

FIGURE 2.54 The Aral Sea in (a) 1973 and (b) 2007

(a)　(b)

Source: australscope

Several strategies have been implemented over time to reduce the level of risk faced by communities and the environment in the region.

- The World Bank's Kok-Aral project in Kazakhstan, completed in 2005, built a dam between the northern and southern parts of the Aral Sea and boosted water levels in the North Aral Sea. The southern part of the sea, however, continued to degrade.
- The United Nations' Multi-Partner Human Security Trust Fund for the Aral Sea Region in Uzbekistan program, which commenced in 2018, focuses on helping communities and reducing poverty in areas affected by the degradation of the Aral Sea.
- Uzbekistan began a 'Green Nation' initiative in 2021 with the aim of planting one billion trees and shrubs across the country by 2026, including in the Aral Sea region.

FIGURE 2.55 Ships on the dry bed of the Aral Sea

- The United States aid agency (USAID) commenced Environmental Restoration of the Aral Sea programs from 2021, which aim to assist with environmental restoration activities. For example, one program in Kazakhstan is creating an 'Oasis' by planting black saxaul trees to help stabilise sandy soils and reduce desertification.

SKILLS ACTIVITY

1. Refer to figures 2.53 and 2.54 and complete the following.
   a. Describe how the size and location of the Aral Sea has changed over time.
   b. Explain how these changes have contributed to ecological hazards in the Aral Sea region.
   c. Create a table to identify social, economic and environmental impacts of these hazards.

2.9 Exercise

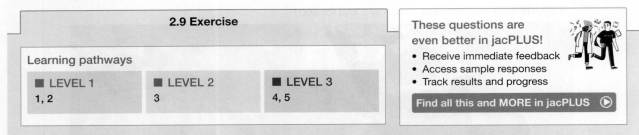
Explain and comprehend

1. **List** three ways in which people can be exposed to lithospheric pollutants.
2. **Describe** the health risks associated with exposure to e-waste.

Analyse and apply

3. **Refer** to figure 2.56.
   a. **Analyse** the graph and describe the patterns of global e-waste generation, collection and recycling.
   b. **Identify** possible implications for people and environments of these patterns.

FIGURE 2.56 E-waste generation and recycling rates per capita in different geographical regions

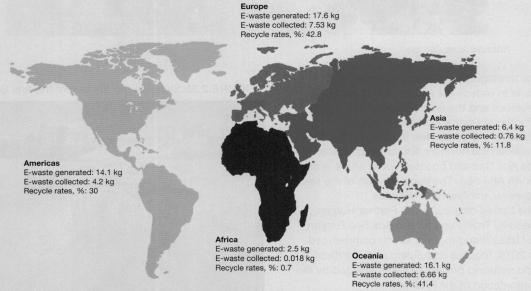

Europe
E-waste generated: 17.6 kg
E-waste collected: 7.53 kg
Recycle rates, %: 42.8

Asia
E-waste generated: 6.4 kg
E-waste collected: 0.76 kg
Recycle rates, %: 11.8

Americas
E-waste generated: 14.1 kg
E-waste collected: 4.2 kg
Recycle rates, %: 30

Africa
E-waste generated: 2.5 kg
E-waste collected: 0.018 kg
Recycle rates, %: 0.7

Oceania
E-waste generated: 16.1 kg
E-waste collected: 6.66 kg
Recycle rates, %: 41.4

Source: Adapted from the Global E-waste Monitor 2024 https://ewastemonitor.info/wp-content/uploads/2024/03/
GEM_2024_18-03_web_page_per_page_web.pdf.

4. Refer to figure 2.57.
   a. **Analyse** the graphs and describe the progress that Ghana has made in achieving its long-term objective for
      the two indicators.
   b. **Explain** possible implications of the trends for people and the environment in Agbogbloshie and the Old
      Fadama informal settlement.
   c. **Identify** ways in which the Ghanaian government might be able to reduce the risk of the hazards involved
      with e-waste in these two places.

FIGURE 2.57 Indicators for Ghana: (a) proportion of urban population living in slums and (b) poverty headcount ratio at $3.65/day

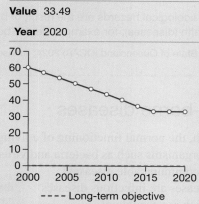

(a) Proportion of urban population living in slums %

Value 33.49

Year 2020

---- Long-term objective

Source: Housing, slums and informal settlements. UN-Habitat Indicators Database. United Nations Human Settlements Programme. Retrieved from https://data.unhabitat.org/pages/housing-slums-and-informal-settlements.

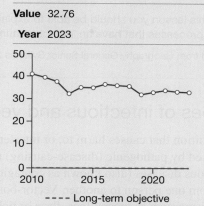

(b) Poverty headcount ratio at $3.65/day 2017 PPP, %

Value 32.76

Year 2023

---- Long-term objective

Source: WORLD POVERTY CLOCK by World Data Lab, https://worldpoverty.io/

5. Select one action that has been taken to manage the impact of lithospheric hazards in the Aral Sea region. **Explain** how this action is helping to reduce the effects of the hazards.

Sample responses are available in your digital formats.

LESSON
2.10 Infectious and vector-borne diseases

LEARNING INTENTION

By the end of this lesson you should be able to explain how ecological hazards are the result of biological and anthropogenic processes that have an impact on human health (diseases); for example, vector-borne diseases.

Source: Adapted from Geography General Senior Syllabus 2024 © State of Queensland (QCAA) 2024; licensed under CC BY 4.0.

2.10.1 Types of infectious and vector-borne diseases

Disease is a condition that causes harm to, or interferes with, the normal functioning of a living thing. Many diseases are caused by pathogenic (disease-causing) microorganisms such as bacteria and viruses and by parasites. Infectious diseases (also known as contagious or communicable diseases) can be passed from one person to another. Vector-borne diseases are infectious diseases carried by organisms such as mosquitoes, fleas or ticks, which are capable of transmitting pathogens from one person to another.

disease a condition that causes harm to, or interferes with, the normal functioning of a living thing

FIGURE 2.58 Polluted drinking water is a common source of infectious disease for poor communities.

FIGURE 2.59 Mosquitoes can act as vectors for many diseases.

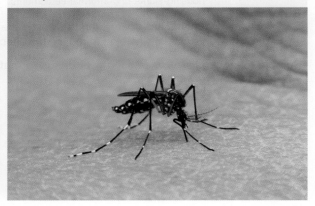

Infectious diseases are usually transmitted via water, air or food, or by vectors that carry or contain the disease-causing organism.

- Water-borne infectious diseases include cholera, typhoid, botulism, polio and giardia.
- Air-borne infectious diseases include COVID-19, influenza, tuberculosis, smallpox and chickenpox.
- Food-borne diseases include those caused by salmonella, campylobacter, norovirus and E. coli.
- Vector-borne diseases include malaria, dengue fever, yellow fever and Lyme disease.

More than 90 per cent of all infectious disease deaths are the result of a small number of diseases, predominantly pneumonia, diarrhoeal diseases and tuberculosis (see table 2.15).

TABLE 2.15 Global Health Observatory data (WHO), 2015

Disease	Deaths	DALYs (disability-adjusted life year)*
Lower respiratory infections	3 190 300	142 384 000
Diarrhoeal diseases	1 388 600	84 928 000
Tuberculosis	1 373 200	56 037 000
HIV/AIDS	1 000 000	62 759 000
Malaria	429 000	38 520 000
Measles	89 780	12 278 708

Note: *DALY is the disability-adjusted life year, a measure of overall disease burden.
Source: ©WHO 2018 http://www.who.int/sustainable-development/news-events/breath-life/air-pollution-by-numbers.jpg

2.10.2 Disease hazard zones

Infectious and vector-borne disease hazard zones can range in scale, from a single village to worldwide. When a disease is prevalent across a wide geographic area at a particular time, it is known as a **pandemic**. COVID-19, the influenza outbreak in 1918, and the Black Death, which swept across Asia and Europe in the 14th century, are examples of pandemics.

When a disease is common in or specific to a particular place, it is known as **endemic**. For example, in many tropical countries, malaria is endemic. **Epidemics** are sudden outbreaks of a disease that spreads rapidly among many people in a community at the same time. An outbreak of influenza is an example of an epidemic.

FIGURE 2.60 The outbreak of COVID-19 became a global pandemic.

2.10.3 Factors contributing to the risk of disease

Climate conditions and climate change

Many infectious diseases are climate sensitive. This is especially the case for vector-borne diseases. Mosquitoes, for example, are much more prevalent in warmer, wetter climate regions, so diseases the mosquitoes spread, such as malaria, yellow fever and dengue, are often endemic in many tropical countries (see figure 2.61).

Water-borne diarrhoeal diseases such as cholera can also be climate-related; cholera outbreaks occur most commonly in wet seasons. Often, diarrhoeal disease outbreaks are a secondary impact of seasonal natural hazard events such as tropical cyclones. Some infectious diseases, such as influenza, are more common in winter.

Climate change is likely to increase the risk of exposure to infectious diseases worldwide. For example, the geographic range and transmission seasons for vector-borne diseases are likely to grow as the climate warms. Temperate climate zones currently at less risk may experience more disease outbreaks, (see figure 2.61). The risk of water-borne diarrhoeal diseases may also increase with increased climate variability in the future.

pandemic a disease outbreak that is prevalent across a wide geographic area, including beyond the region in which the outbreak began

endemic located in a specific area

epidemic a disease outbreak that affects many people in a specific region

FIGURE 2.61 Re-emergence of significant mosquito-borne diseases, 2017–2019

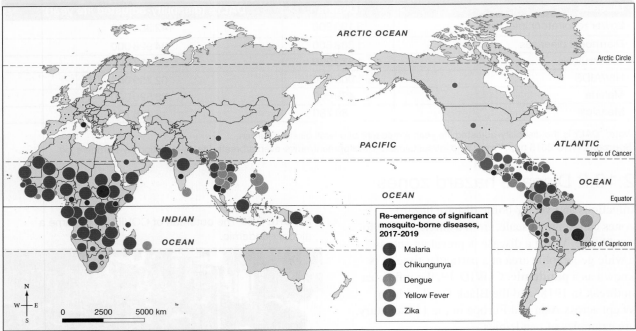

Re-emergence of significant mosquito-borne diseases, 2017-2019
- Malaria
- Chikungunya
- Dengue
- Yellow Fever
- Zika

Source: Based on data from Dahmana H, Mediannikov O. Mosquito-Borne Diseases Emergence/Resurgence and How to Effectively Control It Biologically. *Pathogens*. 2020; 9(4):310. https://doi.org/10.3390/pathogens9040310. Map redrawn by Spatial Vision.

FIGURE 2.62 Projected change in malaria prevalence by 2080

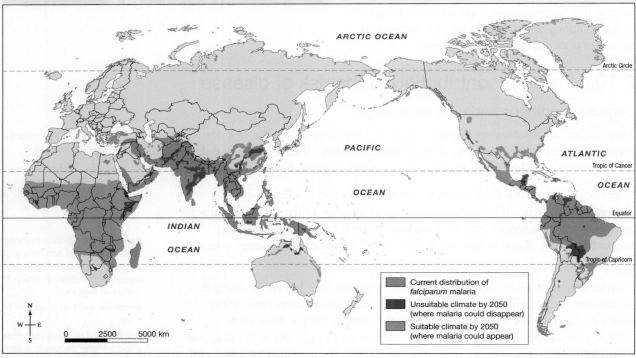

- Current distribution of *falciparum* malaria
- Unsuitable climate by 2050 (where malaria could disappear)
- Suitable climate by 2050 (where malaria could appear)

Source: David J. Rogers, Sarah E. Randolph, *The Global Spread of Malaria in a Future, Warmer World*. Ahlenius H. 2005, *UNEP/GRIO Arendal Maps and Graphics Library*. © European Environment Agency (EEA). Map redrawn by Spatial Vision.

Levels of economic development

Poverty is a major factor affecting the risk of infectious diseases; poorer countries are much more vulnerable to infectious and vector-borne diseases than wealthier countries. The main factors that make people in low-income countries more vulnerable to health problems and disease include the following.

- Physical factors, such as extremes of climate, natural disasters and limited access to safe water
- Socioeconomic factors, such as lack of adequate sanitation, health infrastructure, education, skills and technology, and urban overcrowding.
- Political factors, such as poor governance, corruption, limited human rights (especially for women and children), and civil unrest.

A combination of these factors means that more than half of all deaths in low-income countries are caused by infectious diseases, nutritional deficiencies and maternal causes (conditions arising during pregnancy and childbirth). In wealthier countries, fewer than 7 per cent of deaths are a result of these causes. (See figure 2.63 for an outline of the top ten causes of death in Nigeria versus Australia in 2019, and figure 2.64 for global deaths from communicable maternal, neonatal and nutritional diseases.)

FIGURE 2.63 Disease burden versus GDP per capita

Disease burden to communicable, material, neonatal and nutritional diseases, measured in DALYs (disability-adjusted life years) per 100,000 individuals versus gross domestic product (GDP) per capita, measured in constant, international -$

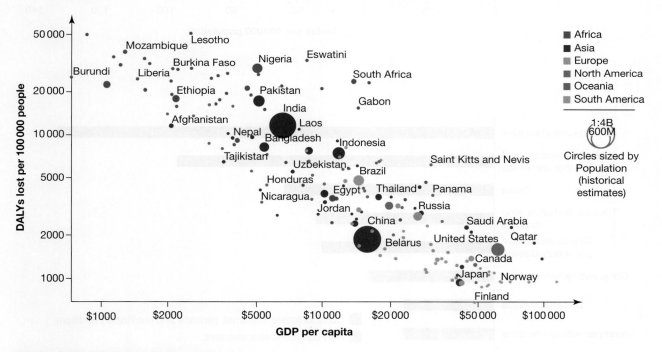

Source: OurWorldInData.org/burden-of-disease.

Data source: IHME, Global Burden of Disease (2019); data compiled from multiple sources by World Bank. Licensed under CC-BY.

FIGURE 2.64 Top ten causes of death in (a) Nigeria (a low-income country) and (b) Australia (a high-income country) in 2019

(a)

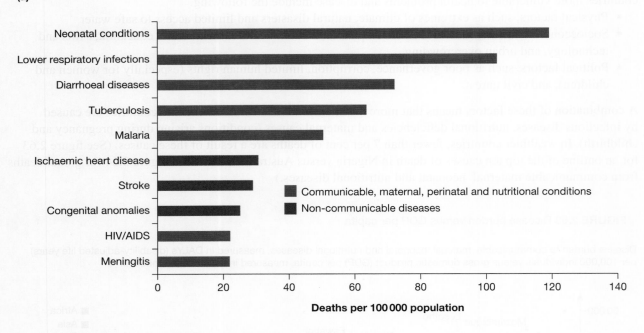

(b)

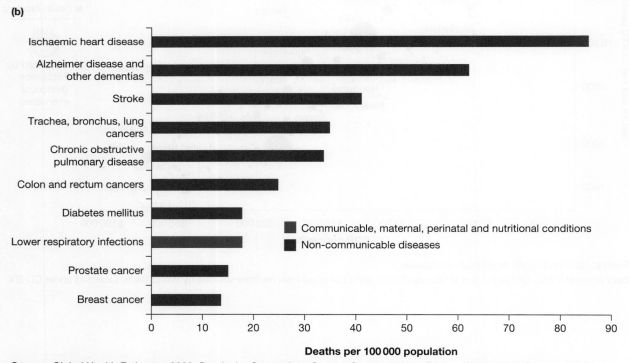

Source: Global Health Estimates 2020: Deaths by Cause, Age, Sex, by Country and by Region, 2000–2019. Geneva, World Health Organization; 2020.

FIGURE 2.65 Deaths from communicable maternal, neonatal and nutritional diseases

Burden of disease, 2019

Disability-Adjusted Life Years (DALYs) per 100,000 individuals from all causes. DALYs measure the total burden of disease – both from years of life lost due to premature death and years lived with a disability. One DALY equals one lost year of healthy life.

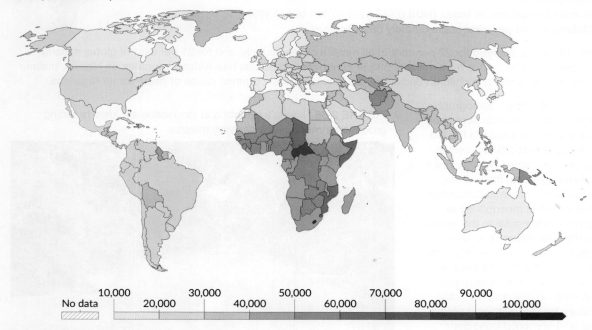

| No data | 10,000 | 30,000 | 50,000 | 70,000 | 90,000 |
| | 20,000 | 40,000 | 60,000 | 80,000 | 100,000 |

Data source: IHME, Global Burden of Disease (2019)   OurWorldInData.org/burden-of-disease | CC BY

Note: To allow for comparisons between countries and over time, this metric is age-standardized[1].

---

1. Age standardization: Age standardization is an adjustment that makes it possible to compare populations with different age structures by standardizing them to a common reference population. 📄 Read more: How does age standardization make health metrics comparable?

Source: OurWorldInData.org| burden-of-disease, licensed under CC BY. Data source: IHME, Global Burden of Disease (2019).

SKILLS ACTIVITY

1. Refer to figure 2.61.
   a. Describe the spatial pattern of mosquito-borne diseases illustrated on the map.
   b. Identify possible risk factors responsible for the patterns identified.
2. Refer to figure 2.62.
   a. Describe the pattern of change in malaria prevalence illustrated on the map. Refer to specific places (countries and regions) in your answer.
   b. Suggest possible factors responsible for the patterns identified.
   c. Describe and explain the likely changes in malaria prevalence in Australia by 2080.
3. Refer to figure 2.63.
   a. Analyse the graph and describe the relationship between disease burden due to communicable diseases and GDP per capita. Refer to specific places (regions and countries) in your answer.
   b. Suggest possible factors responsible for the patterns identified.
   c. Identify two country outliers. Explain what makes these two countries outliers.
   d. Provide possible reasons for these countries not fitting the general pattern.
4. Refer to figure 2.65.
   a. Describe the spatial pattern of deaths from communicable maternal, neonatal and nutritional diseases.
   b. Suggest possible reasons for the patterns that you have identified.

CASE STUDY: Nigeria and malaria

Quick facts: Nigeria
- **Location** West Africa
- **Population** 220 million
- **Capital:** Abuja
- **Human Development Index (HDI)** 0.548 (ranked 161 out of 193)
- **Malaria cases per year** approximately 68 million

In 2022, Nigeria accounted for 27 per cent of all malaria cases globally, and for 31 per cent of global malaria-related deaths. Of all deaths from malaria, 53 per cent occurred across four African countries. In Nigeria, malaria is the top cause of death in children under five, and the fifth most common cause of death for all Nigerians.

Environmental, social, demographic, economic and political factors make Nigeria a significant malaria hazard zone. Because malaria is a vector-borne disease carried by *Anopheles* species mosquitoes, the climatic and other environmental conditions needed for these mosquitoes to breed are fundamental to the spread of the disease. Nigeria's location between 3°N and 14°N results in the country's tropical and subtropical climate, with warm temperatures, and the high rainfall and humidity necessary for mosquitoes. Another critical environmental condition is the presence of water bodies, including temporary ponds after heavy rainfall, where mosquitoes can breed. Climate change may increase the risk of malaria, especially in currently drier northern Nigeria.

FIGURE 2.66 Nigeria's climate and landscape are ideal breeding grounds for mosquitoes carrying malaria.

Social, demographic and economic factors such as low levels of education, poor health education, population growth, poverty, land use changes and deforestation all help to account for malaria risk in Nigeria. Nigeria has relatively low levels of human development, ranking 161st out of 193 countries in the 2023 Human Development Index. This means that Nigeria faces major challenges in the prevention, treatment and control of malaria.

In addition to these factors, the political situation in Nigeria also presents challenges. There are, for example, major problems with terrorist and bandit groups in the north-west of the country, and secessionist pressures in the south-east Biafran region. This can make it even more difficult for aid to reach people in these areas.

Despite the country's ongoing challenges, the Nigerian government has committed to the goals of reducing deaths from malaria over time and eliminating the disease. Nigeria has a National Malaria Elimination Program (NMEP), and in 2022 established a Nigeria End Malaria Council to assist in achieving its goals. Existing methods to mitigate the risk of malaria include insecticide-treated nets (ITNs), indoor residual spraying (IRS) for mosquitoes, and artemisinin-combination therapy (ACT), These methods, together with the 2021 World Health Organization–approved vaccine for children against malaria, the Mosquirix RTS,S malaria vaccine, should help reduce the risk of malaria in Nigeria. The country will, however, also have to deal with the underlying social, demographic, economic and political causes of malaria.

SKILLS ACTIVITY

1. Refer to table 2.16 and complete the following.

TABLE 2.16 Malaria cases in four countries, 2001–2021 (000s)

Year	Nigeria	Democratic Republic of Congo (DRC)	Mozambique	Uganda
2001	51 229	23 541	8997	12 579
2003	52 423	25 246	9278	12 758
2005	57 411	27 239	8517	12 118
2007	61 400	27 977	8256	12 308
2009	63 192	26 605	8469	13 289
2011	58 184	25 108	8783	13 141
2013	53 921	23 984	9291	11 722
2015	54 115	23 263	9403	9501
2017	57 869	25 976	9422	11 833
2019	61 379	28 830	9421	11 282
2021	65 399	30 518	10 283	13 023

Source: Data based on estimated number of malaria cases, Global Health Observatory data. © 2014 WHO. Retrieved from https://www.who.int/data/gho/data/indicators/indicator-details/GHO/estimated-number-of-malaria-cases

a. Construct a multiple line graph to illustrate the changes over time in malaria cases in the four countries.
b. Discuss what might explain the similarities and/or differences in the patterns for the four countries?
c. Compare the patterns of malaria cases shown on your graph with the patterns of malaria fatality rates shown on figure 2.67.

FIGURE 2.67 Malaria death rates, Nigeria, Mozambique, Democratic Republic of Congo and Uganda

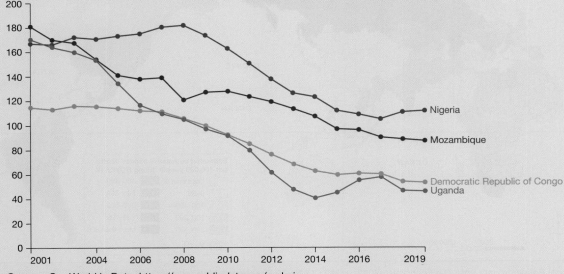

The number of deaths from malaria per 100 000 people.

Source: Our World in Data, https://ourworldindata.org/malaria

CASE STUDY: COVID-19 pandemic

Quick facts: COVID-19
- **Disease virus** SARS-CoV-2
- **Origin** Wuhan, China
- **First Australian case** 25 January 2020 in Melbourne
- **Total deaths** more than 24 000 in Australia, over 7 million worldwide

COVID-19 is an air-borne, viral disease that most likely had its origins in an animal host (known as a zoonotic disease). The SARS-CoV-2 virus caused the first major global pandemic since the outbreak and global spread of an influenza virus between 1918 and 1920.

The COVID-19 pandemic began with an outbreak of pneumonia in Wuhan in mid-December 2019. From there, the disease spread across the world, and affected almost every country. Over the course of the pandemic, more than 700 million people were infected and, by March 2024, over 7 million people had died. In contrast, experts estimated that over 50 million people died during the 1918–20 influenza pandemic.

As is the case when responding to all hazardous events, the COVID-19 response had several stages. In Australia, the initial emergency response involved medical treatment of people who had contracted COVID-19 and the banning of travel — firstly from China, and then from South Korea and Italy. In an attempt to reduce the risk of the disease spreading, large gatherings of people, both indoors and outdoors, were also banned. Subsequently, people were banned from travelling overseas, and a 14-day quarantine was put in place for overseas arrivals.

As the COVID-19 pandemic continued, and before the introduction of vaccinations, state governments began to implement lockdowns and the closure of schools and workplaces, other than for essential services. This meant that people who were able to began to work from home and schools introduced online classes. Once vaccinations were available and lockdowns ended, most states began to use check-in apps and encouraged or mandated mask-wearing.

FIGURE 2.68 Estimated cumulative excess deaths per 100 000 people during COVID-19

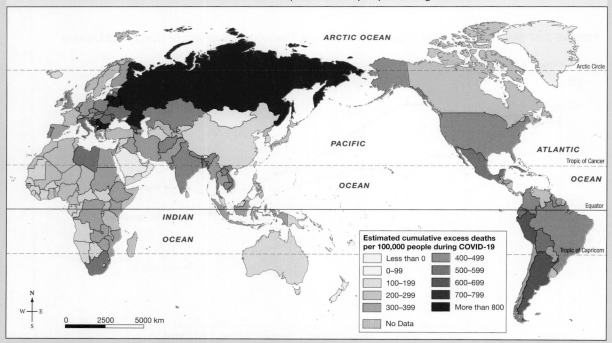

Source: Based on data from *The Economist* (2022); WHO COVID-19 Dashboard. Map redrawn by Spatial Vision.

COVID-19 had wide-ranging health, social and economic impacts. The effects of the disease varied from very mild to extremely severe. People with very severe cases often ended up in hospital and some needed a ventilator to help them breathe. Older people with severe cases often died from the disease. Social impacts were often a secondary effect of lockdowns, and included restrictions on recreation and sporting activities, and social isolation. Widespread economic costs to governments, businesses and individuals emerged, although costs to businesses and individuals were partly offset by the federal government's income support scheme (called JobKeeper).

Resources

🔗 **Weblinks**   Current malaria data
                 Reducing malaria cases

2.10 Exercise

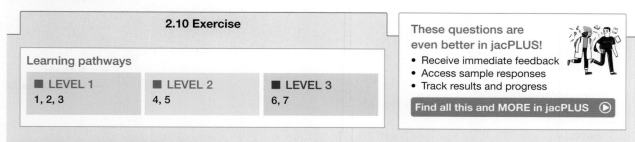

2.10 Exercise
Learning pathways
■ LEVEL 1 — 1, 2, 3   ■ LEVEL 2 — 4, 5   ■ LEVEL 3 — 6, 7

These questions are even better in jacPLUS!
- Receive immediate feedback
- Access sample responses
- Track results and progress

Find all this and MORE in jacPLUS ▶

Explain and comprehend

1. **Define** the term 'pandemic' and provide an example from the text.
2. **Describe** the factors that make people in low-income countries more vulnerable to infectious diseases.
3. **Explain** the role of vaccination programs in preventing and mitigating the effects of infectious diseases.

Analyse and apply

4. Refer to figure 2.68.
   a. **Describe** the spatial patterns of COVID-19 deaths illustrated by the choropleth map.
   b. **Suggest** possible reasons for the patterns identified.
5. Refer to figure 2.69.
   a. **Compare** the patterns of deaths illustrated by the graphs for Australia and South Korea.
   b. **Suggest** what might explain the similarities and/or differences in the patterns for the two countries.

FIGURE 2.69 Daily new confirmed COVID-19 deaths per million people, Australia and South Korea

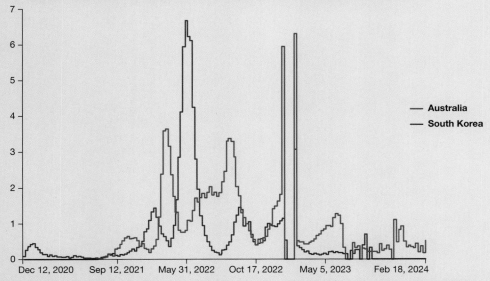

7-day rolling average. Due to varying protocols and challenges in the attribution of the cause of death, the number of confirmed deaths may not accurately represent the true number of deaths caused by COVID-19.

Source: OurWorldInData.org | malaria. Licensed under CC BY. Data source: IHME, Global Burden of Disease

Propose and communicate

6. Refer to figure 2.70.
   a. **Compare** the patterns of COVID-19 cases illustrated by the graph for Australia and the United States.
   b. **Propose** a possible explanation for the similarities and/or differences in the patterns for the two countries.

FIGURE 2.70 Daily new confirmed COVID-19 cases, Australia and the United States of America

7-day rolling average. Due to limited testing, the number of confirmed cases is lower than the true number of infections.

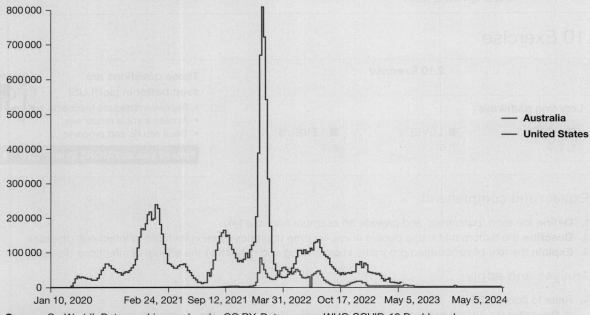

Source: OurWorldInData.org. Licensed under CC BY. Data source: WHO COVID-19 Dashboard.

7. **Evaluate** the effectiveness of strategies implemented by the Nigerian government to reduce the prevalence of malaria, considering environmental, social, economic and political factors.

Sample responses are available in your digital formats.

2.11 APPLY YOUR SKILLS — Analysing spatial and statistical data

2.11.1 Analysing spatial and statistical data

Countries with high levels of economic and social development almost always have universal access to safe drinking water services, so diseases associated with polluted drinking water are extremely rare. In contrast, access to safe drinking water is often far from universal in poorer, less economically developed countries, so water-borne diseases are much more common. Such water-borne diseases can be a major contributor to the deaths of babies and young children, resulting in high infant and under-five years mortality rates.

TABLE 2.17 Scattergraph data set

Country	People using at least basic drinking water services, 2022 (% of population)	Under five mortality rate, 2022 (per 1000 population)
Angola	58	66.8
Benin	67	80.7
Burkina Faso	50	78.7
Cambodia	78	23.7
Central African Republic	36	96.8
Colombia	98	12.4
Comoros	80	48.2
Congo Republic	74	41.6
Fiji	95	28.2
Gambia	86	45.6
Ghana	88	42.3
Guinea	71	95.9

Source: People using at least basic drinking water services (% of population), World Bank data from WHO/UNICEF Joint Monitoring Programme (JMP) for Water Supply, Sanitation and Hygiene. The World Bank Group. All rights reserved. Retrieved from https://data.worldbank.org/indicator/SH.H2O.BASW.ZS

SKILL ACTIVITY

Constructing a scattergraph

A scattergraph is used to plot and compare two sets of data and investigate whether a link exists between them. The idea is to plot two sets of data — one using the horizontal or *x*-axis and the other using the vertical or *y*-axis, and then determine if the distribution of points shows a pattern. If a patten exists, you can draw a trend line or 'line of best fit' on the graph to illustrate the trend.

If a link can be shown between the two sets of data, this is referred to as a correlation. It may be positive or negative, depending on the slope of the pattern of plotted points. Important things about interpreting scattergraphs are as follows.
- If data on the *y*-axis increases as data on the *x*-axis increases, the correlation is positive.
- If data on the *y*-axis decreases as data on the *x*-axis increases, the correlation is negative.

▶

- If points are scattered randomly with no clear trend, no correlation exists.
- If a plotted point is well away from the line of best fit, it is called an anomaly.

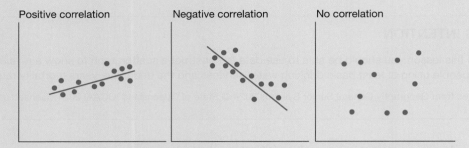

Positive correlation   Negative correlation   No correlation

Create a scattergraph

1. What patterns or trends can you identify just by looking at the data in table 2.17?
2. Rank the ten countries in the table by:
   a. the percentage of people using at least basic drinking water services in 2022.
   b. the under-five mortality rate in 2022.
3. Does this ranking list now show a more evident pattern or trend?
4. Construct a scattergraph by plotting the percentage of people using at least basic drinking water services (this is the independent variable) on the *x*-axis and the under-five mortality rate (the dependent variable) on the *y*-axis.*
5. If you can see a pattern or trend of plotted points on the completed graph, identify the correlation (positive or negative), and draw in a line of best fit. Identify any anomalies.
6. Describe the pattern illustrated by the completed scattergraph. Suggest reasons to explain the correlation and any anomalies.

Remember that changes in the independent variable determine changes in the dependent variable, so your reasons should refer to the percentage of people using at least basic drinking water services first.

*Note: when comparing two sets of data to determine whether or not they are linked, you should identify which of the two is causing, or leading to, the other. In this activity, the lack of safe water is likely to lead to higher under-five mortality rates. This means that the percentage of people using at least basic drinking water services is the independent variable, and should be plotted on the *x*-axis.

Sample responses are available in your digital formats.

 Resources

▶ **Video eLesson**   SkillBuilders: Constructing and interpreting a scattergraph (eles-1756)

🧩 **Interactivity**   SkillBuilders: Constructing and interpreting a scattergraph (int-3374)

LESSON
2.12 Review

2.12.1 Summary

Ecological hazards

- Ecology studies the relationships between organisms and their environments. Ecological hazards are sources of danger that can cause harm to people and environments, such as diseases, pollutants and invasive species.
- Ecological hazards can result from biological processes, such as diseases caused by bacteria and viruses, or anthropogenic processes, such as pollution from industrial activities. These hazards can lead to health issues, environmental damage, and social and economic disruption.
- Infectious diseases, pollution (such as air and water pollution), and invasive species (such as rabbits and cane toads in Australia) are examples of ecological hazards. These hazards affect human health, the physical environment, and can disrupt livelihoods and ecosystems. Ecological hazard zones are areas at risk, such as schools during flu season, or even an entire country or region experiencing a disease outbreak.

The impacts of ecological hazards

- The severity of ecological hazards depends on various factors, including speed of onset, magnitude, frequency, duration, temporal spacing, mobility, location and climate change. For example, rapid-onset hazards can be unpredictable and more dangerous, while slow-onset hazards allow for better preparation.
- The magnitude of hazards varies; a small oil spill from a boat is less severe than a large tanker spill. Frequency also affects impact; small, frequent events are usually less severe than rare, large-scale events such as the 1918 influenza pandemic. Duration and temporal spacing of hazards, such as seasonal flu outbreaks, influence their severity.
- Managing ecological hazards involves identifying, preventing (if possible) and preparing for hazards to reduce damage. Measures vary by hazard type; for example, managing invasive plants differs from handling disease outbreaks. Effective risk management includes early detection, mitigation strategies and sometimes vaccinations.

Plant and animal invasions

- Invasive plants and animals are species introduced by human activities beyond their normal distribution, posing threats to environmental, agricultural or social resources. Invasive plants and animals are not native, result from human actions and are hazardous. They often grow rapidly, spread easily, thrive in diverse habitats, outcompete native species and are difficult to control.
- Invasive plants and animals have significant environmental, economic and social impacts. Environmentally, they threaten biodiversity, leading to species extinction and habitat degradation. Economically, they incur direct costs (monitoring and control) and indirect costs (agricultural damage and reduced production). Socially, they affect amenities, First Nations of Australia practices, health and recreation.
- Invasive species threaten all continents except Antarctica, with island nations such as Australia and New Zealand being particularly vulnerable. European colonisation often introduced invasive species. These species pose the greatest threat to environments similar to their native habitats, and climate change can alter areas at risk of invasion.

Reducing the risk of plant and animal invasions

- Prevention and early eradication are the most cost-effective strategies to reduce the risk of plant and animal invasions. As invasions progress, management becomes more expensive and less beneficial, making early intervention essential.
- Early detection and control are crucial to managing invasive species. Methods include chemical, mechanical and biological controls, though each has its own costs and potential negative impacts.

- Control methods for established invasive species include chemical, mechanical and biological approaches. Chemical control uses herbicides and pesticides but can harm non-target species, be expensive and cause pollution. Mechanical control involves physical removal, trapping, shooting and fencing, and this can be effective in small areas. Biological control uses natural enemies or diseases, such as the cactoblastis moth for prickly pear cactus or viruses for rabbits.

Pollutants

- Pollutants are harmful substances introduced into the environment, mostly resulting from human activities such as industrial processes, transport, agriculture, mining and domestic sources. Pollution can be categorised as point source or non-point source, with the latter being more challenging to manage.
- Pollution affects the atmosphere, hydrosphere, lithosphere and biosphere. Effective risk reduction involves prevention and control, with point sources being easier to manage than non-point sources. Factors such as waste amount, biodegradability and persistence influence the hazard level.
- Pollutants can enter the food chain through bioaccumulation and biomagnification, increasing risks to humans. Some pollutants also create secondary pollutants such as acid rain, formed from reactions of primary pollutants sulphur dioxide and nitrogen oxides in the atmosphere.

Hydrospheric pollutants

- Hydrospheric pollution affects both surface water bodies (rivers, lakes, seas and oceans) and groundwater, with pollutants originating from various sources, including factories, farms, mines, sewage plants, power stations and urban settlements.
- Some pollutants dissolve in water directly from their source, while others are released into the atmosphere first and then deposited into water bodies, such as sulphur dioxide and carbon dioxide, leading to ocean acidification.
- The impact of hydrospheric pollutants includes hazards to human health, with around 1.4 million premature deaths worldwide due to polluted drinking water or unsafe sanitation, particularly affecting populations in developing countries. Other environmental, social and economic impacts are also significant.

Atmospheric pollutants

- Atmospheric pollution involves substances in the air exceeding naturally occurring levels and potentially endangering human health and the environment. Primary pollutants come directly from sources such as motor vehicles, factories and power stations, while secondary pollutants form when primary pollutants combine or undergo chemical reactions.
- Various types of atmospheric pollutants include ozone, which forms through reactions between nitrous oxides and hydrocarbons, and solid particles such as particulate matter (PM2.5 and PM10), which can cause lung and heart diseases due to inhalation.
- Air pollution poses significant health risks, contributing to a substantial portion of global deaths from stroke, lung cancer, respiratory diseases and heart disease, particularly affecting populations in low- and middle-income countries. Additionally, air pollution negatively impacts the natural environment, affecting animals and plants and contributing to phenomena such as acid rain, further exacerbating hydrospheric pollution.

Lithospheric pollutants

- Lithospheric pollution involves substances in the soil exceeding naturally occurring levels, potentially endangering human health and the environment. Pollutants can be organic (e.g. oil-based chemicals, some pesticides) or inorganic (e.g. salt, and metals such as lead, mercury, cadmium and arsenic).
- Human activities such as manufacturing, agriculture, mining and urban waste disposal are major contributors to lithospheric pollutants, which can be transported over long distances from their original source. For instance, electronic waste products are often shipped from wealthier countries to poorer ones, leading to environmental pollution and health risks.

Infectious and vector-borne diseases

- Diseases caused by microorganisms and parasites, including bacteria, viruses and parasites, can be infectious or vector-borne, transmitted via water, air, food or vectors such as mosquitoes.
- Common infectious diseases include cholera, COVID-19, malaria and tuberculosis, with pandemics such as COVID-19 affecting wide geographic areas, while endemic diseases such as malaria are specific to certain regions.
- Climate change influences disease spread, with warmer, wetter climates favouring vector-borne diseases, and economic factors such as poverty exacerbating vulnerability to infectious diseases in low-income countries due to limited access to resources and health care.

2.12.2 Key terms

acidification an increase in hydrogen ions, which lowers the pH of the water

animal invasions the introduction of a non-indigenous animal to an area that negatively impacts the environment

anthropogenic processes processes that involve human activity; for example, the burning of fossil fuels to produce electricity

aquifers layers of rock that can hold large quantities of water in the pore spaces

bioaccumulation the process by which pollutants enter and concentrate through the food chain by passing from one food source to another by being eaten

biodegrade to break down through natural processes

biological processes that are vital for organisms to live, for example, plants require the process of photosynthesis to survive

disease a condition that causes harm to, or interferes with, the normal functioning of a living thing

ecological hazard an interaction between living organisms or between living organisms and their environment that could have a negative effect

endemic located in a specific area

epidemic a disease outbreak that affects many people in a specific region

hazard something that has the potential to cause harm. It may be obvious (for example, a flooded section of road) or not obvious (for example, a damaged hidden electrical wire).

indigenous something that is native to or originates from a specific place; for example, kangaroos are indigenous to Australia. Note that the word often has a different meaning when capitalised. The term 'Indigenous' refers to First Nations Peoples of Australia of Aboriginal or Torres Strait Islander descent.

infectious diseases contagious or communicable diseases that are spread by being passed from one person to another

invasive animal a non-indigenous animal species that has been introduced to a specific area by people (either intentionally or accidentally) and has multiplied to an extent that it threatens to damage or is damaging the economic, environmental or social value of a place

invasive plant a non-indigenous plant species that has been introduced to a specific area by people (either intentionally or accidentally) and has multiplied to an extent that it threatens to damage or is damaging the economic, environmental or social value of a place

non-point sources broad areas a pollution hazard originates from; for example, run-off from city streets

pandemic a disease outbreak that is prevalent across a wide geographic area, including beyond the region in which the outbreak began

plant invasions the introduction of a non-indigenous plant to an area that negatively impacts the environment

point sources a particular location from which a pollution hazard originates; for example, an industrial site or an oil tank

pollutants substances introduced into the environment that are potentially harmful to human health and the natural environment

risk the potential for something to go wrong. This is a subjective assessment about actions that may be predictable or unforeseen.

vector-borne diseases diseases that are carried by organisms, such as mosquitos or fleas, that are capable of transmitting disease-producing bacteria, viruses and parasites from one person to another

vulnerability the degree of risk faced by a place, person or thing, based on an approaching hazard's potential impact, the place's degree of preparedness, and resources available to respond

KEY QUESTIONS REVISITED

1. What is an ecological hazard?
2. Where do ecological hazards occur and why?
3. What biological and anthropogenic factors influence a community's vulnerability to specific ecological hazards?
4. What factors affect the severity of impacts of an ecological hazard?
5. What actions can be taken to reduce risk?
6. What factors affect a community's response to an ecological hazard?
7. How are people in developed and developing communities affected differently by ecological hazards?
8. Why might different communities seek different solutions to the challenges of ecological hazards?
9. How does climate change affect the impact of some ecological hazards?

2.12.3 Exam questions

2.12 Section I – Short answer question

▶ Question 1 (4 marks)

Analyse figures 2.71, 2.72 and 2.73 to explain the pattern of risk from particulate air pollution globally.

FIGURE 2.71 The 21 world regions as defined by the Global Burden of Disease project

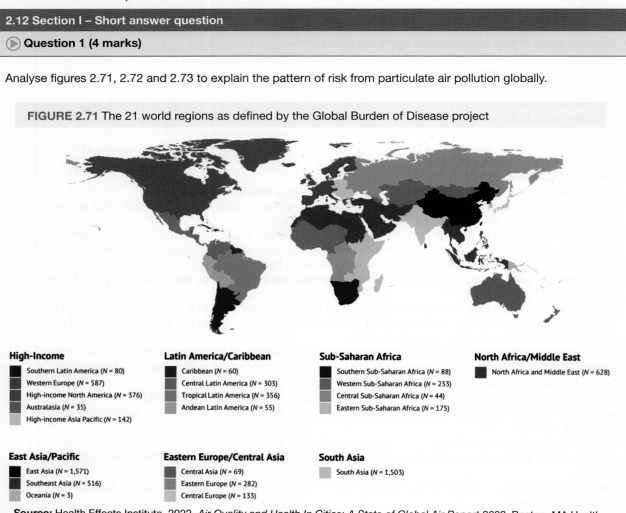

High-Income
- Southern Latin America (N = 80)
- Western Europe (N = 587)
- High-income North America (N = 376)
- Australasia (N = 35)
- High-income Asia Pacific (N = 142)

Latin America/Caribbean
- Caribbean (N = 60)
- Central Latin America (N = 303)
- Tropical Latin America (N = 356)
- Andean Latin America (N = 55)

Sub-Saharan Africa
- Southern Sub-Saharan Africa (N = 88)
- Western Sub-Saharan Africa (N = 233)
- Central Sub-Saharan Africa (N = 44)
- Eastern Sub-Saharan Africa (N = 175)

North Africa/Middle East
- North Africa and Middle East (N = 628)

East Asia/Pacific
- East Asia (N = 1,571)
- Southeast Asia (N = 516)
- Oceania (N = 3)

Eastern Europe/Central Asia
- Central Asia (N = 69)
- Eastern Europe (N = 282)
- Central Europe (N = 133)

South Asia
- South Asia (N = 1,503)

Source: Health Effects Institute. 2022. *Air Quality and Health In Cities: A State of Global Air Report 2022.* Boston, MA:Health Effects Institute.

FIGURE 2.72 Regional urban population-weighted annual averages of PM2.5 concentrations in 2019 (ugM3)

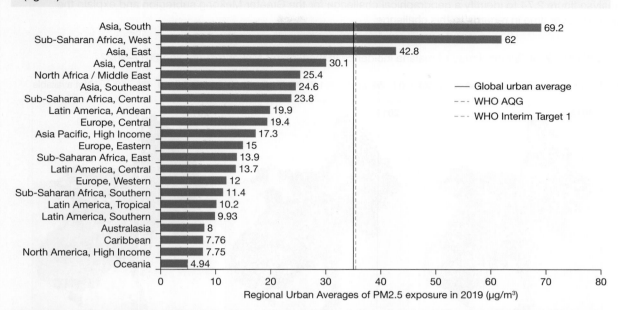

Source: Health Effects Institute. 2022. *Air Quality and Health In Cities: A State of Global Air Report 2022.* Boston, MA: Health Effects Institute.

FIGURE 2.73 Region urban medians of PM2.5 attributable death rates in 2019 (deaths per 100 000)

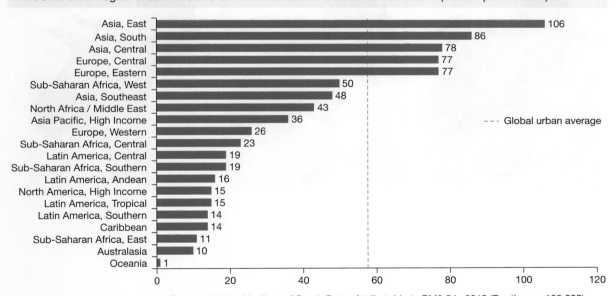

Source: Health Effects Institute. 2022. *Air Quality and Health In Cities: A State of Global Air Report 2022.* Boston, MA: Health Effects Institute.

Analyse figure 2.74 to identify a geographical challenge for the Greater Mekong subregion and explain the changes over time in regards to this challenge.

FIGURE 2.74 Regional map of malaria incidence in the Greater Mekong subregion 2012–2022

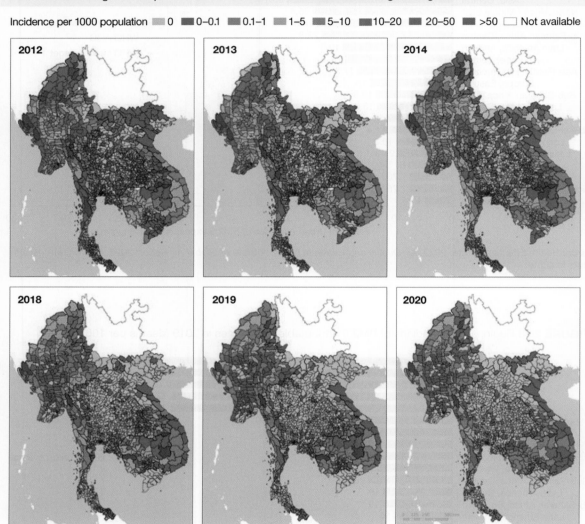

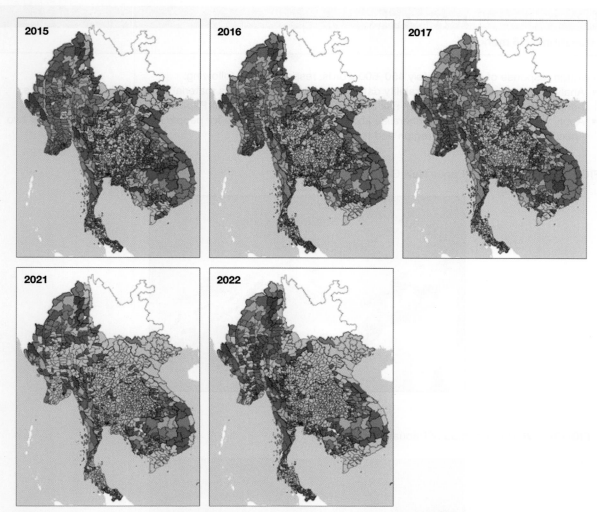

Source: World malaria report 2023. Geneva: World Health Organization; 2023. Licence: CC BY-NC-SA 3.0 IGO.

▶ **Question 3 (16 marks)**

In a written response of approximately 450–600 words, respond to the following:

- Analyse and interpret the vulnerability data presented in the stimulus material to explain how the patterns, trends and relationships represent risk for people and environments in Papua New Guinea.
- Apply your understanding to generalise about the potential impacts for people in Papua New Guinea for two specific locations: the Sepik region and the Highlands region.

FIGURE 2.75 Regions of Papua New Guinea

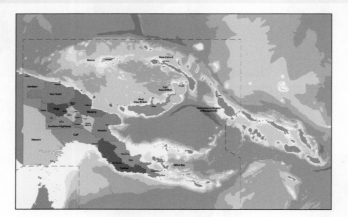

FIGURE 2.76 Physical map of Papua New Guinea

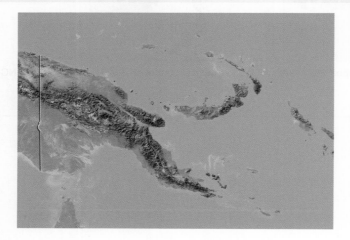

TABLE 2.18 Reported malaria deaths by region, 2010–2022

WHO region Country/area	2010	2011	2012	2013	2014	2015	2016	2017	2018	2019	2020	2021	2022
WESTERN PACIFIC													
Cambodia	151	94	45	12	18	10	3	1	0	0	0	0	0
China	19	33	0	0	0	0	0	0	0	0	0	NA	NA
Lao People's Democratic Republic	24	17	44	28	4	2	1	2	6	0	0	1	1
Malaysia	13	11	11	10	4	4	2	12	12	6	5	13	9
Papua New Guinea	616	523	381	307	203	163	306	273	216	180	188	201	282
Philippines	30	12	16	12	10	20	7	4	2	9	3	3	0
Republic of Korea	1	2	0	0	0	0	0	0	0	0	0	0	0
Solomon Islands	34	19	18	18	23	13	20	27	7	14	3	9	12
Vanuatu	1	1	0	0	0	0	0	0	0	0	0	0	0
Vietnam	21	14	8	6	6	3	3	6	1	0	0	0	0

Source: World Health Organization, Global Health Observatory Data Repository/World Health Statistics. Incidence of malaria (per 1,000 population at risk). (apps.who.int/ghodata). Licensed under CC-BY 4.0.

FIGURE 2.77 Incidence of malaria (per 1000 population at risk), Papua New Guinea

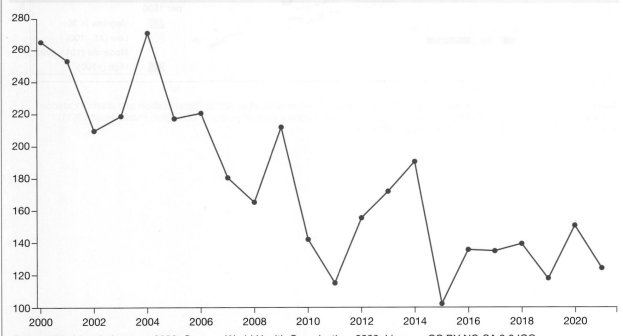

Source: World malaria report 2023. Geneva: World Health Organization; 2023. Licence: CC BY-NC-SA 3.0 IGO.

FIGURE 2.78 Malaria risk strata using the average annual incidence of cases among the general population, Papua New Guinea, 2011–2019

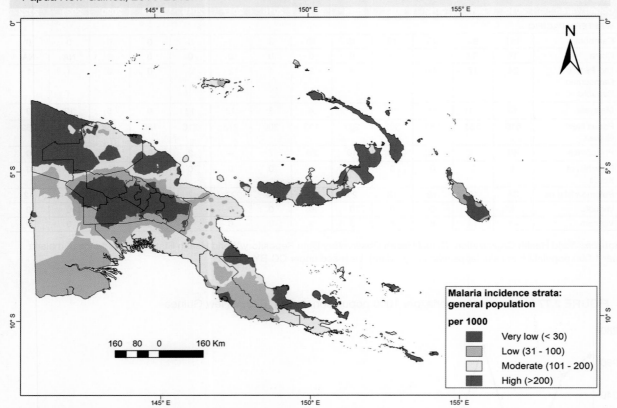

Malaria incidence strata: general population

per 1000

- Very low (< 30)
- Low (31 - 100)
- Moderate (101 - 200)
- High (>200)

Source: Seidahmed O, Jamea S, Kurumop S, Timbi D, Makita L, Ahmed M, et al. (2022) 'Stratification of malaria incidence in Papua New Guinea (2011–2019): Contribution towards a sub-national control policy'. PLOS Glob Public Health 2(11): e0000747. https://doi.org/10.1371/journal.pgph.0000747.

FIGURE 2.79 Prevalence per region of main malaria-causing parasite, Papua New Guinea, 2013 to 2020

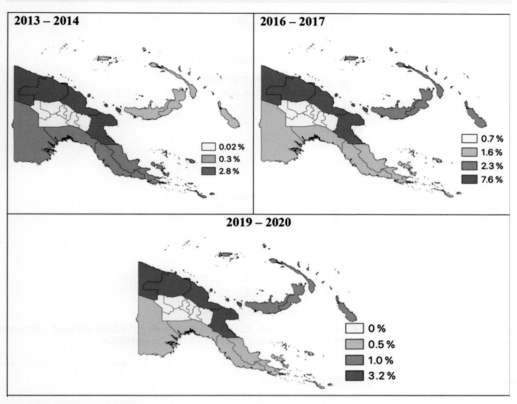

2013 – 2014

☐ 0.02 %
◻ 0.3 %
■ 2.8 %

2016 – 2017

☐ 0.7 %
◻ 1.6 %
▨ 2.3 %
■ 7.6 %

2019 – 2020

☐ 0 %
◻ 0.5 %
▨ 1.0 %
■ 3.2 %

Source: Cleary E, Hetzel MW and Clements ACA (2022) 'A review of malaria epidemiology and control in Papua New Guinea 1900 to 2021: Progress made and future directions'. *Front. Epidemiol*. 2:980795. doi: 10.3389/fepid.2022.980795.

FIGURE 2.80 Under-5 mortality rate, Papua New Guinea, 2000–2021

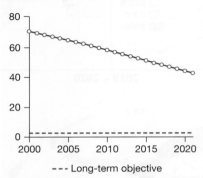

Mortality rate, under-5
per 1000 live births

● Significant challenges remain

↗ Score moderately improving, insufficient to attain goal

Value 42.81

Year 2021

- - - Long-term objective

Source: Data from Sachs, J.D., Lafortune, G., Fuller, G., Drumm, E. (2023). *Implementing the SDG Stimulus. Sustainable Development Report 2023*. Paris: SDSN, Dublin: Dublin University Press, 2023. 10.25546/102924.

Sample responses are available in your digital formats.

3 Responding to challenges facing a place in Australia

UNIT 2 TOPIC 3

SUBJECT MATTER

In chapter 3, students;
- **explain** geographical processes that have shaped the identity of places in Australia
- **recognise** the spatial patterns of these places and the implications for people in remote, rural and urban locations
- **investigate** a specific geographical challenge in a local area using the geographic inquiry model.

LESSON SEQUENCE

Fully worked solutions for this topic are available in the Resources section at www.jacplus.com.au.

LESSON
3.1 Overview

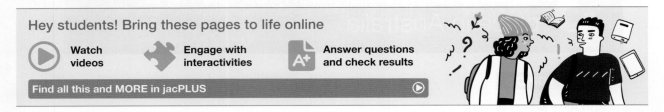
3.1.1 Introduction

In this chapter, you will examine how geographical processes have shaped the identity of places in Australia — the places that you know and visit or get to call 'home'. You will be able to recognise the spatial patterns of these places and the implications for people living in remote, rural and urban locations.

This chapter also provides you with the opportunity to complete fieldwork, to understand the different factors that contribute to liveability in a local area. You will be able to propose action to manage an identified challenge to improve the liveability of a local place.

FIGURE 3.1 Rural Australian landscape

3.1.2 Syllabus links

Syllabus links	Lesson
○ Explain how remote, rural and urban places in Australia are defined by the Australian Bureau of Statistics (ABS).	3.3
○ Explain the processes that shape the identity of remote, rural and urban places, including • urbanisation • suburbanisation • counterurbanisation • population increase • population decline.	3.2 3.3
○ Recognise the spatial patterns of remote, rural and urban places in Australia and represent these on a map, using spatial technologies.	3.4 3.5
○ Explain the factors that have contributed to these patterns (i.e. factors affecting settlement patterns), including physical factors, such as access to fresh water, soil fertility and other natural resource availability; economic factors, such as resource exploitation, employment and affordability; and social factors, such as access to health and education services.	3.4 3.5
○ Explain the implications for people living in remote, rural and urban places; for example, provision of goods and services, transport, housing accessibility and affordability.	3.7
○ Analyse ABS data to explain the changing characteristics of remote, rural and urban places in Australia.	3.8
○ Explain the geographical challenges facing places in Australia as a result of the changing characteristics of places, including • rural and remote places — for example, employment, provision of health and educational services, transportation connections to major centres, expansion or contraction of industry, isolation and remoteness, access to fresh and affordable food, housing availability and affordability, waste management, fresh water quality and availability, and access to communication technology (e.g. NBN) • metropolitan and/or regional cities in Australia — for example, urban sprawl, gentrification, transport options, environmental degradation, land-use zoning, service provision and management, housing availability and affordability, and waste management.	3.6 3.7
○ Conduct a geographic inquiry (e.g. fieldwork) at a local scale to investigate a specific challenge associated with liveability for a place in Australia (remote, rural or urban) and how this challenge might be managed. As part of this investigation, students must • use the geographic inquiry model to investigate a challenge facing a place in Australia • identify and collect the data required and appropriate methods for data collection • analyse data to describe the nature, location and extent of the selected challenge • apply geographical understanding from their analysis to generalise about the impacts on sustainability and liveability for the place in Australia • propose action/s for managing the identified impacts to improve liveability and sustainability for the place in Australia • transform data and information using cartographic, graphic and mathematical skills, spatial technologies and ICT to communicate understanding.	3.9 3.10

KEY QUESTIONS

1. What defines a place as 'rural', 'remote' or 'urban'?
2. What factors affect the identity of Australian places?
3. How have Australian places changed?
4. How have the population distribution and location of settlements in Australia changed?
5. What physical, social and economic factors affect settlement patterns in Australia?
6. What challenges do people face living in rural, remote or urban areas?
7. What are some of the key challenges where you live, and how might you manage these challenges?

LESSON
3.2 Places in Australia

LEARNING INTENTION

By the end of this lesson you should be able to explain how remote, rural and urban places in Australia are defined by the Australian Bureau of Statistics (ABS).

Source: Adapted from Geography General Senior Syllabus 2024 © State of Queensland (QCAA) 2024; licensed under CC BY 4.0.

3.2.1 Defining place

A place is a location on the Earth's surface. The word 'place' can be used to describe a specific location, a **physical environment**, a building or locality of special significance, or a particular region. The term can be used for locations at almost any **geographic scale**, depending on context.

> **physical environment** natural and built surroundings
>
> **geographic scale** the spatial extent or level of detail at which a geographic area is analysed or represented, ranging from local to global

FIGURE 3.2 What is associated with 'place'?

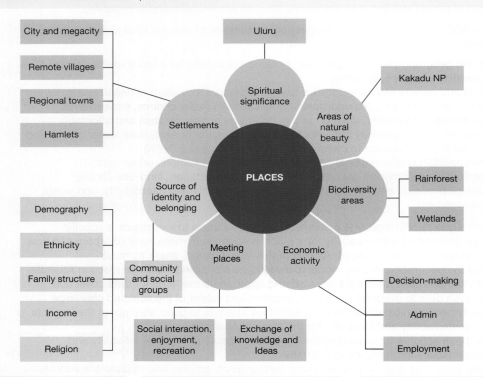

FIGURE 3.3 Mossman Gorge in Queensland is a highly biodiverse place.

FIGURE 3.4 Religious centres, such as churches, are important places to some people.

FIGURE 3.5 Parks are often places where people can practise recreational activities.

One thing we can say with certainty is that the concept of 'places' is a cultural construct, meaning they are places because humans define them as such. Every part of Earth exists in a 'space', which means it can be represented numerically — using latitude and longitude, for instance. However, not every part of Earth is considered a 'place'. It is only those spaces that humans have defined and attached meaning to that become 'places'.

3.2.2 Broad categories of places

Human settlement in different places is not a new phenomenon, but its current size, complexity and rate of growth in the world is. Around 12 500 years ago, humans began to transition from a hunter-gatherer lifestyle to one more settled and sedentary. This period is known as the Neolithic Revolution. This fundamental change in human lifestyle was brought about by the domestication of animals (for food and as a source of power) and the rise of agriculture. This change is still affecting our species and the planet, as people move between **rural** and **urban** places.

rural areas with low population densities, agricultural activities and other characteristics of the countryside

urban areas with high population densities, human development and other characteristics typical of cities and towns

Most urban development in Australia has occurred over the past 200 years, and this development has largely been driven by one or more of the factors shown in table 3.1.

TABLE 3.1 Reasons for settlement in places

Feature	Reason feature may increase settlement
Water supply	Access to freshwater for drinking, while also considering flood avoidance
Resources	Access to building materials, food and fuel
Geographical situation	Considering topography, natural defences, shelter, aspect
Interconnections	Considering nodal points and bridging points such as road crossings, river crossings, connection points on trade routes; any natural or human connection point
Harbours	Access to ocean-based transport was often critical — for example, for the survival of the colonies in Australia as goods and people were transported by sea from places such as Europe
Trade resources	Ability to trade these with other settlements and countries — for example, in areas rich in iron ore, such as Mount Isa, or areas where whaling operations could take place, such as Eden in New South Wales

FIGURE 3.6 Australian towns by population (a) 1911 and (b) 1961

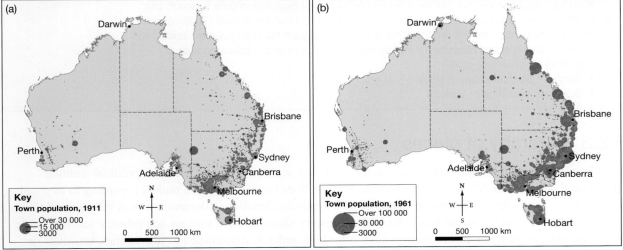

Source: Based on Australian Bureau of Statistics data. © Commonwealth of Australia   *Source:* Based on Australian Bureau of Statistics data. © Commonwealth of Australia

FIGURE 3.7 Australian towns by population, 2023

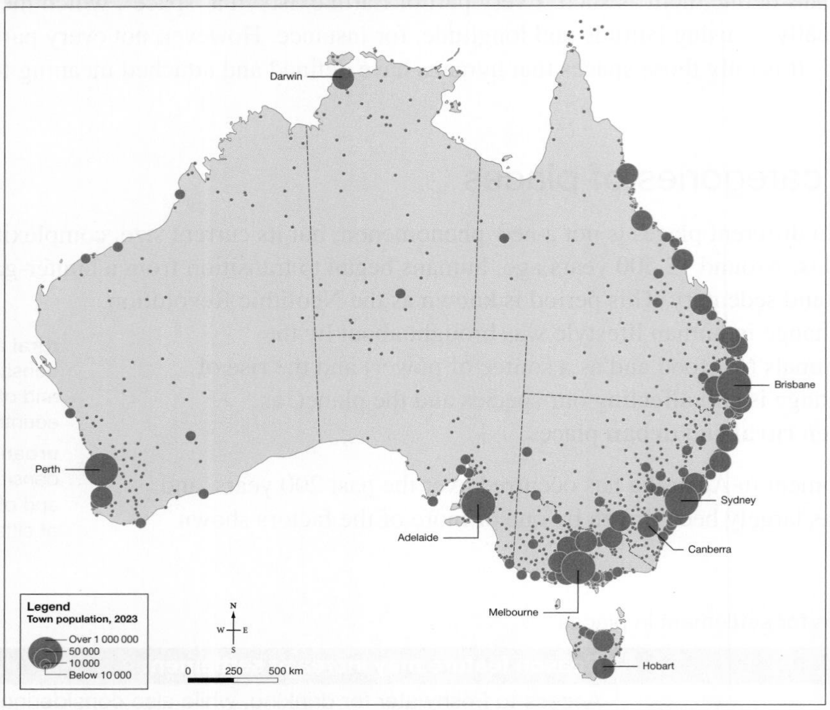

Legend
Town population, 2023
- Over 1 000 000
- 50 000
- 10 000
- Below 10 000

N
W — E
S

0   250   500 km

Source: Based on the data from Australian Bureau of Statistics. (2022–23). Regional population. ABS. https://www.abs.gov.au/statistics/people/population/regional-population/latest-release. © Commonwealth of Australia. Map drawn by Spatial Vision.

SKILLS ACTIVITY

1. Identify some similarities of urban locations across the maps shown in figures 3.6 and 3.7.
2. Identify any differences in the three maps (*Hint:* Don't forget the legend.)
3. Which state shows evidence of newer large centres rather than expansions of existing centres?
4. By 2023, how many cities in Australia had over one million people? Name them.
5. Using figure 3.7 and online sources, outline which category (e.g. over 50 000 but fewer than one million) you would place the following cities.
   - Newcastle (NSW)
   - Rockhampton (Qld)
   - Mandurah (WA)
   - Launceston (Tas)
   - Geelong (Vic)

   You may need to research the location and most recent population data for each city.
6. Identify a centre in each state that has more than 10 000 but fewer than 50 000 people.
7. Identify any centres on the map that have had a significant fall in population since 1961.
8. Consider and discuss reasons for increased urban growth in the following areas.
   - Northwest coastline of WA
   - Northern coastal areas of NSW
   - Southern coastline of Victoria

3.2.3 Hierarchy of places

The places where people settle vary in size and complexity according to their location, political–cultural background, population, levels of infrastructure and functions (services). When ranked according to size and 'importance', this order of settlements is known as a hierarchy. Smaller settlements such as villages and towns provide residents with low-order functions such as grocery shops, service stations and primary schools, while larger settlements such as cities or metropolises offer high-order functions such as universities, hospitals or television stations. The hierarchy of settlements from largest to smallest is shown in table 3.2.

TABLE 3.2 Hierarchy of settlements based on size and importance

Name and description of settlement	Visual
Conurbation Conurbations occur when multiple cities combine into a large urban area. Examples include Tokyo–Yokohama in Japan and the Gold Coast–Brisbane–Caboolture conurbation in South East Queensland. Despite merging, individual cities often maintain their distinct identities. For example, the Gold Coast and Brisbane are essentially connected, but they maintain their individual status.	 **FIGURE 3.8** A conurbation: South East Queensland
Metropolis A metropolis, such as Brisbane, is the largest urban centre in a state or district, often serving as the capital. With large populations, they provide extensive high-level services such as state governance and legal administration.	 **FIGURE 3.9** A metropolis: Brisbane
City A city is a large urban settlement with clear boundaries and municipal functions. Toowoomba (population 173 204 in 2021), Gladstone (population 63 515 in 2021), Townsville (population 234 283 in 2021) and Mount Isa (population 18 727 in 2021) are all cities. Cities are often the centre of local government and regional services.	 **FIGURE 3.10** A city: Townsville
Town A town is an urban settlement. Towns are smaller than cities but larger than a village — for example, Maleny (population 2976 in 2021), Clermont (population 2079 in 2021) and Weipa (population 4100 in 2021). Towns can be large or small, and generally offer residents grocery shops, cafes, hardware stores, service stations and schools.	 **FIGURE 3.11** A town: Maleny

(continued)

Name and description of settlement	Visual
Village A village, like Burketown (population 167 in 2021) or Aratula (population 553 in 2021), is a small settlement with residential dwellings and limited low-level services. Suburban areas such as Samford, located outside the Brisbane CBD, are also referred to as villages.	FIGURE 3.12 A village: Aratula  ***Source:*** © State of Queensland 2018, Qld Globe
Hamlet A hamlet is a tiny settlement with a small number of residential and work buildings, with possibly some other low-order functions, such as a general store, service station or sports oval. On many farm properties, a similar cluster of buildings is referred to as a homestead.	FIGURE 3.13 A hamlet: Cooyar  ***Source:*** © State of Queensland 2018, Qld Globe

SKILLS ACTIVITY: Using GIS

Use an online interactive mapping tool that allows annotation (such as the Queensland Globe, ArcGIS or Google Maps). Individually or in small groups examine a specific suburb or area in your region. Interpret the satellite image layer of data.

1. Identify and mark the different types of land use evident on the map. (Ensure you select appropriate symbols, patterns or colours to represent different land use types.)
2. Adjust transparency to better view the image and the colours.
3. Annotate your map or create a key.
4. Analyse your map to identify the land use patterns. Describe the pattern of land use evident in your area and explain what factors might contribute to those patterns. (Consider a range of factors, including social, physical and economic.)
5. Predict how your area might look in 20 years. Provide justifications for your prediction.
6. Suggest the type of settlement that your area should be categorised as. Explain your answer.

3.2.4 ABS definitions of places

The Australian Bureau of Statistics (ABS) classifies places in Australia into a hierarchy of statistical areas. This is reported in the Australian Statistical Geography Standard (ASGS) — see table 3.3 for some terms used by the ABS and their definitions. The hierarchy used by the ABS is a social geography, developed to reflect the location of people and communities.

Mesh Blocks (MBs) are the smallest geographic units, serving as building blocks for statistical areas in the Australian Statistical Geography Standard (ASGS). There are 368 286 Mesh Blocks covering all of Australia.

TABLE 3.3 ABS definitions of places

Term	Definition	Example
ARIA+ (Accessibility/Remoteness Index of Australia)	Uses GIS data to measure road distance to service centres and provide a value from 0 to 15. The lower the score, the more urban the area.	N/A
Major Cities of Australia	ARIA+ value of 0 – 0.2	Adelaide, South Australia
Inner Regional Australia	ARIA+ value of greater than 0.2 and less than or equal to 2.4	Woodford, Queensland
Outer Regional Australia	ARIA+ value of greater than 2.4 and less than or equal to 5.92	Bright, Victoria
Remote Australia	ARIA+ value of greater than 5.92 and less than or equal to 10.53	Nyngan, New South Wales
Very Remote Australia	ARIA+ value of greater than 10.53	Daly Waters, Northern Territory

Source: Australian Bureau of Statistics. 1217.0.55.001 — Glossary of Statistical Geography Terminology, 2013. © Commonwealth of Australia

 Resources

 Weblink Australian Bureau of Statistics: QuickStats

3.2 Exercise

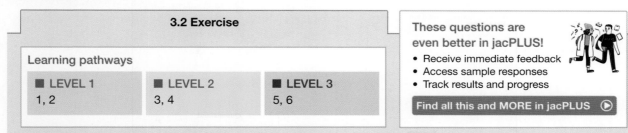

3.2 Exercise

Learning pathways

■ LEVEL 1	■ LEVEL 2	■ LEVEL 3
1, 2	3, 4	5, 6

These questions are even better in jacPLUS!
- Receive immediate feedback
- Access sample responses
- Track results and progress

Find all this and MORE in jacPLUS ▶

Explain and comprehend

1. **Explain** what makes a place different from a space.
2. **Identify** two factors that affected the location of early settlements in Australia and **explain** why these factors were important.
3. **Describe** the factors that make the area where you live an important place. Identify the cultural, economic and physical features.
4. With reference to figures 3.6 and 3.7, **describe** the changes that occurred to the distribution and size of Australia's places between 1961 and 2023.

Analyse and apply

5. Use the **Australian Bureau of Statistics: QuickStats** weblink in the Resources tab to find statistics on the population of your nearest town or suburb.
   Create a table to show how that place **compares** with Queensland and Australia with regards to median age, average number of people per household, full-time workers and median weekly household income.
6. Based on the ABS data for your local area, **suggest** reasons for population changes in your local area: are the causes social, economic, environmental or a combination? Present your ideas as a Venn diagram or table.

Sample responses are available in your digital formats.

LESSON
3.3 Processes that shape places

3.3.1 Urban, rural and remote places

A number of physical differences between urban, rural and remote places can help us understand these places and the challenges they face.

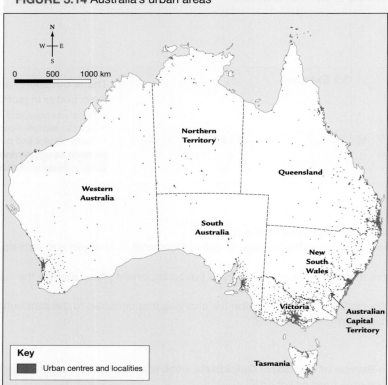

FIGURE 3.14 Australia's urban areas

Key
◼ Urban centres and localities

Source: Based on Australian Bureau of Statistics data. ABS Maps – 2016 Urban Centres and Localities. © Commonwealth of Australia

DID YOU KNOW?

Industries are categorised into three main sectors: primary, secondary and tertiary. The primary sector involves extracting natural resources such as through agriculture and mining. The secondary sector focuses on manufacturing and processing these materials into finished goods. The tertiary sector provides services, including retail, health care and finance. Additionally, some models include a quaternary sector for research and development, and a quinary sector for high-level decision-making.

Table 3.4 highlights four ways to compare urban, rural and remote areas. These four broad areas are also influenced by other factors, such as whether the area is a location for primary, secondary or tertiary industry.

TABLE 3.4 The differences between urban, rural and remote areas

Area of comparison	Urban	Rural	Remote
Population	High	Medium to low	Low
Economic activity	High	Medium	Low
Availability of services	High	Medium to low	Medium to low
Development density	High	Low	Low

Social demography is the statistical study of populations using measures such as age. Urban areas are more likely to have a higher proportion of younger people, while rural areas are more likely to have a higher proportion of people over 65. Younger people tend to leave rural areas to look for work and other opportunities in cities, which has an impact on the demography of these places. As an indication of this, figure 3.15 shows the age distribution by sex of people living in capital cities versus the rest of Australia. Another measure of the social demographics in a community is the total dependent population, or proportion of the community not of typical working age (between 15 and 65).

social demography the statistical study of populations using measures such as age, sex and income

FIGURE 3.15 Age and sex distribution of people living in capital cities and the rest of Australia, 2022

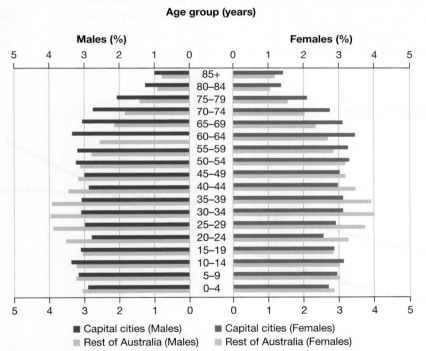

■ Capital cities (Males)   ■ Capital cities (Females)
▨ Rest of Australia (Males)   ▨ Rest of Australia (Females)

Source: Based on Australian Bureau of Statistics. (2022). Regional population by age and sex. ABS. https://www.abs.gov.au/statistics/people/population/regional-population-age-and-sex/latest-release.

DID YOU KNOW?

According to the 2021 census, the median age for capital cities (37.2 years) was younger than the rest of Australia (42.0). The youngest capital was Darwin, with a median age of 34.5 years, while Adelaide was the oldest (39.4). Interestingly, Darwin was the only capital city with more males than females.

3.3.2 Shaping the identity of places

The identity of a place — that is, the meaning and significance of places for the people who live there and use the places — changes over time. These changes are shaped by the following broad processes:

- urbanisation
- suburbanisation
- counterurbanisation
- population increases
- population decline.

Urbanisation

Urbanisation occurs when people move from rural areas to urban areas. It is shown by the growth in the proportion of a population living in urban environments. This phenomenon is occurring across the world, with more people and a greater proportion of the global population living in cities. In 2024, 54 per cent of people on Earth lived in an urban place, and this proportion has been steadily increasing over the last century. The United Nations predicts that, from 2015 to 2030, the global urban population will grow between 1.44 per cent and 1.84 per cent per year. Rates of urbanisation are highest in developing countries, with African and South American cities predicted to see the highest rates of urbanisation in the near future.

In Australia, we have always had a high proportion of our population living in cities. European settlements that began as towns eventually grew, and continued attracting more people. Some people moved into rural areas as the country was opened up to farming, but the trend over the past 100 years has clearly been one of increasing urbanisation. Figure 3.16 shows the change over time in global urban and rural populations, while figure 3.17 shows the change over time of people in Australia living in urban and rural areas.

urbanisation growth in the proportion of a population living in urban environments

FIGURE 3.16 Number of people living in urban and rural areas, world

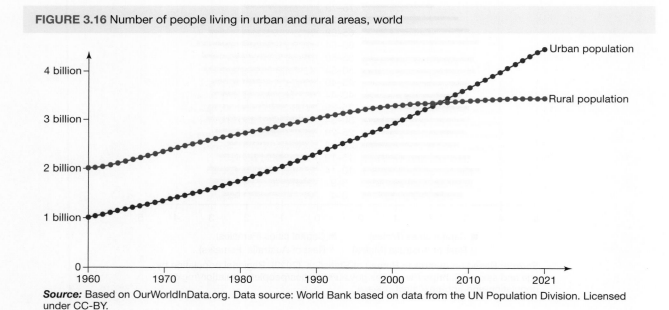

Source: Based on OurWorldInData.org. Data source: World Bank based on data from the UN Population Division. Licensed under CC-BY.

FIGURE 3.17 Number of people living in urban and rural areas, Australia

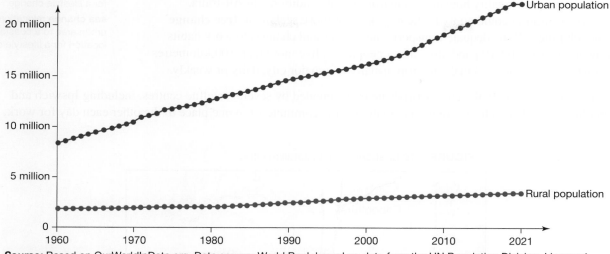

Suburbanisation

Suburbanisation is the development and outward spread of new suburban areas. It includes a population shift from inner-city areas to outer suburban areas, often on the rural fringe, that are accessible to the city centre. In Australian cities, a significant population growth occurred post–World War II, causing our cities to expand and encroach on their rural edges. Many suburban areas surrounding Australia's larger cities were farms or native vegetation not long ago.

Suburbanisation is the key process that leads to **urban sprawl**. Figure 3.18 shows population change at different distances from the Brisbane CBD between 2001 and 2011. The largest proportion of this change has been an increase of people living between 10 and 40 kilometres from the CBD.

Counterurbanisation

Counterurbanisation is a term that describes the migration of people from urban to rural areas. It is a loosely defined term and can occur for a range of reasons.

FIGURE 3.18 Comparison of proportion of population change at various distances from Central Business District, Brisbane, 2001–11

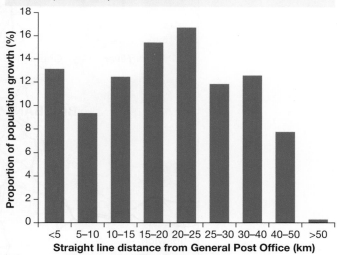

suburbanisation the development and outward spread of new suburban areas

urban sprawl process of outward expansion of urban areas from the CBD into the surrounding countryside and invading adjacent towns, regions and undeveloped land

counterurbanisation the migration of people from urban to rural areas

Counterurbanisation is generally related to push factors in urban areas that are suffering some sort of decline, such as the movement of jobs out of an area, increasing crime rates or poor-quality housing or environmental conditions. In Australia, counterurbanisation could apply to the migration of those seeking a '**tree change**' or '**sea change**'. With adequate transport connections and changes to work habits following the COVID-19 pandemic, many people can live more than 100 kilometres from where they work and make the trip, usually to a major city, daily or weekly.

tree change relocation from a city or suburb to a rural or forested area for a lifestyle change

sea change relocation from an urban area to a coastal or seaside location for a lifestyle change

As shown in figure 3.19, the city of Brisbane is surrounded by several satellite centres, including Ipswich and Caboolture, which are close enough that many people commute from one place to the other each day for work.

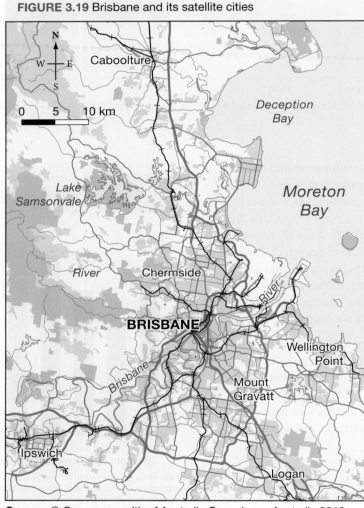

FIGURE 3.19 Brisbane and its satellite cities

Source: © Commonwealth of Australia Geoscience Australia 2018; © Openstreetmap contributors

SKILLS ACTIVITY: Interpreting satellite images

Complete this activity by either referring to figures 3.20 and 3.21, or by using an online interactive mapping tool that allows annotation (such as the Queensland Globe, ArcGIS, ScribbleMaps or Google Maps) to examine an online aerial image or satellite image and topographical map of a place you are familiar with.

FIGURE 3.20 Topographic map extract of the south-eastern section of the Daintree River National Park

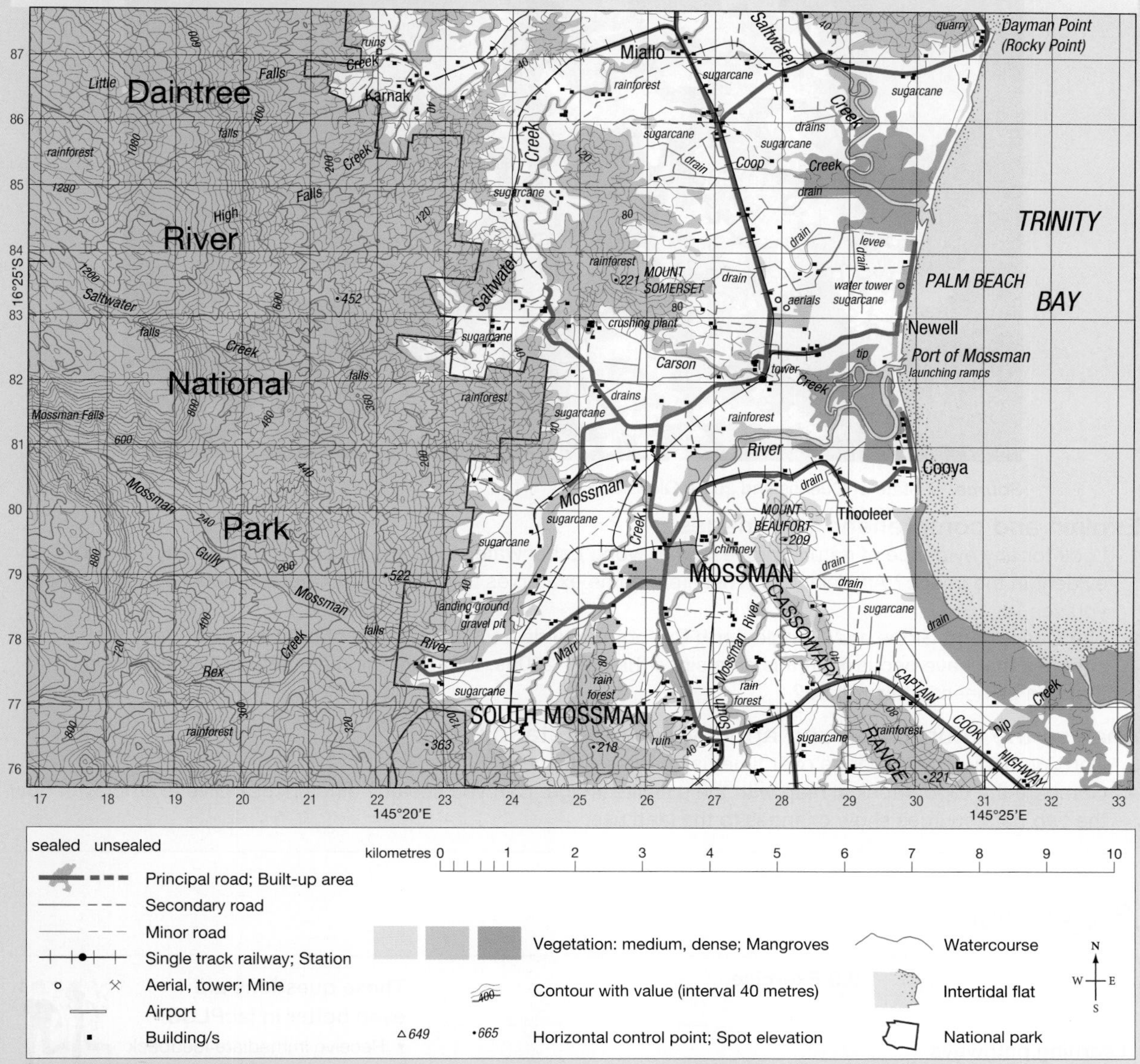

Source: MAPgraphics Pty Ltd, Brisbane

FIGURE 3.21 Examples of rural and urban areas

Source: © State of Queensland 2018, Qld Globe

Explain and comprehend the images

1. Look for any evidence of natural processes at work. Label waterways, mountains or hills. Is any erosion evident in the image? Any islands? (These are some examples of evidence of natural processes; label others you see.)
2. The Queensland Globe and Google Earth Pro both provide access to banks of past imagery. View the changes that have occurred in this area since the early 2000s. What are the main natural and built changes that can be seen in this record?

Analyse the information and apply your understanding

3. Can you deduce any change over time using this evidence?
4. Compare the topographical map with the satellite image. Can you see any differences? Give examples of how the two data sources show changes to the land use.

3.3 Exercise

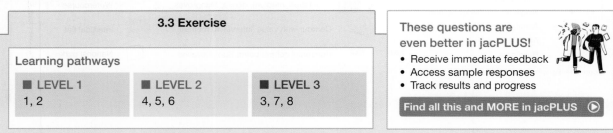

3.3 Exercise

Learning pathways

■ LEVEL 1	■ LEVEL 2	■ LEVEL 3
1, 2	4, 5, 6	3, 7, 8

These questions are even better in jacPLUS!
- Receive immediate feedback
- Access sample responses
- Track results and progress

Find all this and MORE in jacPLUS ▶

Explain and comprehend

1. **Identify** what is meant by the following.
   a. Urban area
   b. Rural area
2. **Outline** the likely demographic of a rural population.

3. **Analyse** the population pyramid in figure 3.15 to explain the impacts of the age and sex distribution on people and environments in both capital cities and the rest of Australia.
4. **Explain** why urban areas have a higher proportion of people in their 20s.
5. **Describe** what is meant by 'urbanisation'.
6. **Describe** what is meant by 'counterurbanisation' and identify reasons this may have increased after COVID.
7. **Categorise** and **describe** the features of each of the places in figure 3.22.

FIGURE 3.22 Examples of rural and urban areas

Analyse and apply

8. **Examine** demographic statistics in your local area using the online ABS Search Census data tool (see the weblink in the Resources tab).
   a. What type of area is it?
   b. What is its population?
   c. What category of settlement would you classify your area of residence?
   d. What are the functions found in your local area that justify the classification you have chosen? **Justify** your decision with statistics.

Sample responses are available in your digital formats.

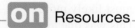 Resources

🔗 **Weblink** ABS Search Census data tool

LESSON
3.4 Where are Australian places?

3.4.1 First Nations places in Australia

Experts estimate that between 300 000 and 750 000 people inhabited the Australian continent prior to European contact in the 18th century. These First Nations Peoples have occupied mainland Australia for at least 65 000 years and were separated into around 250 individual nations. Most nations spoke their own language and had a defined geographic area in which they lived.

The lack of written records means that it is difficult to say with certainty exactly how the population of Australia was distributed at the time of colonisation. First Nations Australians developed ways of managing the natural environment for sustainable use, including sophisticated land management

FIGURE 3.23 Some of the world's oldest rock art was made 20 000 years ago in Burrungkuy (Nourlangie), in Kakadu; however, First Nations Australians have inhabited the continent for much longer.

using fire and other agricultural techniques. The Gunditjmara people of southwestern Victoria, for example, are famous for creating the world's oldest aquaculture infrastructure, traps that corral and catch eels, found in Budj Bim, Victoria.

The First Nations population in Australia, although covering most of the continent, was most likely more populous on the east coast, in what is now Queensland, New South Wales and Victoria. This was due to a favourable biophysical environment, food availability, climate and plentiful water sources from the Great Dividing Range and the Murray–Darling Basin.

FIGURE 3.24 Coastal regions provided favourable conditions for early First Nations populations.

First Nations Australians now

In 2021, First Nations Australians made up 3.2 per cent of the nation's population, with one-third of First Nations Australians living in capital cities. A further 65.6 per cent of First Nations Australians now live in major cities, or inner regional parts of Australia. This is compared to 90.9 per cent of non-Indigenous Australians who live in these places.

Statistically, Furst Nations Australians are significantly younger than the general population, with a median age of 24 compared to Australia's overall median age of 38. Only 5.9 per cent of First Nations Australians are over the age of 65, which is a considerably smaller proportion than for non-Indigenous people at 17 per cent. Figure 3.25 further highlights the differences between First Nations Australians and non-Indigenous Australians by age.

FIGURE 3.25 Aboriginal and Torres Strait Islander and non-Indigenous populations by age groups, 2021

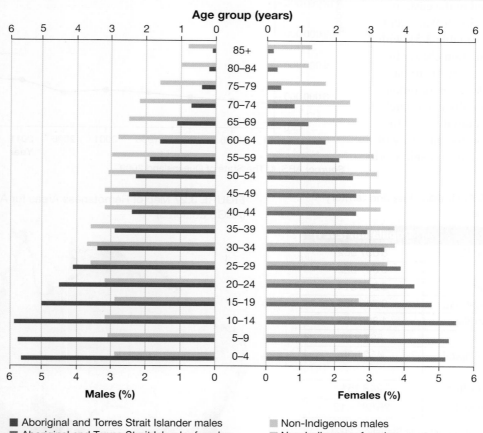

Legend:
- Aboriginal and Torres Strait Islander males
- Aboriginal and Torres Strait Islander females
- Non-Indigenous males
- Non-Indigenous females

Source: Based on the data from Australian Bureau of Statistics. (2021, June 30). Estimates of Aboriginal and Torres Strait Islander Australians. ABS. https://www.abs.gov.au/statistics/people/aboriginal-and-torres-strait-islander-peoples/estimates-aboriginal-and-torres-strait-islander-australians/latest-release.

3.4.2 Australia's population distribution

Australia's population is growing, as shown in figure 3.26. Our population distribution is overwhelmingly urban, with nearly two-thirds of the population living in a capital city. This distribution is heavily influenced by our geography. Most of the major urban settlements can be found on the coastline and, like the pre-European First Nations population in Australia, most of us can be found in the east, in Victoria, New South Wales and Queensland (see table 3.5), with a smaller population cluster in south-west Western Australia. Historically, these major urban settlements, usually each state's capital city, required access to the open seas as the only way of connecting with the outside world.

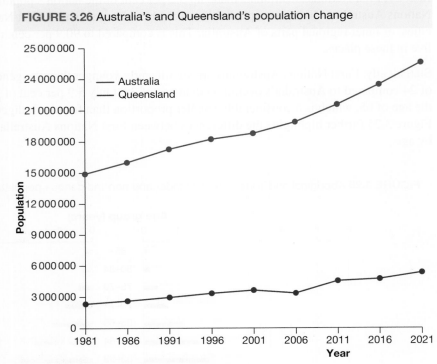

FIGURE 3.26 Australia's and Queensland's population change

Source: ABS, © The State of Queensland 2009

TABLE 3.5 Australian state and territory populations

State	Population (2021)
New South Wales	8 097 062
Victoria	6 547 822
Queensland	5 215 814
Western Australia	2 749 365
South Australia	1 802 601
Tasmania	567 239
Australian Capital Territory	452 508
Northern Territory	248 151
Total (Australia)	25 685 412

Source: Australian Bureau of Statistics, 2021

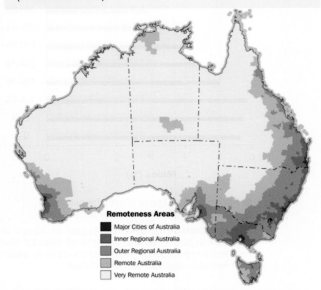

FIGURE 3.27 Map of Remoteness Areas for Australia (ASGS Edition 3)

Remoteness Areas
- Major Cities of Australia
- Inner Regional Australia
- Outer Regional Australia
- Remote Australia
- Very Remote Australia

Source: Australian Bureau of Statistics. (Jul2021–Jun2026). Remoteness Areas. ABS. https://www.abs.gov.au/statistics/standards/australian-statistical-geography-standard-asgs-edition-3/jul2021-jun2026/remoteness-structure/remoteness-areas.

3.4.3 Australia's natural places

As a continent, Australia has many places that haven't been developed or influenced significantly by humans. Australia is known for its natural beauty and much of this has been officially recognised and deemed special or significant by being named parklands, national parks, state forests or recognised through other state, national and international registers. Figure 3.28 shows the main gazetted places of natural and cultural significance in Australia.

Many natural places have special significance to Australia's First Nations population. Some of these places are recognised formally, while knowledge of others is passed from generation to generation through spoken word, stories and songs.

Some natural places are deemed important because of their economic potential, either as forests, dams or areas of resource extraction.

FIGURE 3.28 Australia's sites of natural and cultural significance

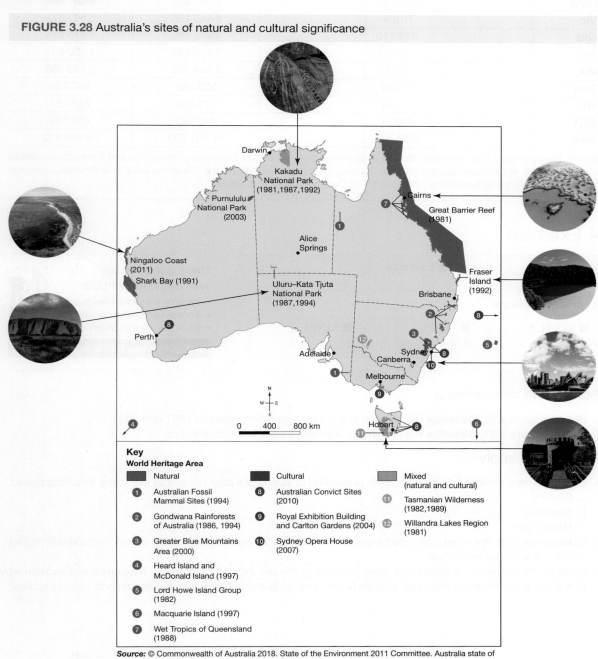

Key

World Heritage Area

■ Natural

■ Cultural

■ Mixed (natural and cultural)

① Australian Fossil Mammal Sites (1994)

② Gondwana Rainforests of Australia (1986, 1994)

③ Greater Blue Mountains Area (2000)

④ Heard Island and McDonald Island (1997)

⑤ Lord Howe Island Group (1982)

⑥ Macquarie Island (1997)

⑦ Wet Tropics of Queensland (1988)

⑧ Australian Convict Sites (2010)

⑨ Royal Exhibition Building and Carlton Gardens (2004)

⑩ Sydney Opera House (2007)

⑪ Tasmanian Wilderness (1982, 1989)

⑫ Willandra Lakes Region (1981)

Source: © Commonwealth of Australia 2018. State of the Environment 2011 Committee. Australia state of the environment 2011. Independent report to the Australian Government Minister for Sustainability, Environment, Water, Population and Communities. Canberra: DSEWPaC, 2011.

SKILLS ACTIVITY

1. Create a compound or multiple bar graph to visually represent the data in table 3.6.
2. Identify the states or territories that have the highest proportion of their total population made up of Aboriginal and/or Torres Strait Islander people.

TABLE 3.6 Aboriginal and/or Torres Strait Islander and non-Indigenous populations by state

	Total Aboriginal and/or Torres Strait Islander	Non-Indigenous	Total
NSW	339 710	7 757 352	8 097 062
Vic.	78 696	6 469 126	6 547 822
Qld	273 119	4 942 695	5 215 814
SA	52 069	1 750 532	1 802 601
WA	120 006	2 629 359	2 749 365
Tas.	33 857	533 382	567 239
NT	76 487	171 664	248 151
ACT	9 525	442 983	452 508
Aust.	983 709	24 701 703	25 685 412

Source: Based on data from Australian Bureau of Statistics. (2021, June 30). Estimates of Aboriginal and Torres Strait Islander Australians. ABS. https://www.abs.gov.au/statistics/people/aboriginal-and-torres-strait-islander-peoples/estimates-aboriginal-and-torres-strait-islander-australians/30-june-2021. © Commonwealth of Australia

3.4 Exercise

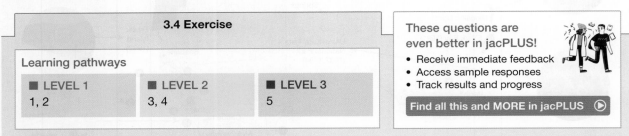

3.4 Exercise

Learning pathways

■ LEVEL 1	■ LEVEL 2	■ LEVEL 3
1, 2	3, 4	5

These questions are even better in jacPLUS!
- Receive immediate feedback
- Access sample responses
- Track results and progress

Find all this and MORE in jacPLUS ▶

Explain and comprehend

1. Refer to figure 3.26. **Describe** the trend in Australia's population between 1981 and 2021.
2. Refer to figure 3.27. **Describe** the patterns in Australia's population distribution.

Analyse and apply

3. Using the data in table 3.7, **calculate** the proportion of Australia's population that live in the following places.
   a. Sydney
   b. Brisbane
   c. Hervey Bay
4. **Discuss** why the five most populated areas shown in table 3.7 have come to be this way. Include reference to some of the important features.
5. Choose one of the lesser populated areas included in the second half of table 3.7. **Research** and **explain** why this place is important to Australia. Include supporting facts about the place, its location and demographics.

TABLE 3.7 Populations for select places in Australia

Place	Population (2021)	Proportion of population, %
Sydney	5 231 147	
Melbourne	4 917 750	
Brisbane	2 526 238	
Perth	2 116 647	
Adelaide	1 387 290	
Wagga Wagga	67 609	
Port Macquarie-Hastings	86 762	
Mildura-Wentworth	64 425	
Hervey Bay	54 649	
Shepparton-Mooroopna	49 862	

Source: Based on data from Australian Bureau of Statistics. (2021). QuickStats. https://www.abs.gov.au/census/find-census-data/search-by-area. © Commonwealth of Australia.

Sample responses are available in your digital formats.

LESSON
3.5 Factors affecting Australia's population distribution

LEARNING INTENTION

By the end of this lesson you should be able to explain the factors that have contributed to these patterns (i.e. factors affecting settlement patterns), including physical factors, such as access to fresh water, soil fertility and other natural resource availability; economic factors, such as resource exploitation, employment and affordability; and social factors, such as access to health and education services.

Source: Adapted from Geography General Senior Syllabus 2024 © State of Queensland (QCAA) 2024; licensed under CC BY 4.0.

3.5.1 Population distribution

Population distribution illustrates the way in which a population is spread across an area. In Queensland, for example, people are not distributed evenly across the state. More people are found along the coastline and in the south-east corner, and these people live in higher densities than in other parts of the state. This pattern is evident in figures 3.30 and 3.31.

Population distribution can be represented visually using a number of methods. The most common methods are dot density maps, where each dot represents a certain number of people, and choropleth maps, where each region is shaded or coloured according to the value in that area (in this case, population). Figures 3.30 and 3.31 both represent the same population data but in two different ways.

FIGURE 3.29 Like most major cities in Australia, Brisbane is close to the coast.

The dot density map of Queensland, figure 3.30, shows more dots towards the coastline and in the south-east corner of the state. It also shows some areas with no dot symbols at all, which doesn't necessarily mean those areas are empty, just that they don't have a large enough population to warrant a dot. Dot density maps should have a key that tells you how many people are represented by each dot. In figure 3.30 it is 1 dot per 100 people.

Figure 3.31 shows a choropleth map of Queensland. Shades of orange and red are used to represent different population densities, with lighter shades indicating lower values (lower density) and darker shades indicating higher values (higher density).

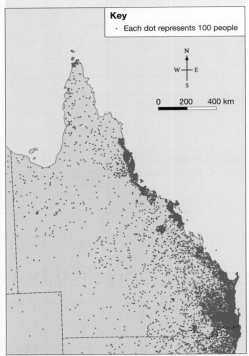

FIGURE 3.30 Queensland's population shown as a dot density map

Key
· Each dot represents 100 people

N
W—E
S

0   200   400 km

Source: Based on data from Australian Bureau of Statistics. (2022-23). Regional population. ABS. https://www.abs.gov.au/statistics/people/population/regional-population/2022-23. © Commonwealth of Australia. Map redrawn by Spatial Vision.

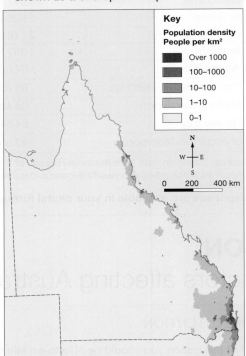

FIGURE 3.31 Queensland's population shown as a choropleth map

Key
Population density People per km²

■	Over 1000
■	100–1000
■	10–100
■	1–10
□	0–1

N
W—E
S

0   200   400 km

Source: Based on data from Australian Bureau of Statistics. (2022-23). Regional population. ABS. https://www.abs.gov.au/statistics/people/population/regional-population/2022-23. © Commonwealth of Australia. Map drawn by Spatial Vision.

Population density for a place can be calculated easily by dividing the total population by the total area in question.

$$\frac{\text{total population}}{\text{area}\left(\text{km}^2\right)} = \text{population density}$$

DID YOU KNOW?

Queensland's population is around 5.2 million and the state's total area is 1.7 million square kilometres. This means that the population density for Queensland as a whole is only about 3 people per square kilometre!

3.5.2 Push and pull factors

When examining changing populations, we need to examine why people might leave one area and move into another. Geographers talk about push and pull factors, and some examples of each of these factors are shown in figure 3.32.

FIGURE 3.32 Push and pull factors affecting population distribution

Push factors

Pull factors

- Population pressure
- Poor infrastructure
- High unemployment
- Limited choice of education
- Poor health care
- Ecological problems
- Natural disasters
- Social compulsions

- Improvement in the standard of living
- Better health care and availability of services
- Varied employment opportunities
- Higher wages
- Quality of education
- Good transport connections
- Future prospects

Push factors are those that encourage people to leave an area. In developed countries, these factors are usually related to job opportunities or lifestyle reasons. For instance, a downturn in a local industry could leave many people unemployed at the same time, so job prospects might be suddenly limited. This would encourage some people to leave the area. Local parkland might be rezoned for medium-density housing, which would remove important recreational facilities for an area and so might push families out. These are examples of push factors. High crime rates, lack of employment opportunities, environmental degradation or removal of services are additional examples of factors that may push people out of an area.

Conversely, **pull factors** are those that attract people to a location. The potential for employment and better job prospects, or the provision of recreational facilities that could encourage families to move into an area are considered pull factors. Other pull factors might include transport connections in an area, service provision, educational and medical facilities or the value of the natural environment.

push factors factors that encourage people to leave an area

pull factors factors that attract people to an area

EXAMPLE: Pull factors of Brisbane

Brisbane attracts migrants from all over Queensland, Australia and even the world because of its many positive features — or its pull factors. However, which factors appeal to different people depends on their current circumstances and preferences.

For example, the cost of living in Brisbane is lower than that of Sydney or Melbourne, and might be a pull factor for people from those cities. On the other hand, someone living in rural Queensland might find that life in Brisbane is more expensive than they're used to, but might be more attracted to the city's education or health facilities.

FIGURE 3.33 Brisbane's many cafes and restaurants work as a pull factor to the city.

3.5.3 Physical factors affecting population distribution

Physical factors of a place include climate, relief (shape of the land), soil, availability of natural resources and energy.

Geography

More physically remote, extreme and harsh areas such as mountains or deserts generally have fewer people. Coastlines and other water sources can boost human settlement because people are attracted to living by the water for a variety of reasons.

FIGURE 3.34 Marre in outback Australia

Physical connectivity is important, too. Post-European contact, most settlements in Australia were founded on the coastline because shipping was vital to maintaining a connection to the other colonies, Britain and the rest of the world. First Nations Australians had well-established trade and transport routes that served as the basis for many of our current highways and roads as they were built on by European settlers. Settlements sprung up along these routes because people moved through them regularly.

Water

People need steady and stable access to fresh water for drinking, cleaning, bathing, industry and agriculture. As early settlement populations increased in Australia, more areas were set aside for permanent water storage, such as Wivenhoe Dam near Brisbane.

FIGURE 3.35 Wivenhoe Dam near Brisbane

The provision of water is one of the fundamental drivers of population — populations will naturally adjust according to the amount of water available for use.

Pipelines and desalination plants can also help to provide water to areas that do not have sufficient natural sources of potable water.

Resources

Resources for building, eating and trading are all predictors of where people will live. Early European settlers in Australia, in particular, required significant resources to establish and grow their communities. As more and larger areas of mineral production were discovered — for example, in central Queensland — populations in those areas increased significantly in order to service the extraction of ore and secondary industries.

CASE STUDY: Natural resources of Mount Isa

Quick facts: Mount Isa
- **Location** North-west Queensland
- **Population** 18 000
- **Median age** 31
- **Most common industry of employment** Copper ore mining

Mount Isa, a city located 1825 km north-west of Brisbane (see figure 3.36), is an example of a settlement that has emerged solely because of the resources found there. Lead, silver, copper and zinc are found in abundance in the region, which drove mining companies to push into the area in the early 20th century. These companies need staff and can offer attractive salary packages, so many people began to reside in Mount Isa. A community has slowly established itself there, and around 18 000 people now call Mount Isa home. The city has most of the same services and features of any other Australian city. Like many mining towns, however, the number of people working in certain industries is very different from the proportion in the wider community (see table 3.8).

FIGURE 3.36 Mount Isa

TABLE 3.8 Top industries of employment for residents of Mount Isa, Queensland and Australia, 2021

Industry of employment	Mt Isa	%	Queensland	%	Australia	%
Copper ore mining	1587	16.9	3145	0.1	8892	0.1
Silver/lead/zinc ore mining	840	9.0	2650	0.1	4697	0.0
Hospitals	519	5.5	122 121	5.0	545 158	4.5
Primary education	339	3.6	59 930	2.5	265 249	2.2
Supermarket/grocery	219	2.3	61 444	2.5	299 810	2.5

Source: Based on data from Australian Bureau of Statistics, ABS: 2021 Census QuickStats Mount Isa.

As table 3.8 shows, mining is a significant area of employment in Mount Isa, with around one quarter of all employed people working in the industry. Meanwhile, for the rest of Queensland and Australia as a whole, mining is a very small employer. However, health and education services are also significant employers in Mount Isa, because those working in mining need hospitals and schools for themselves and their families.

FIGURE 3.37 Mary Kathleen mine, near Mount Isa

SKILLS ACTIVITY

Use table 3.9 to create a population pyramid for Mount Isa.

TABLE 3.9 Age group data for Mount Isa, 2021

Age groups	Males	Females
0–4 years	808	668
5–14 years	1395	1352
15–19 years	619	531
20–24 years	648	652
25–34 years	1597	1647
35–44 years	1269	1293
45–54 years	1100	1064
55–64 years	1009	822
65–74 years	475	417
75–84 years	237	221
85 years and over	50	65

Source: Based on data from Australian Bureau of Statistics. (2021). Mount Isa: 2021 Census Community Profiles. https://www.abs.gov.au/census/find-census-data/community-profiles/2021/UCL313009. © Commonwealth of Australia.

 Resources

 Digital doc Creating a population pyramid (doc-42475)

Soil

The quality of soil in a place and its potential for growing crops attracts people. Higher quality soils maintain environmental quality, make healthier animals and people, and generate greater crop yields. Because of the potential for communities to develop around agricultural practices, areas with high-quality soil attract people, while areas with poor soil quality do not.

EXAMPLE: Soil quality in the Darling Downs

The Darling Downs is a region in southern Queensland with excellent soil quality, which makes it the most fertile farming area in the state. The landscape is dominated by low, undulating hills used to grow different agricultural products, including cotton, wheat, barley, sorghum and different vegetables. Sheep and cattle stations can also be found in the region.

A number of settlements can be found on the Darling Downs, including Toowoomba, Dalby, Warwick, Stanthorpe, Goondiwindi, Oakey, Miles, Pittsworth, Wallangarra, Allora, Clifton, Cecil Plains, Drayton, Millmerran and Chinchilla.

FIGURE 3.38 The Darling Downs provides rich and fertile soil for commercial crop growing.

Climate

People are attracted to areas that experience temperate weather conditions because the temperature is moderate and bearable. Temperate zones are found between the Tropic of Capricorn (which cuts through Australia) and the Antarctic Circle in the southern hemisphere, and between the Tropic of Cancer and the Arctic Circle in the northern hemisphere.

Most of the world's population is found in these zones, particularly in the northern hemisphere because it has more landmass than the temperate zones of the southern hemisphere. These temperate zones then have climate zones within them.

Figure 3.39 shows Australia's climate zones. Also shown is the Tropic of Capricorn, which passes through Queensland, the Northern Territory and Western Australia.

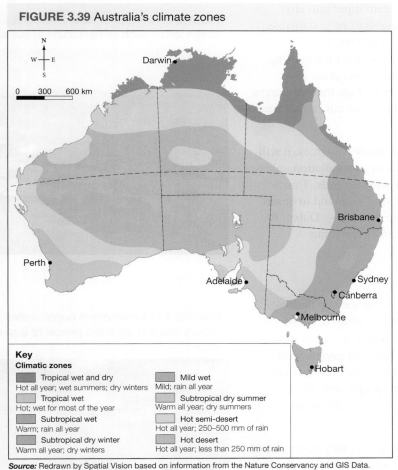

FIGURE 3.39 Australia's climate zones

Key

Climatic zones

Tropical wet and dry
Hot all year; wet summers; dry winters

Tropical wet
Hot; wet for most of the year

Subtropical wet
Warm; rain all year

Subtropical dry winter
Warm all year; dry winters

Mild wet
Mild; rain all year

Subtropical dry summer
Warm all year; dry summers

Hot semi-desert
Hot all year; 250–500 mm of rain

Hot desert
Hot all year; less than 250 mm of rain

Source: Redrawn by Spatial Vision based on information from the Nature Conservancy and GIS Data.

3.5.4 Social factors affecting population distribution

Social factors include education, health care and lifestyle, and these all have an impact on population distribution. In relation to these social factors, people will tend to settle in places that have:

- better educational opportunities for children and young adults
- quality local health services and medical facilities for families and older people
- leisure facilities including parklands and sporting grounds
- community centres such as religious institutions, libraries, sporting clubs or interest groups.

3.5.5 Economic factors affecting population distribution

Economic factors include aspects such as the availability of jobs, wage levels and general affordability of an area.

Resource exploitation and trade

Resources are a significant input into any settlement. For settlements to be sustainable, they require adequate resources. Early settlements need local resources for housing, food and water. As settlements grow and mature, they may begin to trade their resources for other things that they need but cannot access in their region.

If an area doesn't have many resources, it will not be able to sustain a large population unless it can import much of what it needs. This is why very few major cities are found in desert environments but those that are — Dubai, for instance — import nearly everything.

FIGURE 3.40 A lack of natural resources make it hard for desert towns such as Parachilna to sustain large populations.

Employment

Economic conditions influence population distribution significantly. During times of favourable economic conditions, areas can experience a massive influx of people who are looking to profit. However, during times of economic downturn, people will move on to areas that offer better employment opportunities.

Australia famously attracted many international migrants in the period after World War II as many significant infrastructure projects were undertaken, increasing opportunities for skilled and unskilled employment. Many current Australians can trace their ancestry back to these postwar migrants. The recent mining boom has also led to the influx of people to mining communities, both as residents and as FIFO (fly-in fly-out) workers.

FIGURE 3.41 Employment opportunities for miners were key factors that first attracted people to Broken Hill, New South Wales.

Affordability

Affordability is probably the most significant factor in determining where people live. Affordability is influenced by two factors: wages and the cost of living. The benefits of a well-paid job will be undercut if the cost of living is very high. This relationship between wages and living costs can vary considerably across the country.

Figure 3.42 shows the relative socioeconomic disadvantage in Australia. The values are generated from a combination of indicators that take into account the cost of living and wealth of different areas. Figure 3.43 shows the relative median weekly income for households in Queensland — another measure of the economic advantage in a specific area.

FIGURE 3.42 Relative socioeconomic disadvantage in Australia

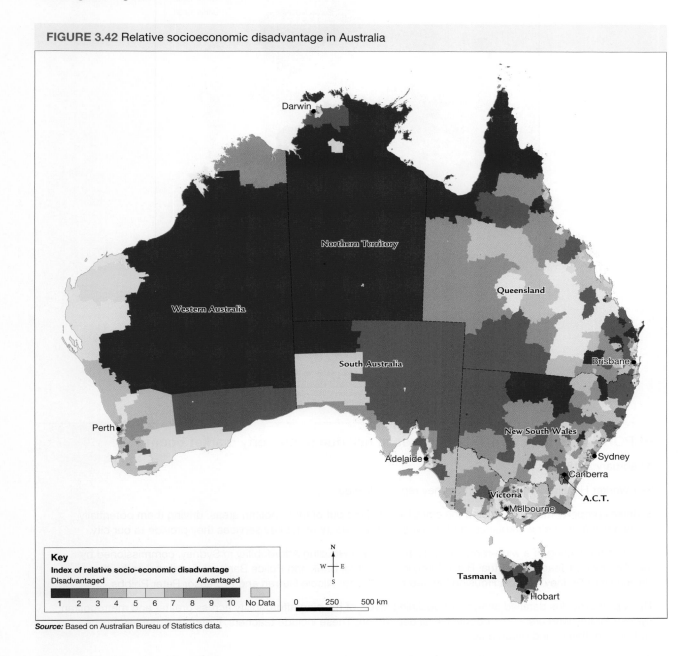

Source: Based on Australian Bureau of Statistics data.

FIGURE 3.43 Median household income per week in South East Queensland

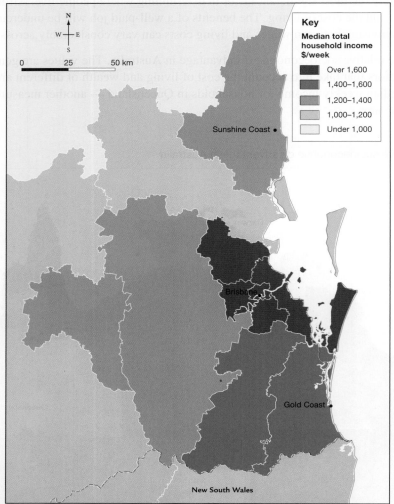

Key

Median total household income $/week

- Over 1,600
- 1,400–1,600
- 1,200–1,400
- 1,000–1,200
- Under 1,000

Sunshine Coast •

Brisbane

Gold Coast •

New South Wales

Source: © The State of Queensland Queensland Treasury 2018. ABS 2033.0.55.001 Census of Population and Housing: Socio-Economic Indexes for Areas SEIFA, Australia, 2016; GBRMPA; Google; ZENRIN.

REPORT: Emergency and key services at risk due to property market boom

5 February 2018

Key Worker Housing Affordability in Sydney report released

Sydney's property market is pricing the city's key workers out of metropolitan areas, driving them potentially hours away from their workplaces, and threatening the viability of the key services they provide to our city.

This was the finding of a wide-ranging study, Key Worker Housing Affordability in Sydney, commissioned by member-owned Teachers Mutual Bank, Firefighters Mutual Bank and Police Bank, and undertaken by the University of Sydney's Urban Housing Lab led by Professor Nicole Gurran and Professor Peter Phibbs.

The report provides detailed analysis of declining levels of housing affordability across greater and metropolitan Sydney for key workers — the people we all depend on. These include teachers, firefighters, nurses, police, ambulance drivers and paramedics.

Based on current trends, the outlook for housing affordability in Sydney's Greater Metropolitan Area is grim.

'Sydney's overstretched housing market is locking teachers, firefighters, nurses, police, ambulance drivers, and paramedics out of home ownership. Our key workers are increasingly being forced to outer metropolitan areas in search of an affordable place to live,' said Professor Gurran.

The report found that in the ten years leading up to 2016, key areas in Sydney lost between 10 and 20 per cent of teachers, nurses, police and emergency service workers to outer and regional areas. Sydney's Inner South West (–14.6%), Inner West (–11.3%), Eastern suburbs (–15.2%), Ryde (–14.2%) and Parramatta (–21.4%) all experienced a net loss of key workers, while areas including the Illawarra (+10.5%), Southern Highlands (+17%) and Hunter Valley (+13.6%) all had net gains.

The nature of this group's shift work combined with living a distance away in areas with unsuitable public transport meant 77.4 per cent of key workers drove their private motor vehicle to work in 2016, compared with just over 43 per cent for the general population.

Only 5 per cent of key workers used public transport to get to work, compared with 12.7 per cent for the general population.

Teachers Mutual Bank CEO Steve James says that the pressure this situation puts on people already working in high-pressure jobs is unfair, whether they're looking to rent or buy.

'Longer commute times, especially in private vehicles, lead to significantly higher financial costs and serious social consequences for key workers and their families, disrupting work–life balance and impacting their lifestyle. Critically, lengthy commute times are also associated with lower rates of workforce participation.'

The report reveals that between 2003 and 2016, the median price of established homes in Sydney more than doubled from $400 000 to around $900 000 — well beyond the reach of many key workers, especially those who are single. Soaring rents have heightened the crisis, making a 20 per cent home loan deposit unattainable for many key workers.

For instance, a single key worker eyeing a property in Sydney's inner ring at the 2016 median price of just over $1 million would need 13 years to save for a deposit. This is a sharp increase from the 8.4 years needed to save a 20 per cent deposit in 2006.

'For a key worker, finding somewhere affordable to live in reasonable proximity to their work is becoming impossible for those not already in the property market,' said Steve James, CEO of Teachers Mutual Bank Limited.

'The report has found that the closest local government area with an affordable median rental price for an entry level enrolled nurse is Cessnock in the Hunter Valley. That's about 150 km from any hospital in Sydney city, making it a 300 km round trip per day,' explains Mr James.

The report also looks at urgent solutions that can be implemented to help key workers buy their own homes both close to their place of work and their established support networks such as family, friends and the wider community, including local groups, clubs and schools.

It identifies five key priorities for policy makers and private sector stakeholders to consider. Professor Gurran emphasised that these approaches can be implemented largely within the current policy framework.

'There are a number of strategies that would improve housing affordability for key workers and their families, with government support. These include reducing the 'deposit gap' to home ownership that key workers now face despite their stable employment; boosting the supply of affordable 'starter homes' for first home buyers; reducing construction costs through design innovation; and looking at alternative forms of housing tenure,' said Professor Gurran.

Teacher's Mutual Bank CEO Steve James continued: 'This study shows that without urgent and genuine intervention on the part of policy makers and other institutions, a growing number of key workers in NSW and Sydney may never be able to afford to own their own home within reasonable distance of Sydney.'

Tony Taylor, Police Bank CEO, said that the five key measures identified in the report to improve housing affordability for key workers were worth serious consideration.

'Without urgent action by all levels of government, in conjunction with the finance and building industries, we could see more key workers pushed further out of the Sydney metropolitan area. The effects of that will be felt by each and every one of us,' he said.

Professor Phibbs added: 'By addressing the key barriers to affordability, residential housing in metropolitan Sydney could be as much as 20 per cent cheaper than today's prices, with vastly more manageable deposits for key workers looking to buy a home.'

In Sydney, 20 per cent of key workers such as teachers, firefighters, nurses, police, ambulance drivers and paramedics left the city in the ten years leading up to 2016. Housing affordability was cited as a key reason. Additionally, 77 per cent of key workers drove their motor vehicle to work while only 5 per cent of key workers took public transport; these are higher and lower than the Australian average respectively.

Source: The University of Sydney, School of Architecture, Design and Planning

3.5 Exercise

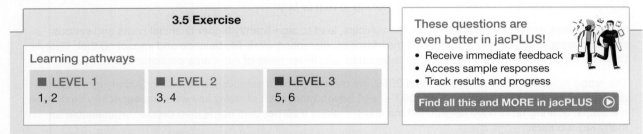

3.5 Exercise
Learning pathways

■ LEVEL 1	■ LEVEL 2	■ LEVEL 3
1, 2	3, 4	5, 6

These questions are even better in jacPLUS!
- Receive immediate feedback
- Access sample responses
- Track results and progress

Find all this and MORE in jacPLUS ▶

Explain and comprehend

1. **Outline** what is meant by push and pull factors.
2. **Define** what is meant by a physical factor affecting a population distribution. Provide an example to support your answer.
3. **Describe** what is meant by a social factor affecting a population distribution.
4. **Describe** what is meant by an economic factor affecting a population distribution.

Analyse and apply

5. Research the population and area of the following places and work out their population density. **Compare** them to Queensland's population density and describe the differences you see.
   a. Your local area, town or suburb
   b. An inner-city Brisbane suburb such as Milton
   c. An outer Brisbane suburb such as Springfield
   d. A regional centre such as Chinchilla
   e. Tokyo or Mumbai

Propose and communicate

6. Go to Google Earth and use the timeline function to examine the following areas from 1990 to today.
   - Your local area
   - Logan
   - Gold Coast
   - Gladstone
   - Beerburrum State Forest
   - Weipa
   a. **Describe** the settlement patterns of each location.
   b. **Describe** the changes you see in both the built and natural environments at each location from 1990 to today.
   c. Find data on the number of government services available at each site. Can you see a pattern in the distribution of services to different areas of the state?
   d. **Identify** what challenges and opportunities residents in each area might have experienced because of the changes you described.
   e. Choose the challenge and the opportunity that you think would have the most impact on your life if you lived in these areas, and **explain** how they would affect you.

Sample responses are available in your digital formats.

LESSON
3.6 Challenges facing remote and rural Australia

LEARNING INTENTION

By the end of this lesson you should be able to explain the geographical challenges facing places in Australia as a result of the changing characteristics of places, including:
- rural and remote places — for example, employment, provision of health and educational services, transportation connections to major centres, expansion or contraction of industry, isolation and remoteness, access to fresh and affordable food, housing availability and affordability, waste management, fresh water quality and availability, and access to communication technology (e.g. NBN).

Source: Adapted from Geography General Senior Syllabus 2024 © State of Queensland (QCAA) 2024; licensed under CC BY 4.0.

3.6.1 Challenges in remote and rural Australia

Australia's rural and remote areas have some unique characteristics, from a geographical, economic and social point of view. This section focuses on some of the challenges faced by residents in these areas, including employment, transport, health care, education, and access to communication technology.

FIGURE 3.44 The Hunter Valley region of New South Wales

Employment

A strong local economy is vital to the survival of a small town. In good economic times, more money is in the area, which encourages employment and drives spending — good things for the economy of a rural town. When times are difficult, however, less money is being spent and businesses are less likely to employ people. These times can see many people leave rural areas to look for employment elsewhere.

Australia experienced a mining boom in the 2000s that added billions to the economy and created jobs and wealth for the hundreds of thousands of Australians who worked in the industry or associated industries. Populations expanded across areas of remote, rural and regional Australia, including in Queensland, where mining activities were based.

FIFO stands for 'fly-in fly-out' and relates to people who work in one place but live in a different place. Most of these workers are employed in the mining industry in remote parts of the country but live in more populated areas, such as South East Queensland. FIFO workers are usually paid very well, due to the remote location in which they work. Additionally, their employer will usually pay for their flights in and out, and provide temporary accommodation.

The practice has positive impacts on people in terms of the money they can earn, but it also has negative impacts on rural communities as they manage increasing costs and an ever-changing community. Individuals suffer due to the regular movement in their lives and the effects of constant travel and change. This makes it an increasingly controversial practice — so much so that the Queensland Government saw fit to address it through a Parliamentary Inquiry in FIFO in 2015.

EXAMPLE: Population change in Moranbah

Table 3.10 shows the population change between 2001 and 2021 for the central Queensland mining town of Moranbah, which is part of the Isaac region. Although Moranbah doesn't have a large population, and many mine site workers are FIFO workers and live elsewhere, the population still expanded by more than 25 per cent from 2001 to 2011. From 2010, the mining boom started to wind down and Moranbah's population contracted slightly. As mining in the area slowed, fewer mine site employees were required and this had a knock-on effect on secondary businesses that fed into the mining industry, with these businesses slowing as well. However, between 2016 and 2021, Moranbah saw an increase in population again, to a new high of 8899.

FIGURE 3.45 Peak Downs coal mine, near Moranbah

TABLE 3.10 Population for Moranbah, Queensland, 2001 to 2021

Year	Population
2001	6124
2006	7133
2011	8626
2016	8333
2021	8899

Source: Based on data from Australian Bureau of Statistics. (2021). Moranbah: 2021 Census All persons QuickStats. https://www.abs.gov.au/census/find-census-data/quickstats/2021/UCL314017 © Commonwealth of Australia.

Housing

Encouraging people to buy into an area is a good way to get them to stay. The mining boom of the early 21st century saw many rural and remote areas change dramatically, with property values often skyrocketing as soon as mining or mineral projects were announced. This brought with it developers and investors, and soon housing prices and rents were so unaffordable in the area that local residents were suddenly priced out of the market.

In contrast, when mining and other projects finish, the reverse can happen: many property investors and buyers pull out of the area, which suddenly guts the market and leaves many people with overpriced property and no interested buyers.

Managing the transition between different phases of the economic cycle is important for economic stability and sustainability in these areas.

FIGURE 3.46 Much of Silverton, New South Wales, was abandoned after local silver mines closed.

Health care and education

According the 2021 census, around 7 million people — or 28 per cent of the Australian population — live in rural and remote areas. These Australians face the following unique challenges due to their geographic location:
- generally poorer health outcomes than people living in urban areas
- higher rates of hospitalisations, deaths, injury
- reduced access to, and lower use of, primary healthcare services, compared to people living in urban areas.

Young people growing up in rural and remote places can have reduced access to education services compared to students living in urban areas. This includes:
- fewer choices of school
- limited access to university or TAFE
- long travel times to school or further education.

Transport

Rural and remote areas often have a lack of transport options, with people being forced to rely on their own vehicles. The public transport network, such as buses and trains, is rarely extended beyond major urban areas. When trains or buses are available, the stations or stops are limited and services less frequent. This is a challenge for many rural and remote communities, making access to employment and amenities difficult. It also contributes to social isolation — in other words, loneliness and a lack of connection to others.

FIGURE 3.47 Rural and remote communities often have limited access to transport.

Waste management

Managing waste and wastewater in small, regional areas can be challenging, depending on the needs of the community. All towns and cities in Queensland have some form of wastewater treatment facility in the town. Smaller or more dispersed areas may still be seeking an option. Remote and rural areas often have their own landfill or waste management processes.

In some rural and regional areas that rely on one industry — for instance, mining or agriculture — waste management is controlled by that industry. This can lead to issues where residents may be left out of the decision-making process and waste is managed in an inappropriate or unsustainable manner for that area.

 on Resources

 Weblink Australian National Waste Report

Fresh water quality and availability

Water is the key resource required in any settlement. If there is no regular supply of water, there will be no settlement. Careful management of fresh water is crucial, especially in smaller towns and settlements, to ensuring a stable and healthy population.

In the mid-2000s, South East Queensland experienced a prolonged drought that heavily impacted on the region's ability to supply water to the population. Severe water restrictions were put in place on residents as water storage dropped to critical levels. Water delivery infrastructure was also upgraded, which allowed water to be moved around the region more easily.

FIGURE 3.48 Glenlyon Dam, Queensland

SKILLS ACTIVITY

Consider figure 3.49.

FIGURE 3.49 Average annual rainfall in Australia (1991–2020)

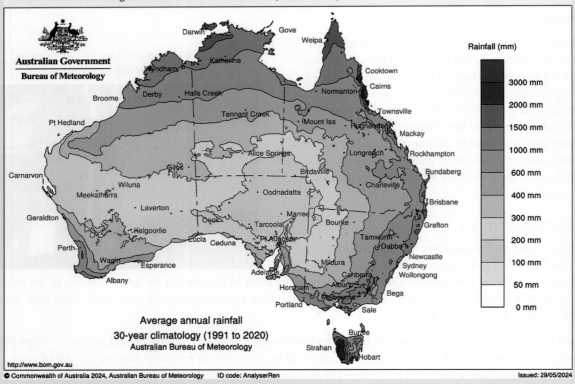

Source: Commonwealth of Australia 2021, Average annual, seasonal and monthly rainfall maps. Retrieved from http://www.bom.gov.au/climate/maps/averages/rainfall/?period=an&region=aus.

1. Describe the spatial association between rainfall and major population centres in Australia.
2. Suggest why this spatial association exists.
3. Identify one challenge that this spatial association implies for rural parts of Australia.

Access to communication technology

The National Broadband Network (NBN) is a nationwide piece of infrastructure designed to modernise Australians' access to the internet. Faster internet speeds will allow users to do more online, including watching more television shows or movies, creating more digital content such as videos, 3D models and games, conducting medical procedures in real time using remote access, and educating people in rural and remote areas who rely on distance education more effectively.

Over time, the scope and technology proposed by the NBN has been reduced by the government. Critics say these changes were made to minimise the financial impacts on private corporations offering products that rely on downloads, such as pay television channels and online movie distributors, although the official reason is saving costs.

Many parts of rural and regional Australia have very poor telephone and internet coverage, and this absolutely hampers the ability of people in those areas to participate in online communities and the online economy. It also affects whether someone is able to work from home or run a home business. When initially announced, the NBN was promoted as a 'nation building infrastructure project' by then-prime minister Kevin Rudd. Sadly, the NBN has been used as a political football by successive governments and its technological scope has been eroded over time. Those charged with delivering the NBN would argue this has been done to curb its cost.

CASE STUDY: Life in Coober Pedy

Quick facts: Coober Pedy
- **Location** South Australia
- **Population** 1566
- **Median age** 50
- **Most common industry of employment** Accommodation (tourism)

Nestled in the arid landscape of South Australia, Coober Pedy stands as a testament to the resilience of rural communities in Australia. Home to approximately 1500 people, its remote location presents unique challenges for its residents.

Employment opportunities in Coober Pedy primarily revolve around opal mining and tourism. However, the seasonal nature of these industries and the limited job opportunities often lead to high levels of unemployment and underemployment within the community. One of the reasons that Coober Pedy is popular with tourists is that it offers the chance to see the unique housing in the town. This housing is characterised by its underground dwellings, known as dugouts, which offer relief from the extreme desert temperatures. These homes are unique to the region but can pose challenges in terms of maintenance and ventilation.

FIGURE 3.50 The town of Coober Pedy is located in the South Australian outback.

Healthcare services in Coober Pedy are provided by the Coober Pedy Hospital and Health Service, which offers a range of medical facilities, including emergency care, general practice and community health services. Meanwhile, education is supported by local primary and secondary schools, providing education for the town's children and young adults. However, the distance to major educational facilities in larger towns remains a challenge for some residents seeking higher education opportunities.

FIGURE 3.51 Bathroom of an underground dugout in Coober Pedy

Transportation in and out of Coober Pedy primarily relies on the Stuart Highway, connecting the town to major cities such as Adelaide (846 kilometres away) and Alice Springs (688 kilometres away). However, the long distances and isolation can make travel difficult, especially during severe weather events.

Despite these challenges, Coober Pedy residents demonstrate remarkable resilience, innovation and community spirit as they navigate life in the outback.

3.6 Exercise

3.6 Exercise

Learning pathways

■ LEVEL 1	■ LEVEL 2	■ LEVEL 3
1, 2	3, 4	5, 6

These questions are even better in jacPLUS!
- Receive immediate feedback
- Access sample responses
- Track results and progress

Find all this and MORE in jacPLUS ⊙

Explain and comprehend

1. **Outline** why rural and remote communities can lack public transportation options.
2. **Explain** how employment can be a challenge faced by remote and rural areas.
3. **Examine** the challenges that poor mobile phone and internet coverage in rural and remote areas might present.

Analyse and apply

4. Research the history of Emerald, a mining town in central Queensland.

FIGURE 3.52 Lake in Emerald, Queensland, Australia

a. **Provide** some basic facts about the place, such as when it was established, why it was established, the main industry in the area, and the economy at the time of establishment.
b. What challenges would Emerald face from a declining population?
c. What could the people of Emerald do to mitigate any negative impacts of a contracting economy? (For some possible strategies, research how Newcastle in New South Wales has managed their changing economy after the mining boom.)
5. Recent studies have suggested that life expectancy in regional areas of Australia is one to two years lower than in Australia's major cities, while in remote areas it is up to seven years lower. **Discuss** possible reasons for this.

Propose and communicate

6. Some have argued that more schools and hospitals should be built in rural and remote areas of Australia. **Evaluate** whether or not you believe this should occur.

Sample responses are available in your digital formats.

LESSON
3.7 Challenges facing urban Australia

3.7.1 Challenges in metropolitan and regional cities in Australia

A **metropolitan area** is a densely populated urban area with significant economic activity, cultural diversity and social interaction.

Such urban places in Australia face a spectrum of challenges, reflective of the complexities of modern urbanisation. These challenges include managing urban sprawl, issues associated with gentrification and pressure on transport options.

3.7.2 Urban sprawl

As urban populations in Australia steadily increase, so does the demand for residential space. At times, this demand may be at a rate faster than that of natural population increase and immigration. Many cities progressively grow outwards, spreading into the surrounding countryside and invading adjacent towns, regions and undeveloped land.

FIGURE 3.53 Morning rush hour in Brisbane, Queensland

This process of outward expansion of urban areas from the city centre is known as urban sprawl. The term is imprecise but it is accepted that urban sprawl:
- lacks planning (in that each housing estate is clearly planned but how they tend to work together is generally not)
- encourages **monofunctional development**
- usually consists of large areas of residential development
- is low density
- includes commercial strip development
- has fewer public transport options
- is dominated by motor vehicles.

metropolitan area a densely populated urban area with significant economic activity, cultural diversity and social interaction

monofunctional development developments that have a single function or purpose, such as a development that is solely for residential buildings

Urban sprawl occurs as a direct result of increasing population in urban areas. The 'Great Australian Dream' of a quarter-acre block (roughly 1000 m^2) has been a driving factor as urban populations and the demand for large residential blocks of land have increased while the availability of land has decreased. This has pushed developers to build large, low-density residential developments on the edges of our cities, causing these urban areas to expand rapidly (see figure 3.54). As an example of this, in 1990, Brisbane's outer northern suburbs were largely undeveloped — the area was effectively farmland. Over the next 25 years, development slowly expanded into these northern suburbs and, by 2018, North Lakes in the outer-north was a vibrant suburb housing more than 22 000 people.

FIGURE 3.54 Low-density suburbs, such as these in Brisbane, contribute to urban sprawl.

Impacts of urban sprawl

Urbanisation could be argued to have positive and negative consequences. Increasingly, however, geographers are associating the negative impacts with urban sprawl, rather than urbanisation itself. As a result, city planners are required to consider ways of providing housing and services to growing populations without spreading the city further outwards.

Loss of biodiversity and reduction of agricultural land

Environmentally, urban sprawl leads to a number of negative impacts. Environmental degradation occurs as land is taken for residential use. In some cities, native vegetation is replaced with housing, roads and the utilities required to sustain a modern community. This results in a loss of habitat, such as the loss of koala habitat in Brisbane's southern suburbs, a loss of animal species and an increase in fragmentation that results in the loss of wildlife corridors.

EXAMPLE: Environmental impacts of Brisbane's sprawl

Development across Brisbane has resulted in the loss of wildlife corridors. This restricts the movement and migration of many species, which, in turn, impacts their breeding ability.

The additional loss of their habitats means they increasingly become displaced and begin to mingle with people in residential areas. This is often seen as an issue by residents, as snakes and possums are seen more often in people's backyards. Such animals are often removed and relocated, which also impacts on their breeding ability. Other species such as koalas move further out and away from residential areas, but the availability of their native environment is decreasing, leaving them with fewer areas to move to.

FIGURE 3.55 Residential development in Brisbane is disturbing natural habitats.

The reduction of agricultural land on the edges of cities also impacts Australia's food security, both now and into the future.

Other impacts of urban sprawl

Table 3.11 outlines further impacts of urban sprawl.

TABLE 3.11 Consequences of urban sprawl

Features of urban sprawl	Impact
Disrupted waterways and increased flooding	Replacement of natural land and soil with impermeable bitumen and development increases surface run-off. Fewer areas for water to be absorbed leads to flooding and often flash flooding because water has nowhere to go.
High energy use	Increased energy use, waste and air pollution increases the occurrence of urban heat islands. This happens due to the increase in dark surfaces in the city compared to rural areas. These areas can be up to 4 °C higher than surrounding areas. (See chapter 4 for more information on urban heat islands.)
Limited access to services and facilities and high vehicle dependency	Areas of urban sprawl often lack functional public spaces and public transport options, and rely on road infrastructure. Strip development with fast-food outlets, service stations, grocery stores and retail chains are all found in these areas. These areas also increase vehicle dependency because public transport often doesn't keep up with the speed of urban sprawl, even though road networks may be in place. The further from the CBD people live, the more cars are needed to access schools, medical facilities and larger grocery stores.
Poor cost efficiency	As urban sprawl increases and moves further from the CBD, providing services to these areas becomes more inefficient and costly, which can lead to increasing inequality.

FIGURE 3.56 Flooding in an urban centre in Queensland

FIGURE 3.57 Many of these car drivers are likely to be going to, or coming from, the suburbs that sprawl outwards from Brisbane.

Factors affecting urban sprawl

The factors that influence urban sprawl are complex and will differ from place to place. However, urban sprawl can generally be attributed to a combination of:

- cost of land
- desire for low-density living
- access to motor vehicles
- safety
- population increase
- development guidelines
- housing policy.

Affordability and access

The price of land usually drops the further it is from a city's CBD. Exceptions exist to this rule but, generally, those on lower incomes or who are less wealthy are attracted to these places because they can afford to buy or rent a home there.

The Great Australian Dream has pushed many people into these areas as they look for affordable, low-density living (see figure 3.58). Families with children often prefer to have their own backyard, especially if they are moving from a part of the city with little green space or infrastructure for children, such as playgrounds or skate parks. Urban sprawl can be driven, in part, by this consumer attitude that prefers the quarter-acre block over inner-city apartment living.

FIGURE 3.58 Single-storey homes with a backyard are part of the Great Australian Dream.

As more households drive more motor vehicles, the ability for people to travel long distances, especially for work, has also increased. This has meant that suburban development, which used to only occur along transport corridors (such as rail lines or highways), can move further away from these corridors and more land can be developed.

Safety

Outer suburbs are often perceived as being safer than inner-city areas in terms of individual safety from personal crime but also in terms of quality of air and extent of the natural environment. While this perception is not always accurate, it is a factor in driving people into these outer suburban areas.

Population increase

Interstate and overseas migration help to increase the population in many urban areas and this also contributes to urban sprawl. Families that expand tend to stay within the same city, but as children move out of home, more homes are required.

FIGURE 3.59 Newly constructed suburb on the edge of Melbourne

Development restrictions

In the past, outer-fringe subdivisions were allowed to occur with few restrictions. Ill-considered land development and marketing practices on the rural–urban fringe resulted in large areas of land being cleared and quickly developed for commercial gain by real estate developers — sometimes before sufficient infrastructure had been put in place to serve the future residents. Councils also benefit financially from these new suburbs of ratepayers, so approving new developments boosts their income.

Housing policy

In the past, government housing policies have favoured ownership of detached single dwellings as opposed to government-controlled social housing. Direct government assistance, such as home saving grants and tax incentives, are designed to encourage private ownership, while the lending policies of financial institutions have encouraged new home ownership. Australia's political history over the last 20 years has seen house prices skyrocket, but personal income has not grown at the same rate. This has made housing, particularly in urban areas, less affordable for those who are yet to participate in the housing market.

3.7.3 Other challenges for metropolitan and regional areas

Gentrification

Gentrification is the process of older areas, usually of lower socioeconomic status, being slowly bought out and renovated by wealthier people. This is a controversial topic because gentrification usually improves property prices in an area, but at the expense of many of the neighbourhood's original inhabitants and its character.

Usually, poorer groups, such as students or artists, inhabit rundown areas of urban spaces due to the more affordable rent. They bring character and flair to these areas, which increases the demand for housing and causes property prices to rise. Eventually, these 'early gentrifiers' are pushed out of the area because they cannot afford to stay and the character of the area subsequently changes.

gentrification the process of older areas, usually of lower socioeconomic status, being slowly bought out and renovated by wealthier people

---

EXAMPLE: Gentrification in Paddington, Brisbane

Paddington in Brisbane is a great example of an area that has been gentrified. The suburb sits only 3 kilometres west of Brisbane's CBD. In the years following World War II, the area was mainly working class but in the 1980s the demographics shifted as more people bought into the area looking for a home near the CBD.

Increasingly wealthy investors continued to drive up property prices and now Paddington is one of Queensland's most desirable, and expensive, suburbs.

FIGURE 3.60 Paddington's house prices have been driven up by gentrification and the suburb's proximity to the CBD.

 Resources

🔗 **Weblink** Gentrification

Figure 3.61 (a) and (b) show what is referred to as gentrification, with working class areas slowly changing to accommodate professionals and their differing lifestyles.

FIGURE 3.61 Paddington (a) in 1949 and (b) now

Transport

Transport options in urban Australia are very different from those in rural and remote Australia.

As Australia's cities have grown, and continue to grow, higher population densities lead to more traffic congestion. Traffic in these areas is a constant, and increasing, problem, particularly during peak times. Brisbane has witnessed significant investment in roads and tunnels since the early 2000s in an effort to curb increasing traffic congestion. See figure 3.62 for examples of roads in Brisbane and other Queensland locations.

FIGURE 3.62 Roads across different Queensland locations: (a) Brisbane (b) Eula, western Queensland, and (c) northern Queensland

Public transport has also undergone huge investment. Often, one of the challenge for those providing public transport in urban areas is how to manage more people using the system at peak times of the day. Active, sustainable transport options, such as cycling and walking, have also led to changes in the city's infrastructure.

Environmental degradation

Brisbane City Council have a conservation action plan for a large number of plant and animal species, ranked on the basis of concern for their future survival. The ranks include 'least concern', 'near threatened' and 'critically endangered'. For example, the Lewins Rail shown in figure 3.63 is a 'near threatened' species in parts of Brisbane, due to changes in water use and disruption of riverbanks.

FIGURE 3.63 The Lewins Rail

FIGURE 3.64 Areas with Matters of State Environmental Significance (MSES) shown in blue; these are protected under Queensland legislation.

Land-use zoning

Zoning can be a cause of problems in urban areas but also a solution. Zoning is one way governments can manage growing urban environments. Local governments have a system of zoning in place to ensure that only appropriate development occurs in particular places. For instance, placing a waste management facility next to a hospital or in the middle of a residential area would be inappropriate. Appropriate zoning helps ensure this doesn't happen.

Although this system can sometimes be taken advantage of, generally it keeps the built-up areas in cities and towns ordered and logical. The main types of zoning generally correspond to the different land-use types shown in figure 3.65.

Problems can arise when authorities adapt zoning as a means of solving density problems. This can lead to conflict in these areas as residents object to the changes and their negative impacts. For an examination of such a conflict, see lesson 3.8.

FIGURE 3.65 Brisbane zoning

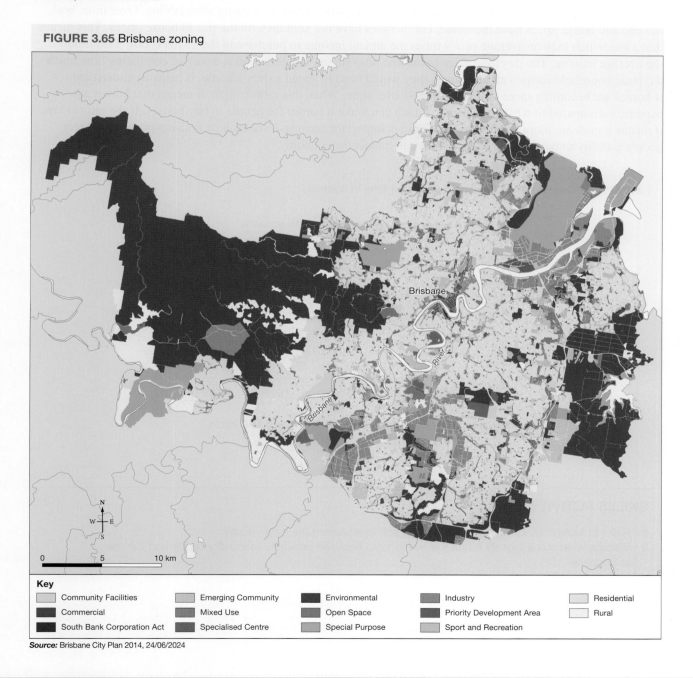

Key

Community Facilities	Emerging Community	Environmental	Industry	Residential
Commercial	Mixed Use	Open Space	Priority Development Area	Rural
South Bank Corporation Act	Specialised Centre	Special Purpose	Sport and Recreation	

Source: Brisbane City Plan 2014, 24/06/2024

Housing availability and affordability

Housing affordability is of particular concern to those in urban areas as populations and demand increase. Everyone needs a place to live, but this goal is becoming harder to achieve for younger Australians as housing costs have increased markedly since the late 1990s and demand has also increased with population increases.

Successive governments have been unable to curb the rapid decrease in housing affordability. Over time, both income and house prices have increased, but incomes have not kept up with the rise in house prices. This means that a house that took an average of 7.9 times the annual income to purchase in the 1970s is now 12.6 times the average income. The dwelling price to income ratio shown in figure 3.66 is a way of comparing how much a typical household earns with how much they would need to spend to buy a home. It helps us understand if homes are becoming more or less affordable over time. When this ratio is high, it means that homes are expensive compared to what people earn, which can make it harder for people to buy houses. If the ratio is low, it means homes are more affordable. This ratio is important because it shows how housing prices can affect people's ability to buy a home and how this can impact the economy.

FIGURE 3.66 Dwelling price to income ratio over time in Australia

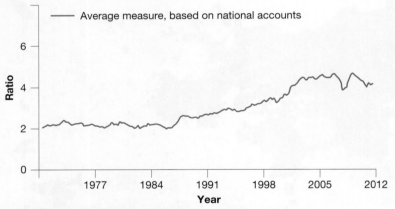

Source: Based on data from Yates 2011, ABS 2017c and ABS 2017d

SKILLS ACTIVITY

1. Refer to table 3.12. Create a multiple bar graph to represent the data visually.
2. Discuss what this suggests about the relationship between proximity to the city centre and car use in Brisbane.

3. Suggest how this information could be used in a debate about urban sprawl.

TABLE 3.12 Car ownership and distance from CBD in Brisbane

	Average motor vehicles per dwelling (2016)	Straight line distance from Brisbane CBD (km)
Spring Hill	1	1
Kedron	1.6	7
North Lakes	1.9	27
Wamuran	2.6	50

3.7 Exercise

3.7 Exercise

Learning pathways

■ LEVEL 1	■ LEVEL 2	■ LEVEL 3
1, 2, 3	4, 5	6, 7, 8

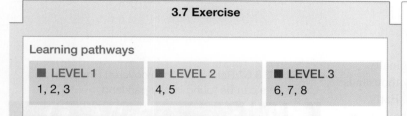

These questions are even better in jacPLUS!
- Receive immediate feedback
- Access sample responses
- Track results and progress

Find all this and MORE in jacPLUS ▶

Explain and comprehend

1. **Define** the term 'urban sprawl'.
2. **Identify** two positive effects of urban sprawl.
3. **Describe** two negative effects of urban sprawl.
4. **Explain** what is meant by gentrification of an area. Provide an example.

Analyse and apply

5. Using table 3.13, **describe** the relationship between vehicle ownership and distance from the CBD in Brisbane.

TABLE 3.13 Car ownership and distance from CBD in Brisbane

	Average motor vehicles per dwelling (2016)	Straight line distance from Brisbane CBD (km)
Spring Hill	1	1
Kedron	1.6	7
North Lakes	1.9	27
Wamuran	2.6	50

6. The suburbs selected in table 3.13 are all north of the CBD. **Analyse** this relationship with suburbs south of Brisbane's CBD. Use an atlas or online map to examine this area.
7. What result would you expect for places that are 75 kilometres and 100 kilometres south of the CBD. **Justify** your hypothesis.
8. Car ownership in cities is discouraged by city councils. **Analyse** why this is the case, applying relevant concepts relating to the challenges facing metropolitan and urban areas.

Sample responses are available in your digital formats.

LESSON
3.8 Understanding geographical challenges — West End, Brisbane

LEARNING INTENTION

By the end of this lesson you should be able to:
- explain the geographical challenges facing places in Australia as a result of the changing characteristics of places
- apply a geographical perspective to understand the challenges to liveability for the local area
- understand the basis for a geographic inquiry (e.g. fieldwork) at a local scale to investigate a specific challenge associated with liveability for a place in Australia (remote, rural or urban) and how this challenge might be managed.

Source: Adapted from Geography General Senior Syllabus 2024 © State of Queensland (QCAA) 2024; licensed under CC BY 4.0.

3.8.1 Development history

The area now known as West End was originally called Kurilpa and was inhabited for many thousands of years by the Turrbal and Jagera nations. Early British colonisers renamed the area after the London borough of the same name.

The Kurilpa peninsula, which covers what is now South Brisbane, West End and Highgate Hill, was once covered in dense rainforest. West End is routinely considered one of Brisbane's most liveable suburbs due to its proximity to the CBD, multiple transportation connections, strong sense of community, safety, recreational facilities, advanced economy, quality local amenities and its proximity to excellent services. However, it hasn't always been seen this way.

FIGURE 3.67 Rainforest once covered Brisbane, and remnants can be found in city parkland.

FIGURE 3.68 Some of the factors that make West End a highly liveable place.

Sense of community	• West End is home to a mix of people, including young families and professionals.
Education	• The suburb has some of Brisbane's top schools.
Location	• West End is close to the city centre, which makes commuting and accessing services and amenities easy.
Local facilities and amenities	• West End has a thriving cultural and social scene. • The area has multiple transport options and is highly walkable.

FIGURE 3.69 West End timeline

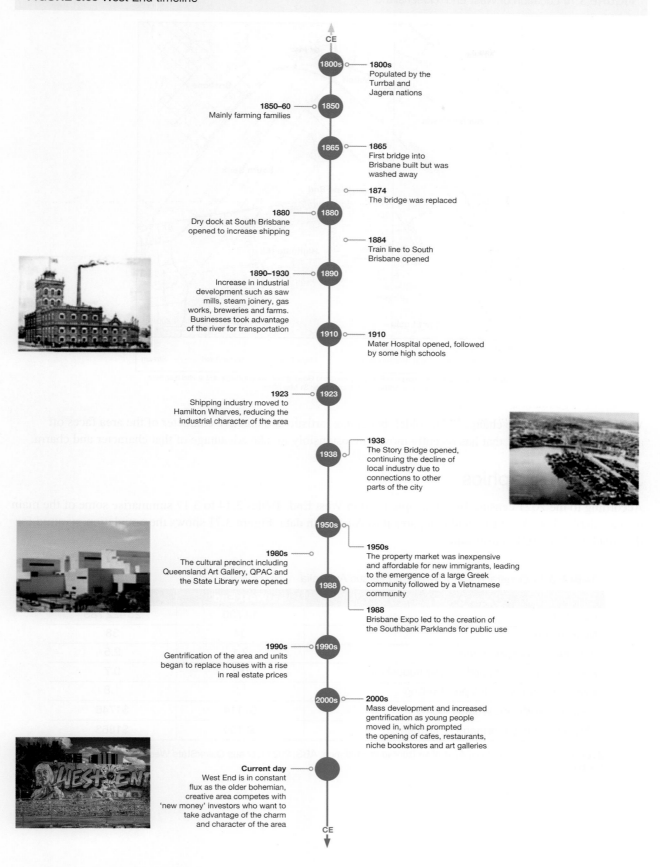

1800s
Populated by the Turrbal and Jagera nations

1850–60
Mainly farming families

1865
First bridge into Brisbane built but was washed away

1874
The bridge was replaced

1880
Dry dock at South Brisbane opened to increase shipping

1884
Train line to South Brisbane opened

1890–1930
Increase in industrial development such as saw mills, steam joinery, gas works, breweries and farms. Businesses took advantage of the river for transportation

1910
Mater Hospital opened, followed by some high schools

1923
Shipping industry moved to Hamilton Wharves, reducing the industrial character of the area

1938
The Story Bridge opened, continuing the decline of local industry due to connections to other parts of the city

1950s
The property market was inexpensive and affordable for new immigrants, leading to the emergence of a large Greek community followed by a Vietnamese community

1980s
The cultural precinct including Queensland Art Gallery, QPAC and the State Library were opened

1988
Brisbane Expo led to the creation of the Southbank Parklands for public use

1990s
Gentrification of the area and units began to replace houses with a rise in real estate prices

2000s
Mass development and increased gentrification as young people moved in, which prompted the opening of cafes, restaurants, niche bookstores and art galleries

Current day
West End is in constant flux as the older bohemian, creative area competes with 'new money' investors who want to take advantage of the charm and character of the area

FIGURE 3.70 Location of West End, Queensland

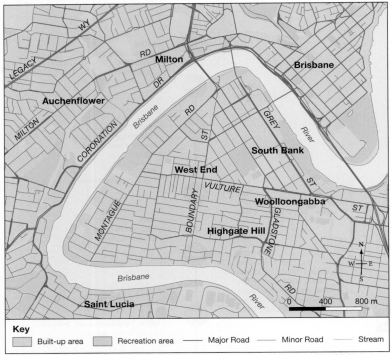

Source: Natural Resources, Mines and Energy, Queensland Government, various maps and spatial data sets, licensed under Creative Commons Attribution 4.0 sourced on 31 May 2018

West End has undergone change as the older, bohemian, artistic and creative character of the area faces off against the 'new money' that has recently moved in, ostensibly to take advantage of that character and charm.

3.8.2 Demographics

According to the 2021 census, 14 730 people lived in West End. Tables 3.14 to 3.17 summarise some of the main data available about West End and compares it to Australian data. Figure 3.71 shows the population pyramid for the area based on 2021 census data.

TABLE 3.14 General data for West End versus Australia

	West End	Australia
Population	14 730	25 422 788
Median age	34	38
Average people per household	2.1	2.5
Average number of children (per household)	0.4	0.7
Average motor vehicles per dwelling	1.3	1.8
Median weekly household income	$2114	$1746
Median monthly mortgage repayments	$2100	$1863

Source: Based on data from Australian Bureau of Statistics, ABS, 2021 Census QuickStats West End Brisbane — Qld

FIGURE 3.71 Population pyramid of West End

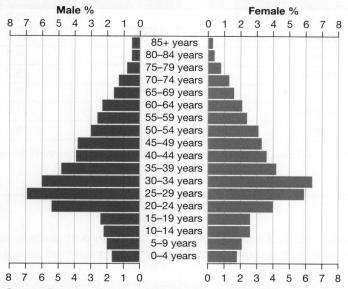

Source: Based on data from Australian Bureau of Statistics. (2022). Regional population by age and sex. ABS. https://www.abs.gov.au/statistics/people/population/regional-population-age-and-sex/latest-release.

TABLE 3.15 Demographic data for West End versus Australia

	West End	Australia
Dependent population (% of total under 15 and over 65)	21.5	35.4
Not married (% of total population)	48.2	41.9
Population with at least a university degree level of education	52.2	26.3
Australian-born population	56.5	66.9
Religious affiliation — No religion or not stated	53.7	38.4
Language other than English — Greek	2.7	0.9
Language other than English — Vietnamese	2.2	0.6

Source: Based on data from Australian Bureau of Statistics, ABS, 2021 Census QuickStats West End Brisbane — Qld, demographic data

TABLE 3.16 Employment data for West End versus Australia

	West End	Australia
Occupation: Professionals	43.4	24
Occupation: Managers	15.2	13.7
Industry: Higher education	4.8	1.3
Industry: Medical	5.3	4.5
Industry: Cafes and restaurants	4	2.2
Travel to work: public transport	12.4	4.6
Travel to work: car	31.7	52.7

Source: Based on data from Australian Bureau of Statistics, ABS, 2021 Census QuickStats West End Brisbane — Qld, occupation stats

TABLE 3.17 Family and home data for West End versus Australia

	West End	Australia
Family composition: Couple family without children	49.9	38.8
Family composition: Couple family with children	34.5	43.7
Dwelling structure: House/townhouse	4.3	12.6
Dwelling structure: Flat/apartment	75.7	14.2
Property tenure: Rented	58.1	30.6
Property tenure: Owned (mortgage)	21.5	35
Property tenure: Owned (outright)	18	31

Source: Based on data from Australian Bureau of Statistics, ABS, 2021 Census QuickStats West End Brisbane — Qld, dwelling and family composition stats

SKILLS ACTIVITY

1. Use the data provided for West End in tables 3.14 to 3.17, and figure 3.71 to answer the following questions.
   a. Select two or three variables that you find most interesting and graph them using a spreadsheet tool such as Microsoft Excel or Google Sheets.
   b. Create a multiple bar graph that shows the rate of your selected data sets for West End and for Australia on the same graph. If you can find further data over a longer period of time, you could use a comparative line graph.

Tensions

Objections to the extent and manner of development in West End are far from new. An article from 1930 describes in detail the native rainforest that was removed to create the modern roads and areas for industry in the 19th century, while lamenting the lack of preservation and foresight of residents and policymakers of the time.

'This jungle was a tangled mass of trees, vines, flowering creepers, staghorns, elkhorns, towering scrub palms, giant ferns and hundreds of other varieties of the fern family, beautiful and rare orchids and the wild passion flower … Here at our very door we had a wealth, a profusion, of botanical beauty which can never be replaced by the hand of man [sic]. Too late have we recognised the desirableness of conserving these glorious works of Nature.' *The Brisbane Courier, March 1930*

By the 1990s, no native vegetation was left in West End and most dwellings were standalone houses. More apartments were developed and this picked up pace in the 2000s, such that by 2001 West End had more apartments than houses. The ratio of apartments to houses has continued to grow ever since, as shown in figure 3.72.

FIGURE 3.72 Dwelling types in West End over time

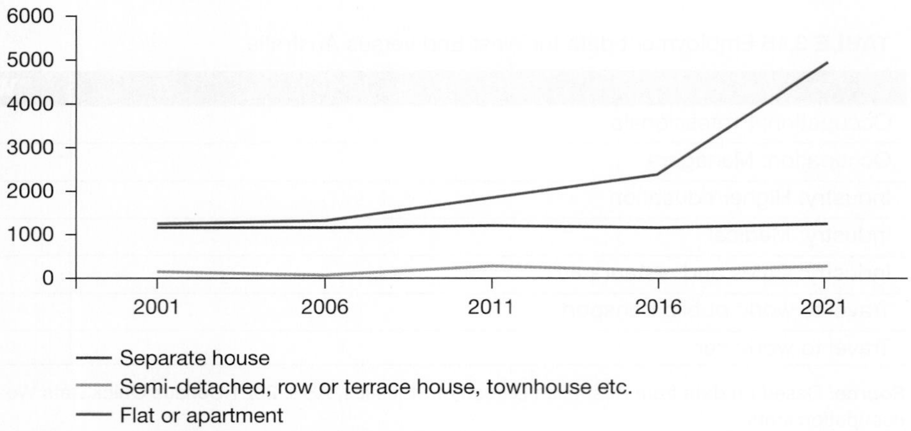

- Separate house
- Semi-detached, row or terrace house, townhouse etc.
- Flat or apartment

Source: Based on Australian Bureau of Statistics data

FIGURE 3.73 Development in West End occurred at pace in the 2000s.

The Brisbane City Council had been encouraging infill development for most of the 2000s and formalised this with the 2006 Brisbane 2026 CityShape initiative. This called for new homes and development to be built around Brisbane's major shopping centres or along major growth corridors, and for jobs to be located close to where residents lived, with better local services and facilities. The Brisbane City Plan 2014 also maintained this outlook.

As the processes of gentrification and urban consolidation have become increasingly visible in West End, with more apartments, building sites, towers and traffic, opposition to developments has increased, and particularly to those that are large in scale and visible in impact. One example of such a development was the West Village development on the former Asboe furniture site. The site was significant because it was very large and also part of the commercial stretch of Boundary Road, the shopping, eating and drinking hub of the area.

The initial development proposal included:
- a 2.6-hectare site
- a 1532 square metre common area
- 1250 apartments
- 7 buildings, each between 8 and 22 storeys high.

In addition, 30 per cent of the site was proposed as a 24-hour accessible open space of laneways and arcades.

The development attracted much controversy in the local community and resulted in many letters to the editor, submissions to Brisbane City Council and the Queensland government over the public consultation period, protests, public meetings, news reports and court challenges. The Queensland Government stepped into the planning process to modify the initial proposal. The project was approved in 2016 and should be completed by 2027.

Right to the City

One notable result of the planning process around the West Village development was the rise of community groups as a vehicle of opposition to significant local developments. At the time of the development proposal, a progressive candidate had recently been elected to the Brisbane City Council representing the area, and another progressive candidate was considered a genuine possibility to win a seat at state government level during the planning process. This meant the government had to take note of community reaction to the proposed development.

These candidates drove much of the opposition, along with the Right to the City movement. Together, they were able to influence the planning process. Right to the City was able to reverse or alter some of the more negative aspects of the development, through changes such as increasing the amount of open space on the site and limiting the height and visual impacts of some of the buildings in the development. However, the developer still managed to force the state government to approve the development, which did not meet the Brisbane City Council planning limits for building height at the time.

FIGURE 3.74 An example of grassroots action in West End

Source: Natalie Osborne Right to the City

3.8 Exercise

Explain and comprehend

1. **Describe** how the demographics of West End have changed over time.
2. **Identify** the negative consequences of development in West End.
3. Using tables 3.14 to 3.17, **outline** five important demographics of people who now live in West End.
4. Figure 3.72 shows that the number of apartments in the West End has increased. **Explain** some of the implications of this.

Analyse and apply

5. **Discuss** some of the impacts of having an oversupply of apartments or units in an area. Consider how might this affect the character of the suburb of West End.
6. **Develop** a set of criteria that could be used for future West End developments. Ensure that your criteria consider all aspects of sustainability, including environmental, economic and social aspects.
7. **Consider** what criteria you would develop to assess the liveability of an area. What factors would be important?

Sample responses are available in your digital formats.

LESSON
3.9 Responding to the challenges facing places in Australia

> **LEARNING INTENTION**
>
> By the end of this lesson you should be able to propose action/s for managing the identified impacts to improve liveability and sustainability for a place in Australia.
>
> **Source:** Adapted from Geography General Senior Syllabus 2024 © State of Queensland (QCAA) 2024; licensed under CC BY 4.0.

3.9.1 Strategies for responding to challenges

The places we live, work and play in have a huge impact on our communities and on us as individuals. They give us identity, affect our health, let us relax, give us opportunity and drive our economy. As outlined through this chapter, the way our places are designed is of vital importance to communities and individuals. So how do we respond to the challenges that arise in the places we live?

Garden cities and green belts

Historically, the first real effort to curtail unhealthy development came in the form of the garden city proposal.

In the early 20th century, governments began to show greater interest in improving the physical and social conditions of cities beyond the bare necessities. This period marks the beginning of modern town planning in Australia and the influence of social reformers such as Ebenezer Howard.

Howard was an English journalist who publicised his ideas in 1898 in a book entitled *Tomorrow: A Peaceful Path to Real Reform*. Influenced by factory towns and suburbs built by industrial philanthropists in the 19th century, he devised the concept of the **garden city**.

FIGURE 3.75 Howard's garden city proposal

Source: Garden Cities of Tomorrow by Ebenezer Howard, 1902

> **garden city** area that contains proportionate areas of residences, industry and agriculture

FIGURE 3.76 Central Adelaide is surrounded by a ring of parkland.

Howard had seen how high land values in cities encouraged high residential densities and discouraged the provision of social amenities such as open space. Therefore, the principles he formulated for the construction of garden cities outlined how the benefits of healthy rural living could be combined with the advantages of city life.

These principles included:
- a population ceiling of 30 000 to ensure that a sense of community was not lost
- all land owned by a public trust run by 'honourable people not driven by the 'pursuit of profit'
- a green core surrounded by two concentric rings of public buildings and housing, both of which would be separated from industry by a **green belt**
- a self-sufficient population that worked in local factories and bought food from surrounding farms.

FIGURE 3.77 Garden city principles are used globally to limit population density and provide green cores.

green belt area of largely undeveloped, wild or agricultural land surrounding or neighbouring urban areas

Inclusive planning and sustainable cities

The environment people live in shapes their behaviour and physical and mental health. We have created places, mainly metropolitan, that encourage a sedentary lifestyle where motor vehicles are almost mandatory for getting to work, shops and schools, and to engage with other members of the community. This reliance on motor vehicles affects rates of obesity and diabetes, incidence of loneliness and overall mental health. Well-designed places, though, can reduce inequality, encourage healthier transport options and impact positively on the physical and mental health of their inhabitants.

FIGURE 3.78 The Gold Coast light rail network is a great example of infrastructure that will enable transport-oriented development.

Essentially, we can respond to the challenges that face our rural, remote and metropolitan places by planning and designing them well. But 'planning' encompasses so much. What is planning? The planning we employ to shape and grow the places we live in needs to be inclusive.

Inclusive planning recognises the importance of the design and sustainability of our places to all who currently inhabit them, and will inhabit them in the future. In many instances, planning has meant that profit has been put before people, which has led to the creation of places that are unhealthy and detrimental to the communities and individuals who inhabit them. Too often, decisions are made that fundamentally change and shape places without consulting the people who live there.

TABLE 3.18 Elements that make up a well-planned, sustainable city

Efficient and affordable public transport	Effective waste management	Green roofs
Walkable neighbourhoods	Water conservation	Access to resources
Vehicle charging stations	Bikeable neighbourhoods	Open green spaces
Renewable energy sources such as solar	Green architecture	Shade in hot cities
The use of urban farming	Energy-efficient buildings	Smart roads (e.g. light grey to reduce urban heat)

Brisbane City Council, similar to many councils across Australia and the world, has an Inclusive Brisbane Plan (2019–2029) for development. This plan aims to put people at the centre of plans for urban development. The planning documents are made accessible, and people from a diverse range of cultures, backgrounds and demographics are invited to have their say and provide input into amenity development, infrastructure and community.

Urban villages

Urban villages are attempts to enact some of the broadly agreed planning concepts that can positively impact communities.

These concepts encompass aspects such as mixed-use development (where residential, retail, commercial and sometimes industrial land use are intermixed), good public transport options, the encouragement of vegetation in the area, medium-density housing, connectivity for pedestrians, and ample public spaces for recreation and community building.

FIGURE 3.79 Modern housing developments often include design features to positively affect health and wellbeing, including communal gardens.

DID YOU KNOW?

Kelvin Grove and Dutton Park in inner Brisbane are two Queensland examples of areas that have employed some or all of the urban village strategies to enhance their social and physical space.

Urban renewal and regeneration

Urban renewal is a philosophy that sees run-down areas of cities repurposed in an effort to revitalise the area. The term has slightly different meanings in different parts of the world, but in Australia urban renewal means increased density in urban areas and more appropriate use of vacant or underused land. This process is driven by local, state and federal governments, which can use their legislative powers to encourage certain types of development in certain places.

EXAMPLE: Times Square, New York City

Times Square in New York is a good example of urban renewal and regeneration of a famous area with pedestrians in mind. The area was closed to cars in 2009 to allow people to move freely, and this has increased the attractiveness of the area for both locals and tourists. Development was completed in 2016.

This transformation involved creating pedestrian plazas, adding seating areas, and enhancing the overall streetscape with lighting and signage. The car-free environment has not only improved safety but also fostered a vibrant street life, with performances, public art installations and outdoor dining contributing to the lively atmosphere.

The redesign prioritised pedestrians, making Times Square more accessible and enjoyable. In turn, this boosted local businesses.

FIGURE 3.80 Times Square and its pedestrianised section

Infill development is also a part of urban renewal and can be employed to help revitalise areas of cities that have lost their charm over time. Infill development aims to increase the population density in an area while minimising impacts on the existing community. This often involves the organic growth of apartment blocks or increased density by way of building works.

infill development construction that occurs on vacant or underused land within an existing urban area

FIGURE 3.81 Examples of infill development in Queensland

SKILLS ACTIVITY: Spatial technologies

Open the **Digital Earth Australia — Land cover over time** weblink and complete the following.
1. Zoom into Brisbane.
2. Describe the change in the size of artificial surface land use from 1988 to 2020.
3. Explain the potential impacts of this change on the natural environment.
4. Propose a strategy to reduce the impact of future population growth in Brisbane. Justify your suggestion.

 Resources

 Weblink Digital Earth Australia — Land cover over time

Smart cities

Smart cities are cities that encourage the collection, analysis and use of data to help make better decisions about development. Smart cities allow data to be collected from a range of sources, including sensors, devices, citizens themselves and existing government sources.

In theory, smart cities make better use of the information that exists in our cities. In practice, the idea of sharing so much personal and intimate data can be off-putting for some people.

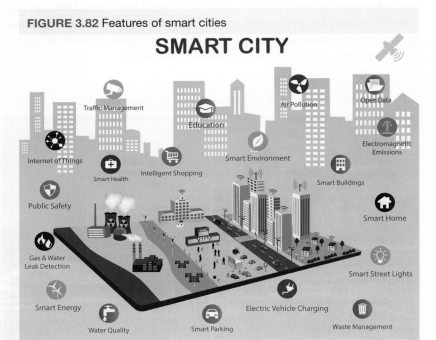

FIGURE 3.82 Features of smart cities

 Resources

 Weblinks Smart Cities: Songdo, South Korea
Critiquing smart cities

3.9 Exercise

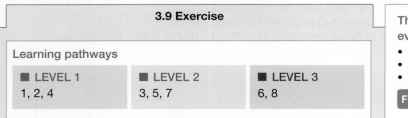

3.9 Exercise
Learning pathways

■ LEVEL 1	■ LEVEL 2	■ LEVEL 3
1, 2, 4	3, 5, 7	6, 8

These questions are even better in jacPLUS!
- Receive immediate feedback
- Access sample responses
- Track results and progress

Find all this and MORE in jacPLUS ⊙

Explain and comprehend

1. **Define** what is meant by a sustainable city and a smart city.
2. **Explain** what is meant by inclusive planning.
3. **Explain** why inclusive planning is important for densely populated areas.
4. **Outline** what is meant by infill developments and give an example from an Australian city.

Analyse and apply

5. **a. List** the features that make a community a good place to live. Categorise each feature as a social/cultural feature, environmental feature or economic feature. (If you can, compare your list to other groups or people's lists.) Adjust yours if you think you need to.
   b. Discuss the extent to which your local community includes these features.

Propose and communicate

6. What changes have you seen in your local area that have been designed to make the place a better place for people to live? **Propose** strategies to improve your community.
7. If you were given the power, what changes would you make to your local area to increase its sense of community and to make it a better place to live? **Prioritise** your ideas according to what would have the greatest positive impact.
8. Write an extended response to explain how you would achieve the change that you have given the highest priority and **justify** your response by outlining how this action will make your community more liveable.

Sample responses are available in your digital formats.

LESSON
3.10 APPLY YOUR SKILLS — Conducting a geographical inquiry

LEARNING INTENTION

By the end of this lesson you should be able to
- collect and present quantitative and qualitative primary data using field techniques such as observing and recording, interviews and questionnaires, photographing, sketching and annotating, measuring and surveying and using GNSS location data (e.g. GPS)
- analyse and interpret primary data collected in the field to explain geographical processes and recognise spatial distribution, geographic patterns and trends
- use fieldwork data to propose responses and predict outcomes.

Source: Adapted from Geography General Senior Syllabus 2024 © State of Queensland (QCAA) 2024; licensed under CC BY 4.0.

3.10.1 Using the geographic inquiry model

The geographic inquiry model is a framework used by geographers to help structure an inquiry. It is a great way to help you organise your thinking as you undertake planning an investigation, and can be used to address issues from local to global scales. The geographic inquiry process is shown in figure 3.83.

FIGURE 3.83 The geographic inquiry process

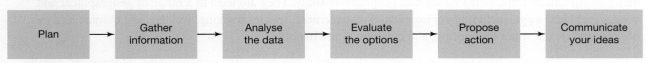

Plan → Gather information → Analyse the data → Evaluate the options → Propose action → Communicate your ideas

Plan

Every inquiry should start with a question or problem. Try to frame your question in such a way as to allow you scope to collect data, analyse it, make conclusions and respond. This might mean expanding your geographic area or the scope of your problem.

Develop key questions and focus questions that help you to organise your research and data collection. These key questions usually take the form of the following.

- What and where is the issue or problem at hand?
- How and why does it occur?
- What are the environmental, social and economic impacts?
- What are the solutions?

The key questions can then be narrowed down through the use of focus questions that focus your research. These focus questions can be constantly updated and can change as you gather more information through your research and field work. Once you have established your key questions, write a plan outlining how you will undertake your research to find answers to these questions — your methodology.

Figure 3.84 outlines how you might plan an inquiry into development in a coastal, urban suburb. As you begin to research your focus questions, you will find more information and uncover more questions to ask, which means your investigation should be reviewed and will be constantly evolving.

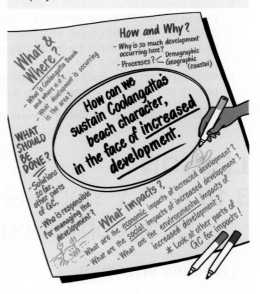

FIGURE 3.84 Planning a geographic inquiry

Gather information

Here you carry out the inquiry by collecting your **qualitative** and/or **quantitative** data and analysing it. This also includes conducting any research to prepare your understanding of the issue at hand.

Primary and secondary data

Primary data is data that is collected first-hand. In research, primary data holds the most weight because it is data collected by the author to help them answer specific research questions — it is data collected for that purpose. As you collect your data, reflect on whether you are collecting the right data to answer your questions, and whether you could improve your collection methods in any way.

Secondary data is data collected by someone else and can include census data, government data and information, as well as research findings. Your analysis, and subsequently your solutions or proposals, will be evidence-based provided you have collected appropriate primary data, and sourced reliable and relevant secondary data.

Spatial technologies

Spatial technologies are one of the tools of geographers. They help you answer 'where' questions and can help you illuminate some of the issues that geographers study. They are used in a range of geographical and nongeographical ways, including environmental management, species tracking, weather and climate analysis, oceanic and inland water monitoring, and economics and history.

Geospatial technologies help you to measure, analyse and represent spatial or geographic information, usually digitally. For your field study, you could use spatial tools to:

- examine land use and/or land cover from a satellite image in the Queensland Globe
- investigate public transport and cycling infrastructure in the area around your place of study using the Queensland Globe
- visualise census and other demographic data using Google Maps or ABS online mapping tools such as TableBuilder
- visualise environmental data, such as geology, vegetation or species extent using online digital tools such as the Queensland Globe, ARCGIS or ESRI
- map data you have collected in the field, such as water quality data or survey results using Google MyMaps or QGIS.

qualitative data based on personal experience
quantitative data that can be measured as an exact quantity

Table 3.19 provides examples of satellite images showing different land use.

TABLE 3.19 Satellite imagery showing examples of different land use

Land use and example	Land use and example	Land use and example
Residential: land where people live; can be high, medium or low density (Gladstone)	Commercial: land used for commercial purposes, such as a shopping complex or strip (Toowoomba)	Water: land for water delivery and storage in communities (Advancetown Lake and Hinze Dam)
Light industrial: land for industrial activity that uses light-weight materials and produces light goods, such as foodstuffs, arts and crafts, clothing or small electronic items (Townsville)	Recreational: land used for recreational purposes, such as a green space, a park, a playground, sporting field or skate park (Glass House Mountains)	Primary industry/agricultural: land used for primary industries (farming and cropping) (between Cecil Plains and Millmerran)
Heavy industrial: land for industrial activity that uses heavy materials or heavy equipment, such as automotive, aeronautical or ship building, chemical and electronic manufacture, and machine tooling (Rocklea)	Administrative: land used for governmental purposes, such as council offices or courts (Brisbane City Hall)	Military: land used for defence purposes (Gallipoli Barracks, Enoggera)
Educational: land devoted to educational use, such as a kindergarten, school or university (Craigslea State School)	Transport: land that is used to transport people or things, such as airports, ferry terminals, ports, bus corridors, or train lines and stations (Roma airport)	Natural/protected: undeveloped or rehabilitated land in a natural or close to natural state (Daintree National Park)
Medical: land used for the medical industry, such as a hospital (Royal Brisbane and Women's Hospital)	Utility: land for utility services, such as electricity, telephone and internet, power and sewage removal (Stanwell Power Station)	Mining: land used for mining (Mount Isa)

Source (images): © State of Queensland 2018, Qld Globe

Analyse the data

Once you have collected your data you need to use it by analysing it. Data can be analysed using a number of methods. In the main, you are looking for patterns in the data that are related to the topic at hand. Sometimes these patterns are self-evident (e.g. poor water quality near an industrial area) but sometimes additional data needs to be collected or analysed in conjunction with your data to draw out the message in the data.

The primary data that you collect and secondary data that you select and interpret should be able to help you to explain the geographic processes at play in your issue and why they have led to the source of any geographical challenges. Your analysis should also focus on the different impacts that could occur as a result of your challenge continuing. You should be able to extrapolate from your analysis to generalise about the impacts of your issue.

At this stage, you should also consider whether the data you collected and the questions you asked have any flaws or limitations. Do you have the data you need to answer your research questions and suggest ways to combat the problem? Is all of the data you collected accurate?

Evaluating the options

Evaluating is the art of selecting or choosing between options. You should always back up any decision you make with evidence. Think about what your data is telling you about the issue and refer to the specific data that supports your position.

You can use a decision-making matrix to choose between options. First develop criteria and then score each option against the criteria to make your final decision.

Propose actions

Once you have analysed your data, made your decision and/or evaluated your options, you need to propose action(s) to address the problem. This could be selecting one option over another or many, proposing changes to a project, modifying existing suggestions or developing a proposal from scratch. Your proposals should refer back to the evidence you have gathered, demonstrating the issue and using the actual data you have analysed. This then justifies your proposal and shows exactly what, within the data, you are trying to change for the better. Use maps, graphs, charts and annotated images to demonstrate your proposals. You could even use 'before and after' images or maps.

Communicate your ideas

The communication element of the inquiry process is the most important. Geography is a subject with change at its heart, and encouraging change requires effective communication of ideas and prosecution of arguments. Effective communication could be the difference between research going ahead or not, a particular environmentally significant place getting more or less funding, a building being built, a park being revitalised or a patch of forest being cut down.

Information can be communicated to an audience in many ways apart from speaking — for example, using a presentation aid such as Microsoft PowerPoint, promotional videos, and visual methods such as billboards, posters or flyers. However, the field report is still the most common method of reporting findings in the professional and academic worlds because field reports are easy to catalogue and retrieve for later use.

3.10.2 Fieldwork example

Sustainable use of public space in Bowman Park, Brisbane

In this fieldwork example, you will consider how a prominent public space is used, and what changes could be made to increase sustainability. You will evaluate a public space and recommend changes that would increase the likelihood of positive sustainability in the area. You could examine any area in your local community that is publicly accessible and used by the community, such as a town square, local park or showground.

In this example, we will examine Bowman Park in the Brisbane suburb of Bardon and ask the question:

How can Bowman Park in Bardon be more sustainable and improve the liveability of the local area?

You will:
- assess the current status of the park: How is it used? Is the site overused or underused by the local and wider community? Does it meet the needs of residents?
- research and make some recommendations as to what improvements could be made to make the space more sustainable and accessible
- communicate your findings using a report format.

The following sections take you through the whole process. You could carry out this investigation in any of your local parks.

FIGURE 3.85 Preparing your field study

Key questions

To develop your key and focus questions, see section 3.10.1.

Use a Plus, Minus, Interesting (PMI) extension activity to flesh out your ideas. See table 3.20 for some example key and focus questions. Remember that these focus questions can change as you begin to investigate your local place — you might find some interesting information that changes your perspective.

TABLE 3.20 Ideas for key and focus questions

	Focus questions	Data collection
What and where?		
What is the nature of the public space?	What is the history of the space? Who is it named after?	History
Where is the public space located?	Map of the area? Where is it in the region/Brisbane? What surrounds it?	Maps, street and satellite
How and why?		
How is the public space used?	What is the land use around the park? What are the public transport links? How do people use the park?	Land use survey Count people using different features of the park (running, sport, playground, BMX track, basketball hoops, outdoor gym, creek tracks, BBQ areas)
Why is it used the way it is?	Who uses the public space? What are the facilities in the park?	Resident/user survey
How else is it used?	What are the types of use at different times of day, week and year?	Resident/user survey
What is the extent of the challenge?		
What are the social impacts?	How does the space impact the social, cultural and political aspects of the area?	Resident/user survey
What are the environmental impacts?	How does the space impact the local environment?	Resident/user survey Water quality assessment
What are the economic impacts?	Does the space encourage or generate economic benefit to the entire surrounding community?	Resident/user survey Research – BCC
What can/should be done?		
What should be done?	What recommendations could be made?	Research
What are other places doing?	What works elsewhere?	Research

Develop criteria

Think about what might be an appropriate use of the space. You will need to develop criteria to determine the nature, location and extent of the challenge. Consider the role of each of these factors in determining your criteria (see table 3.21 for more detail):

- social
- environmental
- economic
- political
- cultural
- historical.

TABLE 3.21 Potential criteria to assess the nature of the challenge of how to improve the sustainability of Bowman Park, Bardon

Factor	Criteria
Social	How well does Bowman Park encourage social interaction and connectivity? *OR* How does the space impact the social, cultural and political aspects of the area?
Economic	Does Bowman Park encourage or generate economic benefit to the entire surrounding community?
Environmental	How does Bowman Park impact the local environment?

Collect data

Go into the field and collect data on:

- *Local history:* notable events and changes in economy or demography that may be relevant. Don't forget to record reference details of any books, articles or websites that you use here as your report will need to reference them.
- *Survey land use:* use spatial technologies to evaluate land use in the area and surrounds. Try the Queensland Government's Queensland Globe or Google Earth. Use the Queensland Government's QImagery tool to examine past aerial photographs of the area.
- *Survey public transport and cycling connections:* include these on the same map as land use. Google Maps and the Queensland Globe has a public transport layer and a cycling layer.

FIGURE 3.86 Collecting local history data

Create a detailed annotated field sketch of surrounding shopfronts and buildings. If you can't map or collect data for the entire area, take a representative sample. Include information on what the buildings are used for (for example, shops, type of shop, office, business or public service) and the age of the buildings.

You could also take a photo with your mobile phone camera to annotate.

FIGURE 3.87 Annotate a photo

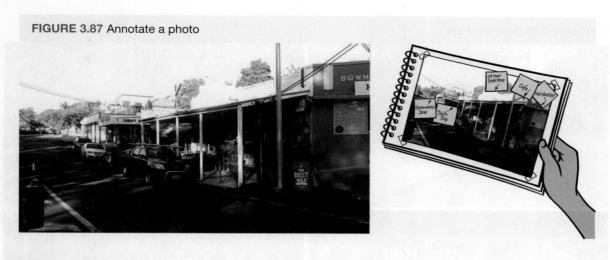

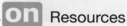 Resources

▶ **Video eLesson**   SkillBuilder: Constructing a field sketch (eles-1650)

✦ **Interactivity**   SkillBuilder: Constructing a field sketch (int-3146)

Survey residents on their views on how the space is used. It would also be useful to ask them about their suggested alternative uses of the space or to offer your proposals. Asking some yes and no questions will allow you to graph some of your responses. Try to survey at least 20 people to make it a fair sample. This can be done as a class and your data could be collated.

Estimate the economic value of the park, and then try to find economic data from the local council or newspaper that outlines the actual value of the space in some way. This might be hard, and you may need to look at other means to determine the economic health of the area, such as the number of shops or spaces for lease, or pedestrian data for the area and surroundings.

FIGURE 3.88 Conduct a resident survey

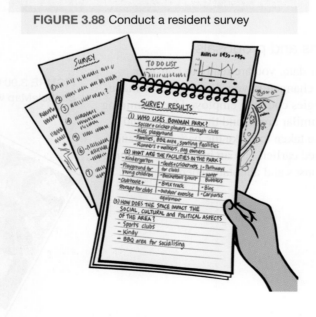

FIGURE 3.89 Bowman Park

Bardon Hall
Community Hall for Hire
247(B) Simpsons Road Bardon

communify

Make recommendations and proposals

Once you have analysed your data, you are ready to consider options for what needs to be changed in your local space to make it better. Find out what strategies could be employed by examining what is successful in other, similar areas, and then propose and justify what actions should be taken to make the park compliant with the criteria you developed earlier. Use an interactive mapping tool to summarise your responses and annotate where required.

FIGURE 3.90 Google My Map or ScribbleMap with annotations

Communicate

You should take some time to write your report. Use the following report structure to create your response to your teacher. Consider presenting your report in digital format. Videos, story maps, Google MyMaps and Google Tour Builder are all appropriate tools that could be used to present your findings and proposed actions in a compelling way.

FIGURE 3.91 The finished product

This is a suggested layout for your report. Sections in bold are not counted towards your overall word count.
Report layout
Title page
Contents page
List of figures
Introduction
Data analysis/main body
Proposals
Conclusion
Bibliography

 Resources

▶ **Video eLesson**   SkillBuilder: Creating a survey (eles-1764)

✦ **Interactivity**   SkillBuilder: Creating a survey (int-3382)

LESSON
3.11 Review

3.11.1 Summary

3.2 Places in Australia

- Places are a cultural construct.
- Places may be, but are not limited to, sites of biodiversity, locations for economic activity, centres for decision-making and administration, meeting places for social interaction, areas of natural beauty and wonder, and areas of spiritual significance.
- People settle in different places for many different reasons; these include water supply, resources, geographical situation and interconnections.
- The Australian Bureau of Statistics (ABS) classifies places in Australia into a hierarchy of statistical areas. This is reported in the Australian Statistical Geography Standard (ASGS).

3.3 Processes that shape places

- Remote, rural and urban places have their own individual characteristics.
- A number of processes shape the identity of remote, rural and urban places, including urbanisation, suburbanisation, counterurbanisation, population increase and population decline.
- The identity of remote, rural and urban places is shaped by population size, economic activity and social influences.

3.4 Where are Australian places?

- Australian places have been developed over time and the Australian continent has been inhabited for at least 65 000 years by First Nations Peoples of Australia.
- Australia's population is growing, and is distributed mostly in urban areas and in coastal clusters.
- Australia has many natural places that have not been developed or influenced by humans.

3.5 Factors affecting Australia's population distribution

- Population distribution illustrates the way in which a population is spread across a geographical area.
- Physical factors such as geography, water, resources and climate affect population distribution.
- Social factors such as the provision of health and education affect population distribution.
- Economic factors such as employment and affordability affect population distribution.

3.6 Challenges facing remote and rural Australia

- Employment can be harder to find in remote or rural places, and access to health care and education can be difficult.
- Living in remote and rural places can contribute to feelings of social isolation, especially if transportation options are limited, preventing people from accessing amenities and social activities.
- Lower levels of connectivity through poor mobile phone and internet coverage can be problematic both socially and economically.

3.7 Challenges facing urban Australia

- Urban areas are challenged by urban sprawl. Urban sprawl occurs as a direct result of the increasing population in urban areas and can also bring opportunities.
- The impacts of urban sprawl are far-reaching and include increased industry and increased populations, but also increased stress on infrastructure and transport, and land-use change.
- Metropolitan and urban areas are also challenged by zoning, housing affordability and gentrification.
- Environmental degradation occurs in urban areas that are growing, and this places stress on plant and animal habitats, leading to loss of species.

3.8 Understanding geographical challenges — West End, Brisbane

- West End, originally called Kurilpa, was inhabited by the Turrbal and Jagera nations for thousands of years. The area, which includes South Brisbane and Highgate Hill, was once a dense rainforest but has undergone significant development since British colonisation.
- West End is known for its liveability due to its proximity to the CBD and community amenities. It has seen a shift from standalone houses to apartments, driven by infill development and gentrification, blending bohemian and affluent residents.
- Large developments such as the West Village project faced strong community resistance. Protests led to some changes, such as increased open space and reduced building heights; however, the development proceeded with modifications.

3.9 Responding to the challenges facing places in Australia

- Modern town planning in Australia began in the early 20th century, influenced by Ebenezer Howard's garden city concept, which combined rural and urban benefits with principles such as population limits and green belts.
- The design of living spaces affects identity, health, relaxation, opportunities and the economy, making good planning crucial.
- Effective planning prioritises community needs, with initiatives such as Brisbane's Inclusive Brisbane Plan and urban villages integrating mixed-use development and public spaces. Urban renewal and smart cities use data and increased density to revitalise areas and improve living conditions.

3.11.2 Key terms

counterurbanisation the migration of people from urban to rural areas

garden city area that contains proportionate areas of residences, industry and agriculture

gentrification the process of older areas, usually of lower socioeconomic status, being slowly bought out and renovated by wealthier people

geographic scale the spatial extent or level of detail at which a geographic area is analysed or represented, ranging from local to global

green belt area of largely undeveloped, wild or agricultural land surrounding or neighbouring urban areas

infill development construction that occurs on vacant or underused land within an existing urban area

metropolitan area a densely populated urban area with significant economic activity, cultural diversity and social interaction

monofunctional development developments that have a single function or purpose, such as a development that is solely for residential buildings

physical environment natural and built surroundings

pull factors factors that attract people to an area

push factors factors that encourage people to leave an area

qualitative data based on personal experience

quantitative data that can be measured as an exact quantity

rural areas with low population densities, agricultural activities and other characteristics of the countryside

sea change relocation from an urban area to a coastal or seaside location for a lifestyle change

social demography the statistical study of populations using measures such as age, sex and income

suburbanisation the development and outward spread of new suburban areas

tree change relocation from a city or suburb to a rural or forested area for a lifestyle change

urban areas with high population densities, human development and other characteristics typical of cities and towns

urban sprawl process of outward expansion of urban areas from the CBD into the surrounding countryside and invading adjacent towns, regions and undeveloped land

urbanisation growth in the proportion of a population living in urban environments

3.11.3 Exam questions

1.12 Section I – Short answer question

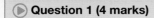 **Question 1 (4 marks)**

Analyse the graphs shown in figures 3.92 and 3.93 to **explain** the different patterns of population change over time.

Identify implications for one of these areas.

FIGURE 3.92 Population trends for Moree, New South Wales, and Mount Isa, Queensland, 2006–2021

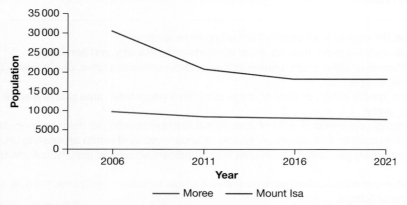

Source: Based on data from Australian Bureau of Statistics (2021) 'Moree 2021 Census Community Profiles and Mount Isa 2021 Census Community Profiles', accessed 12 June 2024.

FIGURE 3.93 Population trends for Sydney, New South Wales, and Brisbane, Queensland, 2006–2021

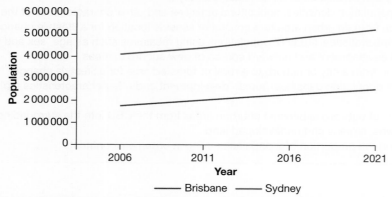

Source: Based on data from Australian Bureau of Statistics (2021) 'Greater Sydney 2021 Census Community Profiles and Greater Brisbane 2021 Census Community Profiles', accessed 12 June 2024.

▶ Question 2 (4 marks)

Analyse the satellite imagery shown in figure 3.94 to **explain** the changes over time to Maroochydore, Queensland, and **identify** a geographical challenge that might arise from this change.

FIGURE 3.94 Satellite imagery showing land cover use, Maroochydore, in (a) 2000, (b) 2010 and (c) 2020

- Water
- Natural Bare Surface
- Artificial Surface
- Natural Aquatic Vegetation
- Natural Terrestrial Vegetation
- Cultivated Terrestrial Vegetation

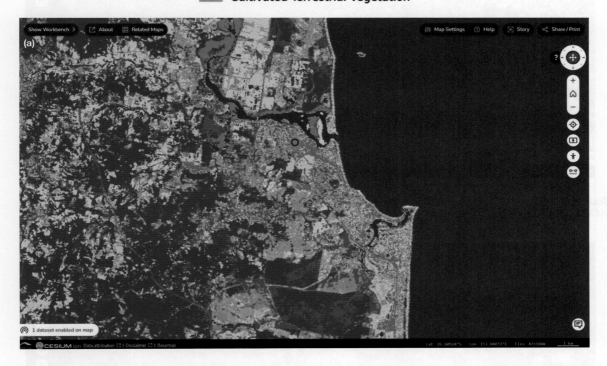

Source: © Geoscience Australia and CSIRO Data 61 2019. Retrieved from https://maps.dea.ga.gov.au/#share=s-2K4ap8F1sfi h2kgfBwOnkV0hdji

FIGURE 3.94 Satellite imagery showing land cover use, Maroochydore, in (a) 2000, (b) 2010 and (c) 2020 *(continued)*

▶ **Question 3 (16 marks)**

In a written response of approximately 450–600 words, respond to the following.

Analyse and interpret the data presented in the stimulus material shown in figures 3.95 to 3.98 to explain how the patterns, trends and relationships represent change for people and environments in Fortitude Valley, Brisbane.

Apply your understanding to generalise about the potential impacts for people living in Fortitude Valley, Brisbane.

FIGURE 3.95 Population Fortitude Valley, 2001–2021

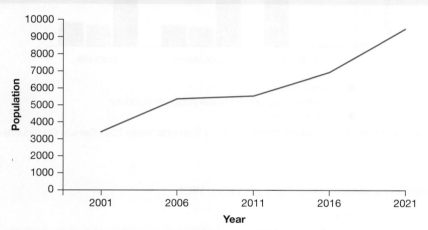

Source: Based on data from Australian Bureau of Statistics (2021) 'Fortitude Valley 2021 Census Community Profiles', accessed 12 June 2024.

FIGURE 3.96 Population pyramid, Fortitude Valley, 2021

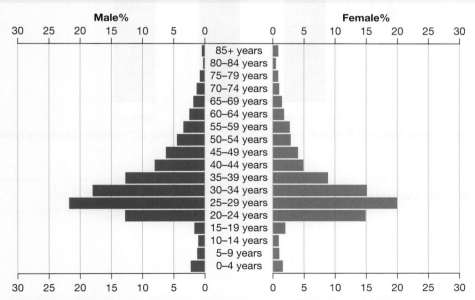

Source: Based on data from Australian Bureau of Statistics (2021) 'Fortitude Valley 2021 Census Community Profiles', accessed 12 June 2024.

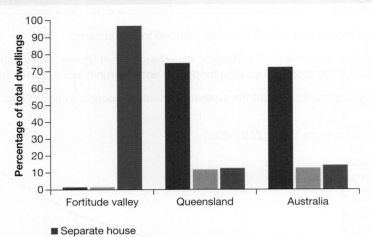

FIGURE 3.97 Dwelling types, Fortitude Valley, 2021

- ■ Separate house
- ■ Semi-detached, row or terrace house, townhouse
- ■ Flat or apartment

Source: Based on data from Australian Bureau of Statistics (2021) 'Fortitude Valley 2021 Census Community Profiles', accessed 12 June 2024.

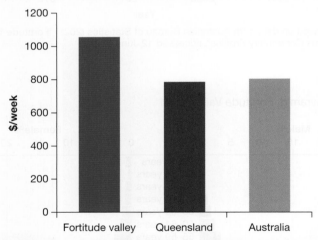

FIGURE 3.98 Median personal income, Fortitude Valley, 2021

Source: Based on data from Australian Bureau of Statistics (2021) 'Fortitude Valley 2021 Census Community Profiles', accessed 12 June 2024.

Sample responses are available in your digital formats.

4 Managing challenges facing megacities

UNIT 2 TOPIC 4

SUBJECT MATTER

In chapter 4, students;

- **explain** the processes of urbanisation that have resulted in the growth of megacities around the world, and how these processes shape the identity of megacities
- **recognise** the spatial patterns of megacities on Earth's surface and the implications for people and environments
- **investigate** a specific geographical challenge by conducting a case study that focuses on one megacity from Africa, Asia or South America
- **understand** the factors that contribute to the growth of an identified megacity
- **understand** how urbanisation and megacities are changing the organisation of the world's populations and the challenges for liveability in a specific place
- **propose** action for managing a geographical challenge in the megacity investigated.

LESSON SEQUENCE

Fully worked solutions for this topic are available in the Resources section at www.jacplus.com.au.

LESSON
4.1 Overview

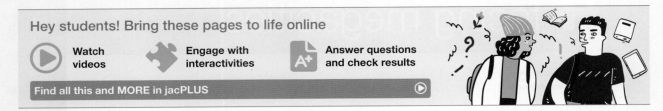
4.1.1 Introduction

In this chapter, you will explore urbanisation, particularly around the growth and challenges of **megacities**: cities with populations of more than 10 million people.

You will learn about where and why megacities have grown in specific places and the factors that have contributed to their development. You will also look at the challenges that growing urban environments present for people and the natural environment.

megacities cities with 10 million or more inhabitants

FIGURE 4.1 Japan's capital, Tokyo, is the world's largest megacity. The greater metropolitan region is home to around 40 million people.

4.1.2 Syllabus links

Syllabus links	Lesson
○ Explain features of global population growth and the processes of urbanisation that have resulted in the rise of megacities.	4.2
○ Recognise and represent the spatial patterns of megacity distribution across the world and the spatial change in this pattern over time.	4.2
○ Explain the impacts of urbanisation and the growth of megacities on human wellbeing and environments using various forms of data and information and spatial technologies. Consider risks and opportunities posed by • rate of urban growth • population density • changing land use and consumption of land • formal and informal economies • settlement infrastructure and land tenure • risk and vulnerability to natural hazards and disasters.	4.3 4.6–4.8
○ Explain the challenges for sustainable development facing megacities in less developed countries compared with challenges facing megacities in more developed countries; for example: • climate change • employment • housing • transport • sanitation • health and education services • provision of fresh water and energy • land availability • waste management.	4.3–4.8
○ Explain how urban planning can be used in the development of resilient cities to mitigate and adapt to the future impacts of urbanisation.	4.9
○ Conduct a case study to investigate a geographical challenge in a megacity in a country in Africa, Asia or South America. As part of this case study, students must • manipulate, adapt, and transform data, using spatial and information and communication technologies, to represent and describe the nature and extent of urban growth for a selected megacity • identify a specific geographical challenge for the selected megacity • analyse data and information to describe the nature and extent of the identified challenge • apply geographical understanding from their analysis to explain the impacts for people and/or environments in the selected megacity • investigate a range of planning strategies that could be used to address the identified challenge for future sustainability and liveability in the selected megacity • propose action that will address the identified challenge and improve sustainability and liveability in the megacity investigated.	4.3–4.8
○ Communicate understanding using appropriate forms of geographical communication.	4.10

KEY QUESTIONS

1. What factors led to the growth of megacities?
2. What patterns or trends can be seen in where and how urban areas and megacities have developed?
3. What predictions can be made about future patterns of megacity development?
4. How do countries manage the challenges presented by urban development? Do responses differ between developed and developing countries?
5. Can megacities be planned in a way that promotes sustainability and the wellbeing of the people who live there?

LESSON
4.2 Global patterns of urbanisation

LEARNING INTENTION

By the end of this lesson you should be able to explain features of global population growth and the processes of urbanisation that have resulted in the rise of megacities, and recognise and represent the spatial patterns of megacity distribution across the world and the spatial change in this pattern over time.

Source: Geography General Senior Syllabus 2024 © State of Queensland (QCAA) 2024; licensed under CC BY 4.0.

4.2.1 The rise of cities

Megacities are a product of both world population growth and **urbanisation**, which is now occurring at a rate faster than ever before in human history. Figure 4.2 shows the percentage of the world's population living in urban areas and the total number of people who reside in urban areas.

> **urbanisation** growth in the proportion of a population living in urban environments

FIGURE 4.2 World urbanisation, 2018

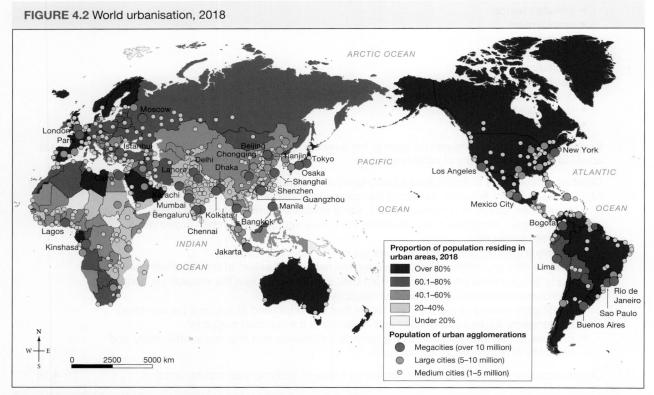

Source: Based on information taken from the United Nations, Department of Economic and Social Affairs, Population Division 2018. World Urbanization Prospects: The 2018 Revision.

Urbanisation has its roots in the ability of early communities to produce reliable food surpluses. These food surpluses could be used to feed the whole community throughout the year, rather than each family or smaller group relying on producing their own food to survive. As crop-growing techniques were developed, people were able to settle in one place, no longer having to move from place to place as the food ran out. Working together to produce food for the broader community also led to the division of labour — different people could complete different jobs to achieve a common purpose. This meant that not everyone was tied to the land, working to produce food to survive. Instead, those with other skills, such as craftspeople, were able to work full-time in their own trade and were, in effect, paid by the food surplus.

Over time, skill specialisation increased and eventually trade took place, thereby increasing the total wealth of the community and the individuals within it. Figure 4.3 shows the correlation between the proportion of the population in different countries living in urban areas, and the percentage of the population working in agriculture. We can see the expected negative correlation between the share of a country's population who live in urban areas and the share who work in the agricultural sector. In other words, the more people live in cities, the fewer who work in agriculture.

FIGURE 4.3 Employment in agriculture versus urban population, 2021

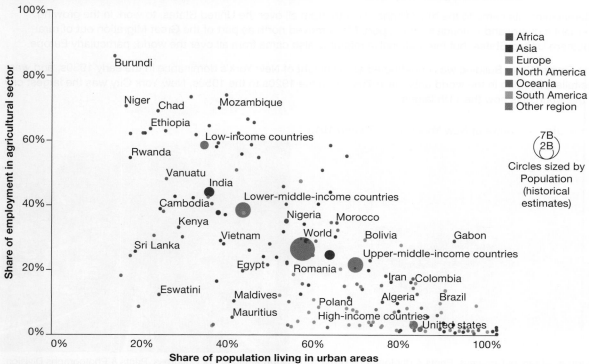

Share of total employment from the agricultural sector, versus the share of population living in urban areas

Source: Based on OurWorldInData.org. Licensed under CC BY. Data source: World Bank (2023); UN Population Division (via World Bank)

EXAMPLE: The city of Ur, 3800 BCE

The ancient Sumerian city of Ur, dating from 3800 BCE, was located in what is now southern Iraq and is a good example of how early communities quickly expanded and thrived. The Sumerians successfully controlled the Tigris and Euphrates rivers to supply water for farming and drinking in a very arid region.

This strategic use of irrigation allowed early settlers to cultivate surplus crops, supporting a growing population and leading to the development of a complex society. Ur became a bustling urban centre, featuring impressive architectural achievements such as the ziggurat, which served as both a religious temple and a symbol of the city's wealth and power. Trade flourished as Ur connected with other cities and regions, exchanging goods such as textiles, metals and grains.

FIGURE 4.4 The Great Ziggurat of Ur

Urbanisation has always been closely linked to economic development, but important advances in manufacturing and production techniques during the Industrial Revolution of the late 1700s and early 1800s spurred the rapid growth of existing cities in Europe and North America. New factories were generally built in or near existing urban areas, accelerating the growth of cities during the late 1800s as people flocked to them looking for work.

EXAMPLE: New York City

New York City's population grew rapidly in the 19th and early 20th centuries as a result of migration, jumping from 696000 in 1850 to 7.892 million in 1950.

Internal migrants came to the main financial centre from all over the United States, to work in the growing garment industry and America's largest port. Many moved north as part of the Great Migration out of rural Southern United States, but international immigrants also came from all over the world, particularly Europe.

The Empire State Building was constructed at the height of New York's dominance in the early 1930s, and was the tallest building in the world until the 1970s. From the 1920s to the 1950s, New York City was the largest city in the world; it is now the 11th largest.

FIGURE 4.5 Views of New York City, 1874 and 1915

Source: Library of Congress, Prints & Photographs Division, [LC-DIG-pga-02708]

Source: Library of Congress, Prints & Photographs Division, [LC-DIG-ds-00182]

As well as the expansion in New York, new industrial cities also emerged across the United States, including Chicago. Developments in the transport network — such as canal construction and, later, railways that allowed for more efficient transportation of goods — made Chicago the fastest growing city in the world. In 1850, the population of Chicago was 29 963, but by 1890 it had risen to 1 099 850. During this period, the proportion of Americans living in towns and cities more than doubled as migrants from the surrounding countryside and abroad moved into the cities in search of work.

Most of these large cities in North America and Europe reached their peak by the mid-1900s; other parts of the world have experienced significant city growth since then. For example, through the 20th century, the economies of Latin America transitioned from agriculture to industry and services as they adopted the import–substitution industrialisation strategy, making more products rather than importing them. In 1950, more than half of the total employment in Brazil was in agriculture but by 2000 this had fallen to just over one-fifth. Today, more than half of all Latin American countries' populations live in urban areas. By 2050, it is estimated that this figure will rise to 90 per cent.

4.2.2 Patterns in urban growth

With the world's population surpassing eight billion people in 2022, the number of people living in both urban and rural areas has continued to grow. This growth is projected to continue until at least 2080, potentially reaching more than 10 billion people. As shown in figure 4.6 and reported by the United Nations, the proportion of the world's population living in urban areas passed the 50 per cent mark in 2007. This proportion is expected to climb to two-thirds by 2050, largely due to increases in urbanisation in Africa and Asia. Experts estimate that India, China and Nigeria alone will account for one-third of this growth in the world's urban population.

FIGURE 4.6 Change in global urban, rural and total population, 1950–2050

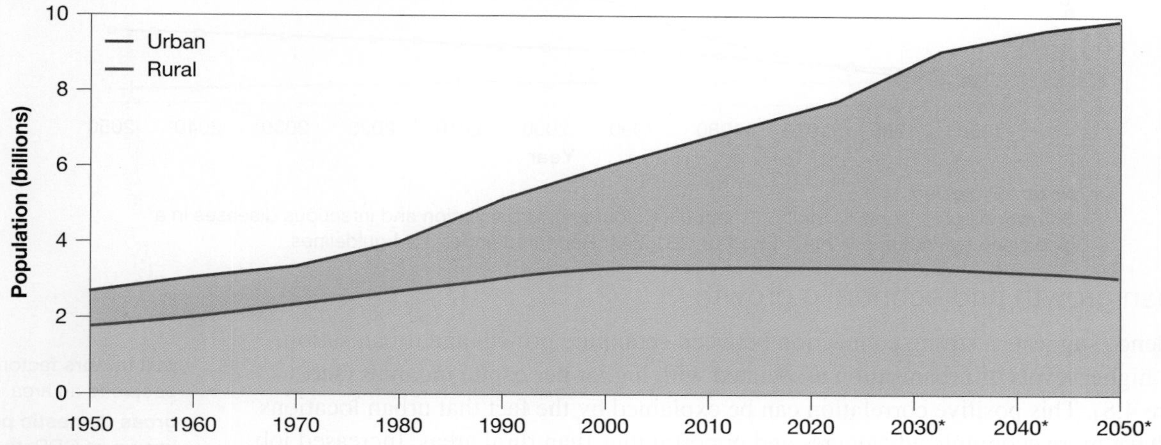

Note: * Projected

Source: United Nations, Department of Economic and Social Affairs, Population Division 2014, Our urbanizing world, Population Facts No. 2014/3

Although urbanisation rates have been slowing in recent decades, the number of people living in urban areas has been increasing steadily. It is important to make the distinction between urbanisation and population growth in urban areas.

- **Rates of urbanisation** measure the change in the proportion of the total population of an area living in an urban environment, usually expressed as a percentage.
- **Urban population growth** is the increase in the number of people living in an urban area. This can increase without altering the rate of urbanisation if the rural areas are also growing at the same rate.

The United Nation's urban population projection for 2015–50 is for more than 2 billion urban dwellers in Africa and Asia alone. Experts also estimate that two-thirds of the world's population will be calling cities home by 2050. These population increases are the result of natural increases (from births) and migration from rural areas.

Figure 4.7 shows the considerable differences in the population trends for developed and developing countries. While developed regions have seen a slight and consistently gradual rise in urban populations since the 1950s, the growth of urban centres in less developed regions has been far more rapid, and is projected to continue as their economic structures continue to shift.

FIGURE 4.7 Global urban and rural population growth in developed and developing regions, 1950–2050

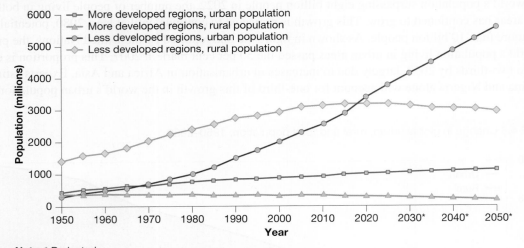

Note: * Projected
Source: Alirol E, Getaz L, Stoll B, Chappuis F, Loutan L, Urbanisation and infectious diseases in a globalised world, Lancet Infect Dis, 112, 131–241. Reprinted under STM guidelines.

Urban growth and economic growth

Evidence suggests a strong connection between economic growth and urbanisation, with higher levels of urbanisation associated with higher per capita incomes (see figure 4.8). This positive correlation can be explained by the fact that urban locations can have more economic advantages and opportunities than rural areas. Increased job opportunities and greater wealth expectations have long been major **pull factors** that have caused cities to grow from rural migration. According to the World Bank, cities now account for more than 80 per cent of the world's **gross domestic product** (GDP).

> **pull factors** factors that attract people to an area
>
> **gross domestic product** also known as GDP; the total value of all goods and services produced within a country, usually over a period of one year

FIGURE 4.8 Share of the population living in urbanised areas versus GDP per capita, 2022

GDP per capita data is adjusted for inflation and for differences in the cost of living between countries.

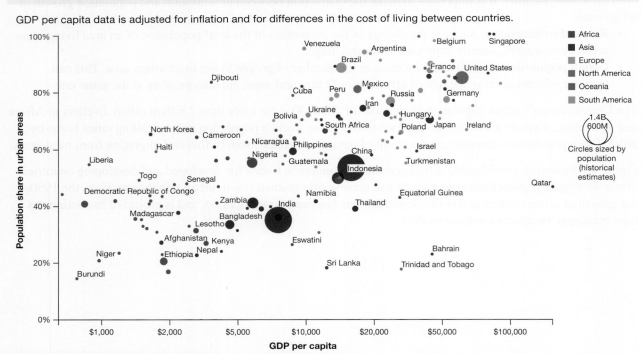

Source: Based on OurWorldInData.org. Licensed under CC-BY. Data source: HYDE (2023); Bolt and van Zanden - Maddison Project Database 2023 (2024)

Evidence also supports the view that **globalisation** and urbanisation are inextricably connected. Globalisation manifests itself in the marketing and sale of identical goods and services, such as technology, clothing and fast-food 'superbrands', simultaneously around the world. Trade between countries provides employment opportunities, often in manufacturing and transport industries. This, in turn, leads to urbanisation where these industries are based, often on coastal areas with access to shipping routes for international trade.

Today, most urban growth in the developed world is a result of a natural increase among those already dwelling in cities. Additionally, formerly small settlements are being reclassified as urban areas as the populace living there grows from within.

globalisation the process by which the world has become increasingly inter-connected through freer movement of capital, goods and services. It is reflected in the value of cross-border world trade expressed as a percentage of total global GDP.

FIGURE 4.9 The rise of globalisation has led to the widespread availability of some brands around the world.

SKILLS ACTIVITY

1. Using table 4.1, construct ten sets or a selection of clustered column graphs for each of the decades to show contrasting patterns of urbanisation. (You could also construct multiple line graphs to show these patterns.)
2. Describe the differences between the regions, noting those that have experienced broadly similar or very different rates of urbanisation.

TABLE 4.1 Average annual change in urbanisation (%), 1950–2050

	1950–60–	1960–70	1970–80	1980–90	1990–2000	2000–10	2010–20	2020–30*	2030–40*	2040–50*
Asia	1.9	1.2	1.4	1.7	1.5	1.8	1.3	1.0	0.8	0.7
South-East Asia	1.8	1.5	1.7	2.1	1.9	1.6	1.2	1.1	0.9	0.8
Europe	1.0	1.0	0.7	0.4	0.1	0.2	0.3	0.3	0.4	0.4
North America	0.9	0.5	0.0	0.2	0.5	0.2	0.3	0.3	0.3	0.2
South America	1.9	1.5	1.2	0.9	0.7	0.3	0.3	0.2	0.2	0.2
Latin America and the Caribbean	1.8	1.5	1.2	0.9	0.7	0.4	0.3	0.3	0.3	0.2
Africa	2.6	1.9	1.7	1.6	1.0	1.1	1.1	1.1	1.0	0.9
North Africa	2.0	1.6	1.1	1.0	0.6	0.4	0.4	0.5	0.7	0.7
Sub-Saharan Africa	3.3	2.1	2.1	1.9	1.3	1.4	1.4	1.3	1.1	1.0
Oceania	0.7	0.5	0.1	−0.1	−0.3	0	0	0.1	0.2	0.3

Note: * Projected
Source: From World Urbanization Prospects, 2014 Revision, by Department of Economic and Social Affairs, Population Division. © United Nations 2015 Reprinted with the permission of the United Nations

4.2 Exercise

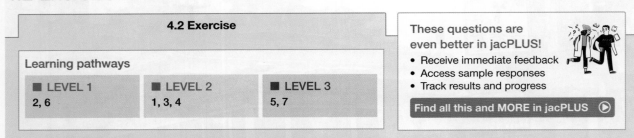

4.2 Exercise

Learning pathways		
■ **LEVEL 1** 2, 6	■ **LEVEL 2** 1, 3, 4	■ **LEVEL 3** 5, 7

These questions are even better in jacPLUS!
• Receive immediate feedback
• Access sample responses
• Track results and progress

Find all this and MORE in jacPLUS ▶

Explain and comprehend

1. a. **Compare** the levels of urbanisation shown in figure 4.10 for 2015 between Africa and Asia, and the rest of the world.
   b. **Describe** how the differences are expected to change by 2050. Consider the gap between their respective levels of urbanisation.

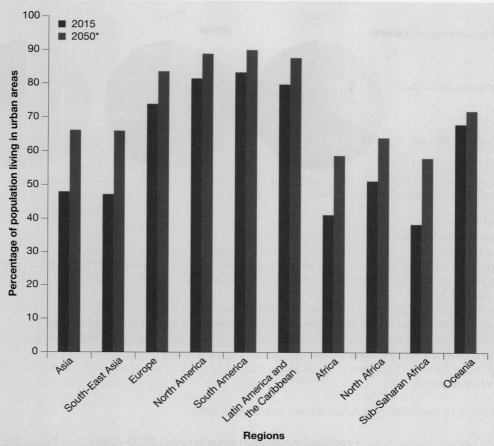

FIGURE 4.10 Percentage of people living in urban areas in 2015 and 2050*

Source: United Nations, Department of Economic and Social Affairs, Population Division 2015. World Urbanization Prospects: The 2014 Revision, ST/ESA/SER.A/366

2. a. **Identify** the two main processes contributing to the growth of megacities.
   b. **Outline** the main changes in early communities that led them to be able to settle down in one place, resulting in the start of urbanisation.
   c. **Identify** which event in the 1700 and 1800s led to rapid urbanisation in Europe and America.
   d. **Identify** which of the following are examples of pull factors.
      i. Poor climate
      ii. Employment opportunities
      iii. Education opportunities
      iv. Food insecurity
   e. **Identify** the year the proportion of the world's population living in urban areas exceed 50 per cent.

Analyse and apply

3. a. With reference to figure 4.3, **describe** the relationship between the number of people employed in agriculture and the share of the population living in urban areas (the urban population).
   b. Give possible reasons for this relationship.
4. Refer to figure 4.11.
   a. **Explain** how changes in economic structure can lead to megacities, particularly in China.
   b. **Suggest** one reason Australia, despite having a low proportion of its GDP generated by the agricultural sector, doesn't have any megacities.

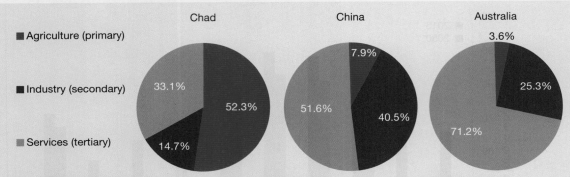

FIGURE 4.11 GDP composition by sector of origin: Chad, China and Australia

- Agriculture (primary)
- Industry (secondary)
- Services (tertiary)

Chad

33.1%
52.3%
14.7%

China

7.9%
51.6%
40.5%

Australia

3.6%
25.3%
71.2%

Source: The World Factbook. Central Intelligence Agency. Retrieved from https://www.cia.gov/the-world-factbook/field/gdp-composition-by-sector-of-origin/

5. Refer to figure 4.2.
   a. **Describe** the location of the megacities in the world in 2018.
   b. **Describe** the pattern of the proportion of the urban population across the world.
   c. What is the relationship between the location of megacities and the areas with higher urban population?
   d. **Suggest** a reason India has a lower urban population but still has many megacities.

6. Study figure 4.12.
   a. Which region has the highest predicted population growth rate?
   b. What is predicted to happen to the growth rate between 2010 and 2050 for all regions?
   c. Which regions have the biggest change in population growth rates, and what is the change?
   d. What is different about Europe's predicted population growth rate?
   e. What does figure 4.12 indicate is predicted to happen to urban areas in Africa and Oceania compared to the other regions?

FIGURE 4.12 Predicted population growth rates, 1950–2050

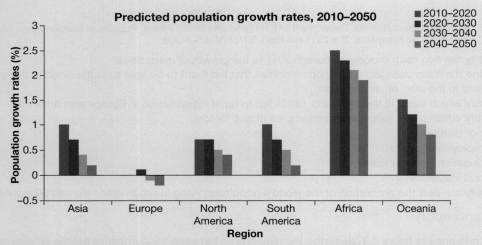

Source: From World Urbanization Prospects, 2014 Revision, by Department of Economic and Social Affairs, Population Division. © United Nations 2015 Reprinted with the permission of the United Nations

7. a. With reference to figure 4.12, **describe** the predicted population growth rates for the different regions of the world.
   b. What can you **infer** from the patterns you have described for the urbanisation of the different regions, and in particular Africa and Europe?

Sample responses are available in your digital formats.

LESSON
4.3 Vulnerability to natural hazards in megacities

LEARNING INTENTION

By the end of this lesson you should be able to explain the impacts of urbanisation and the growth of megacities for human wellbeing and environments using various forms of data and information and spatial technologies, considering risks and opportunities posed by the rate of urban growth, population density, and risk and vulnerability to natural hazards and disasters, and explain the challenges for sustainable development facing megacities; for example, climate change.

Source: Geography General Senior Syllabus 2024 © State of Queensland (QCAA) 2024; licensed under CC BY 4.0.

4.3.1 Factors affecting vulnerability

With the world's population becoming increasingly urbanised, more people are working and living in cities than ever before. The result of this is the concentration of wealth in large urban areas. This excessive concentration makes national and global economies more exposed to **anthropogenic** and natural disasters.

DID YOU KNOW?

The Lloyd's City Risk Index 2015–25 quantifies this exposure to disasters by calculating the potential GDP@Risk in more than 300 cities from threats such as tropical windstorms, floods and earthquakes. GDP@Risk models the expected loss to GDP from a significant natural or anthropogenic disaster. Eight cities in the top 10 of this list are megacities, including Tokyo, New York, Manila, Istanbul, Osaka, Los Angeles, Shanghai and London.

Cities, especially in developing economies, struggle to manage rapid population growth. Invariably, poor migrants from rural areas settle illegally in hazardous areas with poorly constructed housing and lack of infrastructure, leaving themselves vulnerable to the impact of natural hazards.

Asia has the largest number of people exposed to natural disasters. Verisk Maplecroft, a UK-based risk management company, estimates that nearly 1.5 billion people in South Asia are exposed to at least one natural hazard, the most common being flooding. However, they found that African countries are the most vulnerable to risk of poor governance and the lack of preparedness for natural disasters.

Urban researchers are predicting that many of the world's largest cities can expect to be flooded as climate change and rapid urban expansion combine.

FIGURE 4.13 Flooding is a common natural hazard for residents of Dhaka, Bangladesh.

anthropogenic change caused or influenced by people, either directly or indirectly

CASE STUDY: Flood risk in the Pearl River Delta

Quick facts: Pearl River Delta
- **Location** Southern China
- **Population** 86 million
- **Also known as** Guangdong–Hong Kong–Macao Greater Bay Area
- **Includes** Hong Kong, Shenzhen, Dongguan, Guangzhou, Foshan, Zhongshan, Jiangmen, Zhuhai, Huizhou and Macau
- **GDP per capita** US$22 500

One example of how rapid urbanisation can make a place more vulnerable to natural hazards can be seen in the Pearl River Delta economic zone in China. The Pearl River Delta economic zone was one of the three major engines propelling the rapid economic development of China. During the last 30 years, the Pearl River Delta has experienced rapid economic growth and urban expansion. The result was an increase in run-off and abnormal rising water levels during the high-rainfall season, from June to September. These combined effects overloaded the limited drainage systems and increased the risk of floods.

FIGURE 4.14 Canton Tower, next to the Pearl River, Guangzhou

In this region, 75 per cent of the population now live in urban areas. The city of Foshan is the third most populous city in the region, with a population of about 7.5 million. Since 1990, the municipal government has been spending 0.5 per cent of the GDP each year on flood mitigation infrastructure projects. Figures 4.15 and 4.16 indicate how river discharge at Macau has increased since 1994 but also the maximum river levels have been reduced, indicating the success of measures designed to reduce the flood risk.

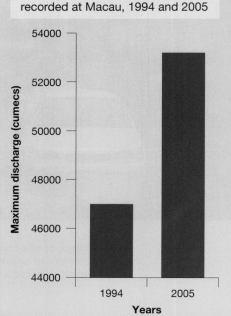

FIGURE 4.15 Maximum discharge recorded at Macau, 1994 and 2005

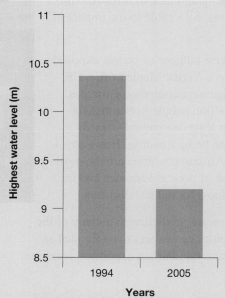

FIGURE 4.16 Highest water level recorded at Macau, 1994 and 2005

Source: Zhang H, Ma W, Wang X. Rapid Urbanization and Implications for Flood Risk Management in Hinterland of the Pearl River Delta, China: The Foshan Study. Sensors Basel, Switzerland. 2008;84:2223-2239.

Urban land use and catchments

FIGURE 4.17 Drainage basin (arrows indicate volume of water)

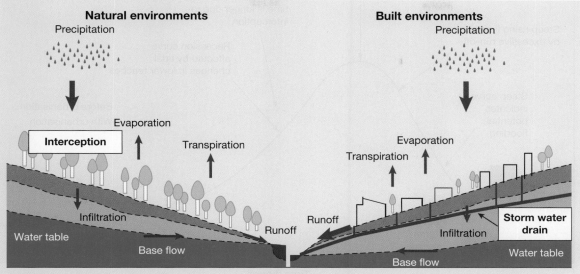

Source: © Melbourne Water. Retrieved from https://www.melbournewater.com.au/building-and-works/stormwater-management/introduction-wsud.

A catchment in its natural state will act like a sponge after a rainstorm. Rain is initially intercepted by trees and shrubs before it hits the ground. It can then soak into the soil layer and later percolate down into the rock layer, if it is permeable, before making its way slowly to the river. However, if the rainfall is intense or the catchment is saturated with water from previous rainfall, the water will simply run over the surface and reach the river more quickly. This may cause the river to flood.

In contrast, urban land use contributes to flood risk because urban areas contain many impermeable surfaces, such as roads, car parks and roofs. This creates widespread water run-off after heavy rain. Furthermore, urban drainage systems are designed to remove rainwater quickly into neighbouring creeks, but in their natural state these creeks cannot cope with such large volumes of water arriving so quickly.

Storm hydrographs are used to measure the discharge of a creek over time as a result of a rainstorm. The **discharge**, normally measured in cubic meters per second (**cumecs**), is the volume of water passing a point, usually a gauge station, during a specific period of time. Discharge may also be measured by the height of the creek water above a measuring weir. Data is plotted on a storm hydrograph as a line, whereas rainfall is plotted as a histogram (see figure 4.18).

Creeks with a high flood risk have a flash response hydrograph. Discharge rises rapidly after a storm, creating a high peak discharge. Creeks with a low flood risk have a much lower peak discharge and longer **lag time,** meaning the difference in time between peak rainfall and peak discharge is much longer.

discharge the volume of water flowing through a river channel, measured in cumecs

cumecs an abbreviation of 'cubic metres per second'

lag time the time between the peak rainfall and the peak river discharge

FIGURE 4.18 Model of a storm hydrograph

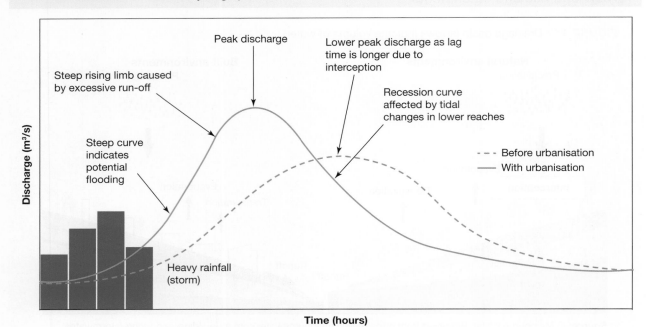

Peak discharge

Lower peak discharge as lag time is longer due to interception

Steep rising limb caused by excessive run-off

Recession curve affected by tidal changes in lower reaches

Steep curve indicates potential flooding

- - - Before urbanisation
—— With urbanisation

Heavy rainfall (storm)

Discharge (m³/s)

Time (hours)

CASE STUDY: Natural hazards in Manila, Philippines

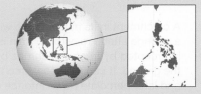

Quick facts: Manila
- **Population (metro)** 13.3 million
- **Average annual rainfall** (approximately) 2200 mm
- **Typhoon frequency** 6 to 9 per year
- **Earthquake risk** 20 to 30 per year

Metropolitan Manila, with a population of more than 13 million, is one of the megacities most exposed to risk. In a study of megacities at risk, the Economist Intelligence Unit ranked Manila at 55 out of 60 based on 49 risk indicators, including digital, health, personal and infrastructure security. Manila was also ranked the highest for rates of death caused by natural disasters, with tropical cyclones (typhoons) and earthquakes accounting for 80 per cent of the share of total GDP at risk, according to the Lloyd's City Risk Index. Manila is in the bottom five in infrastructure security, and is second last in digital security.

Manila is located in a highly active cyclone area and has a high risk of flooding. Because it is adjacent to the Eurasian, Philippines and Sunda plate boundaries, Manila has high earthquake risk too. The Marikina River, which flows through Manila, runs in a rift valley along two parallel fault lines: the West Valley Fault and the East Valley Fault.

The Philippine Institute of Volcanology and Seismology believes that rupturing along the West Valley Fault has a recurrence interval of 200 to 400 years and could result in a 7.2 magnitude earthquake. A seismic event like this in the vicinity of Manila could result in more than 35 000 fatalities. Manila is now inside the recurrence interval for another earthquake. A buffer zone on each side of the fault line, where no buildings should be established, was recommended, but with such rapid urban development such a buffer has been impossible to enforce. Poor administration of building standards and the absence of clear planning directions also mean that local governments have allowed crowding of limited urban space.

FIGURE 4.19 The Manila Cathedral after an earthquake in 1880. Experts believe Manila is due for another major earthquake soon.

Predictably, the poorer residents live in these hazardous locations, and are vulnerable to flooding, tropical cyclones and earthquakes without adequate protection.

4.3.2 The effects of climate change

As the sizes of cities and the density of populations increase, natural river systems are less able to cope with the levels of run-off from non-permeable surfaces of the growing urban environment. However, this is complicated further by the impacts of climate change. Climate change presents a multifaceted challenge to megacities worldwide, by amplifying the risk of natural hazards.

One significant impact is the intensification of extreme weather events. For instance, in recent years, megacities such as Tokyo have experienced record-breaking heatwaves, attributed in part to climate change. These heatwaves strain infrastructure and public health systems, posing significant risks to vulnerable populations.

FIGURE 4.20 Heatwave in Tokyo

Climate change also alters precipitation patterns, leading to more frequent and intense rainfall events or prolonged droughts, both of which heighten the risk of flooding and water scarcity in megacities. For instance, megacities such as Jakarta and Manila face recurrent flooding due to heavy rainfall and inadequate drainage systems. These floods disrupt daily life, damage infrastructure, and threaten public health by promoting the spread of waterborne diseases.

Additionally, sea level rise exacerbates the vulnerability of coastal megacities to storm surges and coastal erosion. Cities such as Miami and Shanghai face the prospect of inundation and saltwater intrusion into freshwater supplies due to rising sea levels. The loss of coastal land and infrastructure threatens the economic viability and livability of these cities, necessitating adaptive strategies such as seawalls and coastal retreat measures.

FIGURE 4.21 Flooding in Jakarta

FIGURE 4.22 Hurricane Irma hits Miami

In parts of London, the risk of flash flooding is increasing. British government scientists now believe that as a result of climate change, longer and more intense rain episodes will become more common and will lead to more flooding.

The London Borough of Hammersmith and Fulham has recently been found to be the worst borough for potential flooding, with almost 60 000 homes — 60 per cent of the borough — at risk.

FIGURE 4.23 Flooded suburban road in London

4.3.3 Urban heat management

Large urban environments can present a direct risk to the health and wellbeing of their inhabitants. Hot urban environments can be extremely dangerous because heat has the potential to aggravate pre-existing health conditions, including heart disease, lung disease and asthma. Moreover, very old and very young city dwellers are more susceptible to heat-related illness, especially in poorer neighbourhoods.

An **urban heat island** is a built-up area that is significantly warmer than its surrounding rural areas (see figure 4.24). The phenomenon was first observed by Luke Howard, a pharmacist working in London during the early 1800s. With a keen interest in meteorology, he recorded and published temperature readings for various neighbourhoods from 1800 to 1830, when the London population grew from 1 to 1.5 million people.

urban heat island microclimates created in an urban environment by the hard, dark surfaces that attract and retain heat, such as roads and buildings

FIGURE 4.24 Urban heat island model

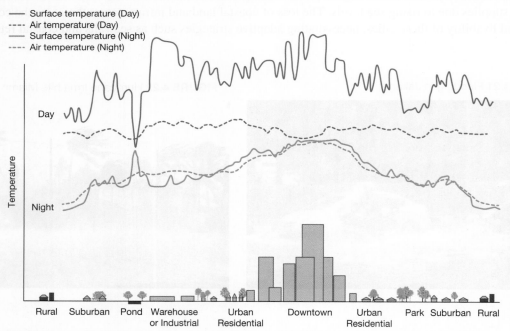

Note: This is a model of how features of urban environments affect the rise and fall of day and night surface and air temperatures.
Source: US Environments Protection Agency, Learn About Heat Islands, https://www.epa.gov/heat-islands/learn-about-heat-islands

As cities spread, the natural cooling elements of a landscape disappear. The result is that less energy is used up by evaporating water and less of the sun's energy is reflected. More heat is also stored by buildings and the ground in urban areas compared to rural areas because materials such as bricks, concrete and bitumen have a high heat capacity, or **thermal mass** — an ability to store heat during sunshine hours and then release it slowly. Heating and cooling equipment and vehicle exhausts can add to the heat island effect.

thermal mass capacity to absorb and retain heat energy

DID YOU KNOW?

Even in areas where urban heat islands are not a major problem now, they could exacerbate heatwaves in the future. Researchers in Hong Kong found that over a 27-year period, the urban heat island duration in a developing urban area of Hong Kong increased from 13.6 hours to 17.5 hours. They suggested that predicted changes in urban heat island duration should be included in planning for large cities.

FIGURE 4.25 Annual mean temperature recorded at the Hong Kong Observatory Headquarters (1885–2023)

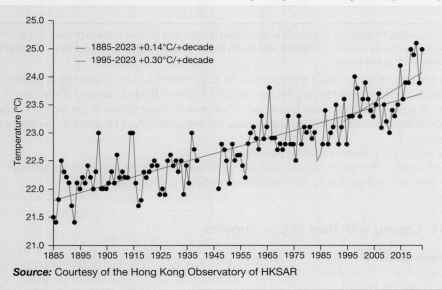

— 1885-2023 +0.14°C/+decade
— 1995-2023 +0.30°C/+decade

Source: Courtesy of the Hong Kong Observatory of HKSAR

EXAMPLE: Delhi, India

With a population of more than 33 million, Delhi, the capital city of India, is one of the world's largest and most extensive megacities. Comprising 11 districts of the National Capital Territory of Delhi, it covers an area of nearly 1500 square kilometres.

The average population density is about 11 320 per square kilometre. Delhi's North district has the highest population density at 36 155 persons per square kilometre whereas New Delhi, the least densely populated district, has 4057 persons per square kilometre.

Researchers at the Centre for Science and Environmen t in New Delhi have identified the heat island phenomenon as a contributing factor to the risk of excessive heat for Delhi's inhabitants. The heat island magnitude in Delhi is high, at around 8 °C and, in 2015, was responsible for accentuating heatwave conditions, resulting in melting tarmac roads and a spike in heat-related deaths. Heatwaves in May 2024 resulted in areas within Delhi setting new temperature records of almost 50 °C.

FIGURE 4.26 New Delhi heat island

SKILLS ACTIVITY

Examine table 4.2 to complete the following questions.

TABLE 4.2 Built-up density of Delhi's administrative districts[*]

Administrative district	1977	2014
North	0.28	0.43
North East	0.13	0.62
East	0.27	0.60
Central	0.38	0.40
New Delhi	0.05	0.35
South	0.08	0.37
South West	0.04	0.39
West	0.12	0.60
North West	0.01	0.35
Delhi	0.10	0.40

Note: * Delhi now has 11 districts. In 2012, the North East district was divided into North East and Shandara, and the South district was divided into South and South East. The old boundaries have been maintained in this table to allow for comparison.

1. Using data from table 4.2, construct a spiderchart to show how the built-up density of Delhi's nine administrative districts changed during the period 1977 to 2014. Built-up density is the ratio of the built-up area to the total area. It indicates how dense the urban area is, irrespective of the location of urban patches within the administrative boundary. (You can access a blank spiderchart template in the Resources tab to complete this task.)
2. Describe the pattern of change shown in your graph.
3. How do the changes in the distribution of built-up land compare with that of built-up density?
4. Explain why the Central district may have remained relatively unchanged.

CASE STUDY: Coping with heat in Los Angeles

Quick facts: Los Angeles
- **Location** California, United States
- **Population** 13 million (metro)
- **GDP per capita** US$86 000

A number of measures can be adopted to mitigate the heat island effect. Los Angeles, for example, was the first city in the United States to make heat reflecting roofs compulsory for new and significantly renovated houses. In 2017, Los Angeles set itself the goal to cool the city by 1.5 °C by 2035. This involves experimenting with painting streets with a sealant that lowers the surface temperature by 5 °C and increasing the urban tree canopy cover by 15 per cent by 2035.

FIGURE 4.27 Los Angeles skyline in heat haze

The Los Angeles Health Department has also flagged urban heat as a serious and growing threat to public health, particularly as the climate warms. The Los Angeles metropolitan area has a population of around 18 million people, and in October–November 2017 experienced a heatwave that broke many existing heat records for the area. Part of the LA County's response to heat in the city is the Emergency Survival Guide, which includes tips for avoiding heat illness. LA County has also established Cooling Centres, which are community facilities opened by local authorities on days of extreme heat for people who have no other means of escaping the heat or might be at greater risk of heat illness, such as the elderly, ill or homeless. These measures may not help to reduce the heat island effect, but they do help to mitigate the impact on the most vulnerable members of the community.

In other cities in the United States, Washington DC has implemented a green area ratio for new developments, while in Seattle a points system is used where property owners select various options, such as tree planting, to meet minimum planning requirements.

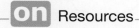 Resources

 Weblinks Evidence on the effectiveness of interventions during heatwaves
Cooling urban heat islands

4.3 Exercise

4.3 Exercise

Learning pathways

■ LEVEL 1	■ LEVEL2	■ LEVEL 3
1, 3, 5	2, 6	4

These questions are even better in jacPLUS!
- Receive immediate feedback
- Access sample responses
- Track results and progress

Find all this and MORE in jacPLUS ▶

Explain and comprehend

1. Examine figure 4.24.
   a. **Describe** the contrasting pattern of day and night surface temperatures.
   b. **Compare** the day air temperature profiles with the night air temperature profiles.
2. **Explain** why the high-density parts of a city experience the highest temperatures.
3. **Explain** the impact that each of the following measures can have on mitigating the heat island effect.
   - Tree planting
   - Vegetated rooftops and wall panels
   - Light coloured rooftops and pavements

Analyse and apply

4. Refer to figures 4.28 and 4.29.

> **FIGURE 4.28** Rainfall and height of Brent River above crest of weir at (a) Wealdstone Brook and (b) Hanwell

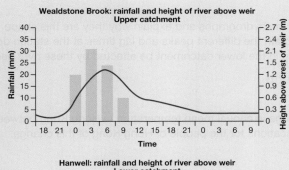

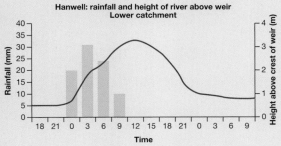

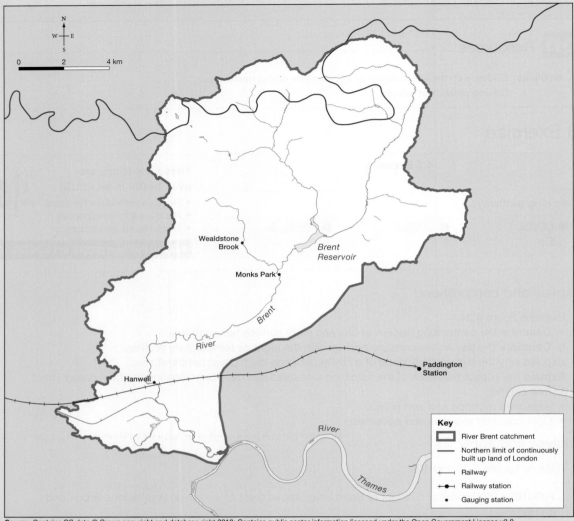

FIGURE 4.29 The Brent River catchment

Key
☐ River Brent catchment
— Northern limit of continuously built up land of London
†—†—† Railway
†—●—† Railway station
● Gauging station

Source: Contains OS data © Crown copyright and database right 2018; Contains public sector information licensed under the Open Government Licence v3.0

a. **Describe** the shape of the hydrographs and explain why they are this shape.
b. What factors might **explain** the different peaks and lag times at the station downstream?
c. How might people living in the lower catchment be affected by these river heights?

Propose and comprehend

5. Study figure 4.25.
    a. What is the overall change in annual mean temperature in Hong Kong between 1885 and 2023?
    b. **Suggest** possible implications for the people of Hong Kong of this change.

6. Study figures 4.30 and 4.31

FIGURE 4.30 Land use in Delhi, 1989 and 2011

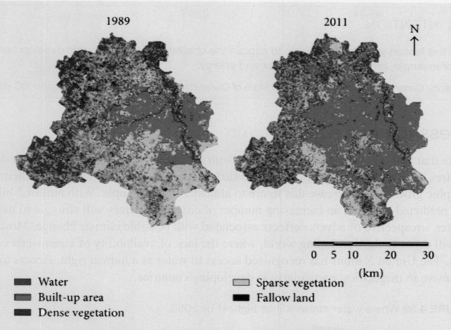

1989   2011

Water

Built-up area

Dense vegetation

Sparse vegetation

Fallow land

0   5   10   20   30

(km)

Source: © 2014 Bablu Kumar et al. Source: Bablu Kumar, Kopal verma, and Umesh Kulshrestra, "Deposition and Minerglogical Characteristics of Atmospheric Dust in relation to Land Use and Land Cover Change in Delhi India," Geography Journal, vol. 2014, Article ID 325612, 11 pages, 2014 https://doi.org/10.1155/2014/325612

FIGURE 4.31 National Capital Territory of Delhi Administrative Districts

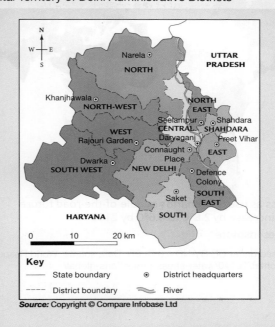

Key

State boundary

District boundary

District headquarters

River

Source: Copyright © Compare Infobase Ltd

a. **Describe** the changes shown in figure 4.30 by referring to figure 4.31 and identifying districts where particular types of land use expanded or contracted between 1989 and 2011.

b. **Explain** the possible effect this may have on the temperatures in Delhi.

Sample responses are available in your digital formats.

LESSON
4.4 Water and energy in megacities

4.4.1 Access to fresh water

Experts estimate that 150 million people live in cities with perennial water shortage, which is defined as having less than 100 litres per person per day of sustainable surface and groundwater flow within their urban extent. By 2050, demographic growth will increase this figure to almost 1 billion people. With almost 3 billion additional urban dwellers predicted by 2050, an increasing number of cities of all sizes will struggle to meet the increasing demand for water, irrespective of adverse effects associated with possible climate change. Most of the future urban growth will occur in the developing world, where the lack of availability of clean water can easily lead to health issues. The United Nations has recognised access to water as a human right. Access to clean water sources is expensive in megacities, particularly in developing countries.

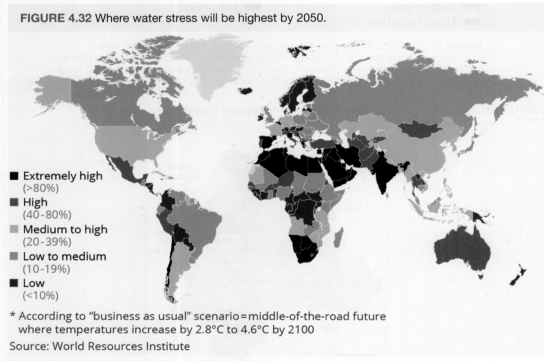

FIGURE 4.32 Where water stress will be highest by 2050.

Extremely high (>80%)

High (40-80%)

Medium to high (20-39%)

Low to medium (10-19%)

Low (<10%)

* According to "business as usual" scenario=middle-of-the-road future where temperatures increase by 2.8°C to 4.6°C by 2100

Source: World Resources Institute

Source: © Statista/ Licensed under CC BY-ND 4.0 DEED. Data source: World Resources Institute. https://www.statista.com/chart/26140/water-stress-projections-global/

Three things are required for urban water supplies to meet the needs of their residents:
- a water supply that is ample and sustainable
- water that is of good quality (potable water; i.e. clean enough for drinking)
- an efficient supply network.

The rapid pace of urbanisation has often far exceeded the capacities of many national and local governments to plan and manage the need for additional water.

In any country, developing or developed, failure to invest in adequate maintenance or increasing capacity of water supply that keeps pace with increasing demand can result in critical water shortage. This risk is heightened during drought conditions.

EXAMPLE: Cape Town's 'Day Zero'

In 2018, a water crisis in Cape Town illustrated the danger of low investment in water supply. Alarm bells began to ring more than 20 years before this, as Cape Town's population grew from 2.4 million in 1996 to more than 4 million in 2017. The city relies almost exclusively on six dams, located in the mountains east of the city, for its water supply. Rather than tackle the supply problem, the focus of action during the water crisis was on curbing demand. Per capita consumption did decline as large water users were required to pay more.

In 2014, the dams surrounding Cape Town were full, but three years of drought thereafter required the introduction of drastic measures to conserve dwindling water supplies. Residents were restricted to 50 litres per person per day and, to heighten people's awareness of the problem, the concept of

FIGURE 4.33 Cape Town, South Africa, came perilously close to completely running out of water.

'Day Zero' was introduced, marking the day when municipal water supplies would be largely switched off. This day was originally predicted to fall in April 2018 but due to the reduced consumption after the restrictions were implemented, it was moved to July 2018.

Fortunately, Cape Town had good winter rains starting in June 2018, and following more rains in 2020 the drought was broken and the water crisis was averted.

on Resources

🔗 **Weblinks**   United Nations water and sanitation programs

We're (not) running out of water — a better way to measure water scarcity

CASE STUDY: Beijing — A thirsty giant

Quick facts: Beijing, China
- **Population** 22 million
- **GDP per capita** US $21 800
- **Annual per capita water availability** 185 cubic metres

With a population of about 22 million, Beijing is struggling to provide an adequate and reliable water supply for its residents. The city's annual per capita water availability is only 120 cubic metres per person, which is well below the United Nations threshold for absolute water scarcity (see table 4.3).

TABLE 4.3 UN Levels of Water Scarcity (UNDESA)

Level of water scarcity	Annual water supplies per person
Water stress	Less than 1700 cubic metres
Water scarcity	Less than 1000 cubic metres
Absolute water scarcity	Less than 500 cubic metres

The total annual water consumption in Beijing is around 4.3 billion cubic metres of water, which is far more than the available, renewable fresh water resources.

Beijing has three water sources:
- surface water captured from five major rivers within the Beijing municipality
- groundwater
- diverted water from the Yangtze River through the South-to-North Water Diversion Project.

Beijing is in one of the driest regions of China. As shown in figure 4.34, for much of the year the city receives little rain. More than 90 per cent of the annual precipitation falls between June and September. Consequently, drought can have a devastating effect on the city's water supply. The drought of 2017 was the worst on record. Less rainfall meant less runoff for sustainable refilling of its surface and groundwater sources.

With 50 000 wells extracting about 1.4 billion cubic metres of water a year, Beijing has a huge groundwater deficit. Between 2000 and 2012, the North China Plain groundwater table, on which Beijing is situated, dropped by 12 metres, a problem that is aggravated by groundwater pollution. In 2014, China's Ministry of Land and Resources sampled groundwater in more than 200 cities. They found that 44 per cent of the samples tested were 'relatively poor' and requiring treatment, and 16 per cent were rated 'very poor' and deemed unsuitable for drinking. This presents a major concern for Beijing, because groundwater currently contributes more than 70 per cent of the city's total water supply.

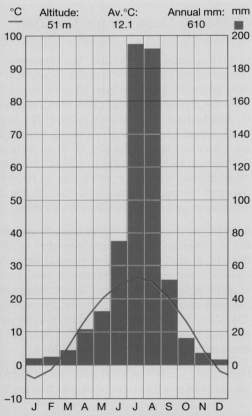

FIGURE 4.34 Beijing climate graph

Despite being short of water, Chinese cities were accused of wastage by selling it at heavily subsidised prices. In 2010, and for the first time in seven years, Shanghai raised residential water prices by 25 per cent, while Beijing increased the price of water for commercial use by nearly 50 per cent. In 2014, Beijing increased residential water prices by 25 per cent and introduced a separate pricing tier for businesses that saw water prices double.

Beijing's water supply has been supplemented by water diverted from the Yangtze River via the South-to-North Water Diversion project (see figure 4.35). This project comprises three routes for diverting water from the Yangtze to the drier north, where other megacities, such as Chongqing and Shenzhen, also face water shortages. However, reduced precipitation, a possible indicator of climate change, has already decreased freshwater reserves in the Yangtze River basin by 17 per cent. The scheme has been heavily criticised for the number of people it has displaced and the potential losses through **evapotranspiration**.

evapotranspiration the process by which water moves into the atmosphere. This occurs through evaporation (from the ground or surfaces) and through the transpiration of plants.

FIGURE 4.35 The South-to-North Water Diversion project in China

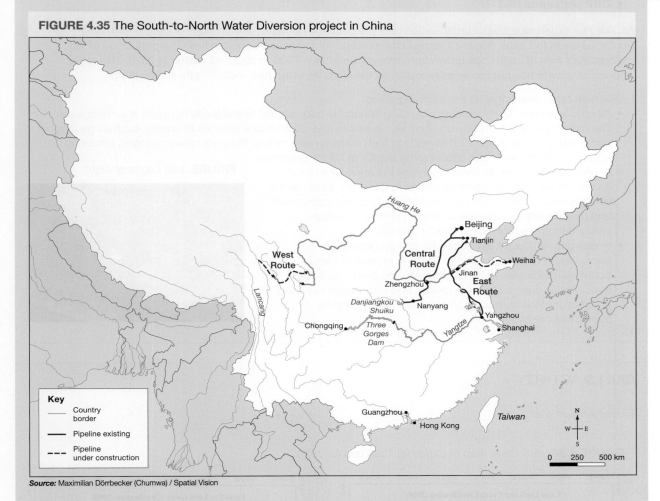

Key
—— Country border
—— Pipeline existing
- - - Pipeline under construction

Source: Maximilian Dörrbecker (Chumwa) / Spatial Vision

China's rapid urbanisation and industrialisation over the past few decades have led to large increases in wastewater discharge. Total wastewater discharge is mainly from industrial and domestic sources. More than 80 per cent of wastewater is treated but only 10 per cent is recycled. In Beijing, a third of wastewater is reused for industrial processing or irrigation.

4.4.2 Providing energy in the megacity

In megacities, providing energy presents a complex and multifaceted challenge. As urban populations continue to grow rapidly, so does the demand for electricity, heating and transportation fuels. Meeting these energy needs in a sustainable and reliable manner is critical for supporting economic growth, enhancing quality of life, and mitigating environmental impacts.

However, energy infrastructure in many megacities, particularly those in the developing world, is often outdated, inefficient and reliant on fossil fuels, leading to air pollution, greenhouse gas emissions and energy insecurity. For megacities that have grown quicker than their economy has, finding the money to invest in sustainable, long-term solutions can be difficult.

CASE STUDY: Lights on in Lagos, Nigeria

Quick facts: Nigeria
- **Capital city** Abuja
- **Population** 230 million
- **Largest city** Lagos
- **GDP per capita** US$1755

Nigeria has experienced rapid urbanisation over the past few decades, with millions of people migrating from rural areas to Lagos, Nigeria's largest city, in search of economic opportunities. The population of the metropolitan area of Lagos has grown from around 8 million in 2006 to about 12 million in 2022. This rapid population growth has put immense pressure on the city's infrastructure, including its energy supply.

Lagos now faces challenges in the following areas:
- *Electricity access:* Despite being a bustling economic hub, access to electricity in Lagos is a major issue. The majority of residents rely on informal, unreliable and often unsafe sources of energy, such as generators powered by diesel or petrol. This leads to high costs for energy and frequent power outages, hindering economic productivity and affecting the quality of life for residents.
- *Energy demand:* The demand for energy in Lagos far exceeds its supply. As the population continues to grow, the strain on the existing energy infrastructure intensifies. Industries, businesses and households all compete for limited energy resources, exacerbating the problem of energy scarcity.
- *Infrastructure:* The infrastructure for energy generation, transmission and distribution in Lagos is inadequate and outdated. Many power plants are operating below capacity, transmission lines suffer from losses, and distribution networks are often inefficient, leading to significant energy losses throughout the system.

FIGURE 4.36 Lagos at night

SKILLS ACTIVITY

Examine figure 4.37.

FIGURE 4.37 Land cover map of Lagos (a) 1990 and (b) 2020.

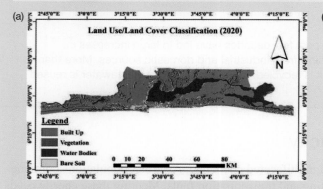
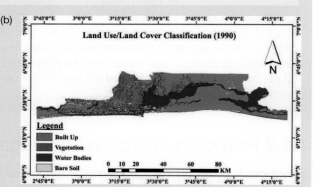

a. **Describe** the changes in urban land cover between 1990 and 2020 in Lagos.
b. **Predict** how these changes might impact on the government's ability to provide electricity to all the residents. **Explain** your answer.

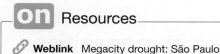

Resources

🔗 **Weblink** Megacity drought: São Paulo

4.4 Exercise

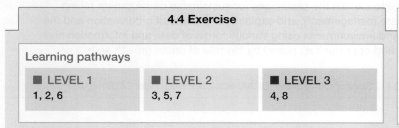

4.4 Exercise

Learning pathways

■ LEVEL 1	■ LEVEL 2	■ LEVEL 3
1, 2, 6	3, 5, 7	4, 8

These questions are even better in jacPLUS!
- Receive immediate feedback
- Access sample responses
- Track results and progress

Find all this and MORE in jacPLUS ▶

Explain and comprehend

1. **List** possible reasons a megacity may face water scarcity in the following situations.
   a. Currently
   b. In the future
2. **Identify** the three things that megacities need to meet the water needs of their residents.
3. Read the Cape Town example and answer the following.
   a. **Identify** and give a brief explanation of the two main reasons that led to the water crisis in Cape Town in 2018.
   b. What was meant by 'Day Zero'?
   c. **Identify** possible implications for the residents of the water shortages.
4. Read the Beijing case study and answer the following.
   a. **Identify** where Beijing currently sources its water.
   b. **Explain** how the South-to-North Water Diversion project will help ensure water security for the residents of Beijing.
   c. **Identify** some of the positives and negatives of this project for the people of China.
5. **Identify** possible reasons people may not have access to clean running water in a city. Categorise your answers into social, environmental and economic.

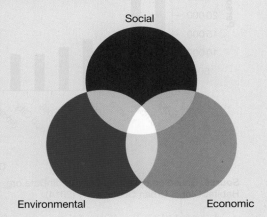

6. Create a list of barriers to consistent energy supply in some megacities.

Analyse and apply

7. Refer to figure 4.32. Which regions appear to be most at risk of water scarcity by 2050?
8. Study figure 4.34.
   a. **Describe** the pattern of annual precipitation in Beijing.
   b. **Explain** how the pattern you have described, combined with the growing urban population, may lead to water scarcity in Beijing.

Sample responses are available in your digital formats.

LESSON
4.5 Sanitation and waste management in megacities

4.5.1 Disease in urban areas

Population density in cities is increasing with continuous urbanisation. Megacities have some of the highest population densities in the world (see figure 4.38). Often, the lack of availability of clean water and sanitation infrastructure in the cities of developing nations can easily lead to outbreaks of vector or waterborne diseases and other health issues. However, as more people live and work closer together, the spread of airborne infectious diseases can be accelerated in highly populated areas, regardless of whether they are in a developing or developed nation.

In December 2019, the disease severe acute respiratory syndrome coronavirus 2 (SARS-CoV-2), known more commonly as COVID-19, spread out from China. In mid-March 2020, the World Health Organization (WHO) called the outbreak a pandemic, and within a month many of the world's largest cities were battling to control the spread of the airborne disease. COVID-19 spread rapidly through cities, and governments implemented lockdowns to prevent people from coming into contact and spreading the virus. The WHO recommended social distancing, and that people stay at least 1 metre from each other,

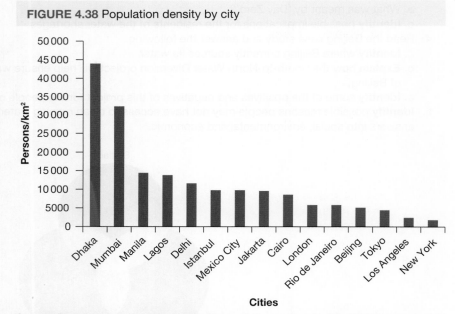

FIGURE 4.38 Population density by city

Source: Based on data from OurWorldInData.org, licensed under CC-BY. Data source: UN Habitat Global Urban Observatory (2014)

which was a great challenge in cities with population densities of thousands of people per square kilometre, and particularly those with densely populated slums. People living in megacities comprise 7 per cent of the global population, yet accounted for approximately 20 per cent of all COVID-19 deaths. As seen in table 4.4, the higher the urban population of an area, the higher the number of excess deaths due to COVID-19. The COVID-19 outbreak was considered by the WHO as a global health emergency for over three years, until May 2023.

FIGURE 4.39 Social distancing sign in the United Kingdom

FIGURE 4.40 Empty streets in New York City, March 2020

TABLE 4.4 Urban population and excess deaths due to COVID-19, 2022

	Urban population %, 2022	Excess deaths per 100 000 population, 2020–2021
World	**57**	**96**
More developed	79	142
Less developed	53	87
Least developed	35	47
High income	81	90
Upper-middle income	70	82
Middle income	54	102
Lower-middle income	43	118
Low income	34	47

Source: Toshiko Kaneda, Charlotte Greenbaum, and Carl Haub, 2022 World Population Data Sheet (Washington, DC: Population Reference Bureau, 2022). Reproduced by permission. All rights reserved.

SKILLS ACTIVITY: Creating a scattergraph

Refer to table 4.4 and create a scattergraph to examine the hypothesis that a link exists between the level of urbanisation and the incidence of excess deaths due to COVID-19. See the Resources tab for a guide to constructing a scattergraph.

4.5.2 The challenge of waste management

As urbanisation increases, so does one of its unpleasant by-products: solid waste. The World Bank estimates that the world generated 2.24 billion tonnes of solid waste in 2020. This is expected to increase to 3.88 billion tonnes by 2050, an increase of 73 per cent. Urban areas, by virtue of their high populations, produce an enormous amount of waste. For cities in many developing countries, this waste is becoming an environmental catastrophe.

Developing nations, urban centres and megacities produce huge amounts of waste, but do not always have the infrastructure to collect or process it. One of the most famous megacity rubbish dumps in the world was

FIGURE 4.41 The mountainous dumps of rubbish in urban areas in Sri Lanka pose a number of health and safety threats to the workers.

the enormous 'Smokey Mountain' in Manila. With approximately 44 per cent of Manila's 10 million residents living in poverty, many of the poorest residents lived near the rubbish dump and scavenged there for food and a living. When Smokey Mountain closed down in 1995, many scavengers migrated to the Payatas dumpsite, where another large scavenging community arose.

Considering the projected growth for cities in Asia, and with many are already struggling to manage their waste, it is easy to see how waste will become an even bigger management issue in the future.

 Resources

 Weblinks   Life after Smokey Mountain
Life scavenging in Manila's garbage mountain
Sri Lankan rubbish dump collapse

EXAMPLE: Vector-borne diseases in Sri Lanka

Garbage can also contribute to the spread of disease. Dengue, a vector-spread disease, is particularly endemic in tropical countries such as Sri Lanka where the Aedes aegypti and Aedes albopictus mosquitoes have widely adapted to urban and suburban environments. (These mosquitoes can spread yellow fever and dengue fever, among other viruses.) They can lay eggs in as little as 5 to 10 ml of water, which easily collects in discarded rubbish as small as a bottle cap.

With an annual rainfall of 2400 mm, the blocked drains and uncollected rubbish in Sri Lanka's urban areas are ideal breeding grounds for mosquito larvae. This is particularly the case in the south-west, where humid and wet conditions allow the mosquito to live longer. The heavy monsoon rain seasons do, however, provide some respite as they flush out the mosquito breeding grounds.

The south-west is also the most urbanised and densely populated part of the country. The broader metropolitan area of Colombo has a population of approximately 5.6 million people and a steady population growth. The population density is particularly high in the poorer districts of the city (approximately 20 000 people per square kilometre). This allows disease to spread relatively quickly. In 2017, Sri Lanka suffered its worst outbreak of dengue in history.

FIGURE 4.42 Colombo, Sri Lanka, under monsoon clouds

4.5.3 Water pollution and subsidence

In megacities, water pollution and subsidence are significant challenges. Rapid urbanisation and industrialisation often lead to untreated wastewater being discharged into rivers, contaminating drinking water sources and harming aquatic ecosystems. Additionally, excessive groundwater extraction for urban development contributes to land subsidence, particularly in coastal megacities such as Jakarta and Bangkok. Subsidence increases the risk of flooding and saltwater intrusion, exacerbating water pollution issues. These interconnected problems threaten public health, damage ecosystems and strain infrastructure in megacities worldwide. Table 4.5 shows the ten coastal cities around the world that are sinking the fastest.

TABLE 4.5 The ten fastest sinking coastal cities

Rank	City	Country	Peak velocity (mm/year)	Median velocity (mm/year)
1	Tianjin	China	43	6
2	Ho Chi Minh City	Vietnam	43	16
3	Chittagong	Bangladesh	37	12
4	Yangon	Myanmar	31	4
5	Jakarta	Indonesia	26	5
6	Ahmedabad	India	23	5
7	Istanbul	Turkey	19	6
8	Houston	USA	17	3
9	Lagos	Nigeria	17	2
10	Manila	Philippines	17	2

Source: Planet Anomaly. Visualized: Which Coastal Cities are Sinking the Fastest?. March 1, 2024. Visual Capitalist.

CASE STUDY: Pollution and subsidence in Jakarta, Indonesia

Quick facts: Jakarta, Indonesia

- **Population** 32 million (greater metropolitan area)
- **GDP per capita** US$9415
- **Subsidence rate** 4.9 cm per year

Only 60 per cent of Jakarta's residents have access to clean, safe, treated water. Unequal access is highlighted by the fact that most homes in Jakarta in the city's low-income areas rely on individual shallow wells to supply household needs.

Because an efficient municipal sewerage system does not exist in Jakarta, 75 per cent of household wastewater is sent to septic tanks. Over 11 per cent of households discharge their wastewater directly into rivers. Off-site sanitation, which includes industry and domestic individual treatment plants, make up the remainder. Septic tanks are a heath concern because many are simply holes dug in the ground, lined with layers of coral rock and fibre at the bottom to filter the water. Often poorly constructed or old, these septic tanks are prone to leak raw sewage, which can then leach into the city's groundwater supply. Experts estimate that more than 50 per cent of the shallow wells (those less than 40 metres deep) are contaminated by bacteria, such as E. coli, that cause pathogenic diseases.

Deeper extraction of groundwater is mostly carried out by industry. This extraction is much more localised, and the rate of extraction is much greater than that for domestic users. Researchers estimate that the level of groundwater in north Jakarta has dropped over 50 metres since 1910.

FIGURE 4.43 Informal housing in outer Jakarta, Indonesia

Source: New Geography, The Evolving Urban Form, Jakarta Jabotabek, by Wendell Cox 05/31/2011, http://www.newgeography.com/content/002255-the-evolving-urban-form-jakarta-jabotabek

FIGURE 4.44 Kampungs in north Jakarta

FIGURE 4.45 Head of groundwater level contours (m) of lower aquifer in Jakarta basin

Lower aquifer (140–250 m) in 1992

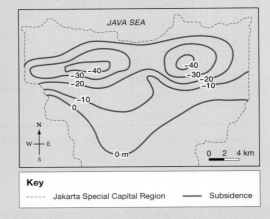

This deep extraction of groundwater is believed to be one of the main factors responsible for the land subsidence observed in the coastal, western and north-eastern parts of Jakarta. Subsidence is also caused by major building projects on soil of high compressibility. A study of land subsidence in Jakarta from 1982 to 2010 found spatial and temporal variations, with the annual rates of subsidence between 1 and 15 centimetres.

4.5 Exercise

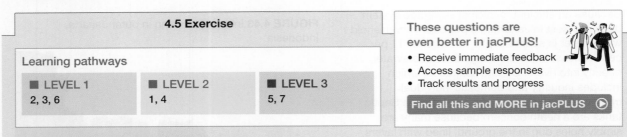

4.5 Exercise

Learning pathways

■ LEVEL 1	■ LEVEL 2	■ LEVEL 3
2, 3, 6	1, 4	5, 7

These questions are even better in jacPLUS!
- Receive immediate feedback
- Access sample responses
- Track results and progress

Find all this and MORE in jacPLUS ▶

Explain and comprehend

1. **Explain** why airborne diseases may be more easily spread in urban areas. Using this explanation, why did governments use lockdowns as a way to control the spread COVID-19?
2. **Explain** how garbage can aid the spread of vector-borne disease.
3. **Identify** reasons that rubbish dumps such as 'Smokey Mountain' develop.

Analyse and apply

4. Read the Jakarta case study and answer the following.
   a. **Define** the term subsidence in your own words.
   b. **Describe** an aquifer.
   c. In a paragraph, **explain** why extracting water from aquifers made areas of Jakarta more prone to flooding.
5. Jakarta's reliance on septic tanks poses a risk of contamination for the city's water supplies. Write a short explanation of why this is the case.

6. Examine table 4.6.

TABLE 4.6 Jakarta special capital territory (660 km^2)

Year	Population ('000)	Mean population density ('000/km^2)
1951	2012	3.0
1961	2973	4.5
1971	4579	6.9
1981	6503	9.9
1991	8259	12.5
2000	8389	12.7
2010	9588	14.5
2015	10 323	15.6
2020	10 770	16.3

Analyse the impact population and population density have on rates of water extraction in Jakarta.

Propose and communicate

7. **Suggest** how each of the following measures could combat Jakarta's land subsidence and flooding problems.
   a. Land use planning
   b. Regulation of groundwater extraction
   c. Sea wall construction
   d. Flood management control
   e. Seawater intrusion control (ways to stop seawater entering aquifers)

Sample responses are available in your digital formats.

LESSON
4.6 Housing people in megacities

LEARNING INTENTION

By the end of this lesson you should be able to explain the impacts of urbanisation and the growth of megacities on human wellbeing and environments using various forms of data and information and spatial technologies, considering risks and opportunities posed by settlement infrastructure and land tenure, and explain the challenges for sustainable development facing megacities; for example, housing and land availability.

Source: Geography General Senior Syllabus 2024 © State of Queensland (QCAA) 2024; licensed under CC BY 4.0.

4.6.1 Informal settlement

For decades, the pull factor of a perceived better life in the city has led many people to leave rural areas. Push factors, such as extreme poverty or lack of work opportunities, can lead rural migrants to see the urban lifestyle as their only means of survival. In some countries, urban life becomes attractive because residents benefit from the competition between businesses and from government assistance. For instance, people in Mexico City receive cheaper education, health care and transport than their rural counterparts. It is, therefore, little wonder that thousands of people from rural settlements have migrated to the fringes of Mexico City to improve their standard of living.

FIGURE 4.46 A favela in Morumbi, on the outskirts of São Paulo

Unfortunately, the hopes of such migrants are not always realised. Often urbanisation is stimulated by industrialisation and economic development; however, these processes can also lead to a phenomenon known as **pseudo-urbanisation**, a type of urbanisation without economic growth that results in large numbers of poverty-stricken residents living in informal settlements. People are forced to live in slum environments, squatter areas and shanty towns, usually found on the outskirts of the city. These environments are overcrowded, and residents lack access to safe drinking water and sanitation. Residents are also subject to insecure **tenure** because the migrants illegally occupy unused land. Globally, about 1 billion people live in slums or squatter settlements, and experts estimate that around 2 billion people will live in urban slums by 2030.

Squatter settlements are also often in areas with high environmental risks, such as on mountainsides prone to landslides or in the vicinity of waste dumps. All of this can negatively affect wellbeing and safety, exacerbating mental health disorders, including depression and anxiety, or leading to greater violence and crime. The UN-Habitat estimates that over half of Africa's urban population lives in slums and faces the aforementioned challenges of poor health and crime. In preparation for the 2016 Olympic Games in Rio de Janeiro, a policy of police occupation of the *favelas* commenced in 2008. The aim was to rid the *favelas* of violent drug traffickers.

pseudo-urbanisation a type of urbanisation without economic growth that results in large numbers of poverty-stricken residents living in informal settlements

tenure the occupancy or lease of an area of land

CASE STUDY: Dharavi, India

Quick facts: Dharavi
- **Located** in Mumbai, India
- **Population** 1 million (approx.)
- **Literacy rate** 69 per cent
- **Toilet availability** 1 per every 1400 people

Mumbai, formerly known as Bombay, is West India's commercial and financial capital. The port city continues to attract migrants from the rural hinterlands of central India. With high birth rates and a large influx of migrants, Mumbai's population continues to grow rapidly. In addition to this, it is estimated that 49 per cent of all urban dwellers in India live in slums.

In the middle of Mumbai is Dharavi, Asia's largest slum. Covering approximately 240 hectares, it is home to more than 1 million people, many of whom are second-generation residents whose parents moved there years ago.

Originally a small fishing village in the mangroves, Dharavi is now a sprawling economic powerhouse of small-scale businesses and light industry. The 'city within a city' produces embroidered garments, leather goods, pottery and plastic, which are sold both locally and on international markets. Annual turnover is up to US$1 billion.

The slums have a severe shortage of living accommodation. Almost none of the people who live in Dharavi own the land, but many own their homes and businesses, some of which they rent out. Rents are very low compared with nearby Mumbai. Most houses have electricity, and some have running water, but there is little other infrastructure.

A shortage of facilities means existing ones must be shared. On average, residents have access to only one toilet for every 1400 people, 78 per cent of which do not have running water. The squalor of open sewer drains and piles of garbage make the slums a haven for disease, particularly during the monsoon season. Despite the poverty, Dharavi is the most literate area of India with a literacy rate of 69 per cent. Ironically, crime is much lower than in other urban parts of the country.

FIGURE 4.47 Dharavi, Asia's second largest slum, is located in central Mumbai and is home to more than 1 million people.

FIGURE 4.48 Shanty houses in Dharavi

Despite providing some economic and social benefits to residents, the Indian government has often seen the area as a social embarrassment to progress. Plans to renew and improve the living standards of slum dwellers have been proposed since 2004, and in 2023 Adani Realty won the bid to redevelop Dharavi at an estimated cost of US$3 billion.

 Resources

 Weblinks 'Slum' is a loaded term. They are homegrown neighbourhoods.
Adani plans Dharavi redevelopment

SKILLS ACTIVITY

Go to the Resources tab to find a downloadable map of Jakarta's districts.
1. Using the data in table 4.7, construct a choropleth map to show the population density across Jakarta. The number of each district in the table corresponds to the numbers on the map.
2. Describe the distribution of high population density across Jakarta.
3. Research the location of slum settlements in Jakarta and mark them on your map. Discuss how well this information correlates with your population density map.

TABLE 4.7 Districts of Jakarta, Indonesia and their population density (people per square km)

District	Population density	District	Population density	District	Population density
35. Cakung	13 381	25. Kebayoran Baru	10 797	29. Pancoran	19 766
20. Cempaka Putih	21 119	24. Kebayoran Lama	18 451	31. Pasar Minggu	14 030
9. Cengkareng	22 031	11. Kebon Jeruk	20 347	40. Pasar Rebo	17 364
30. Cilandak	11 101	6. Kelapa Gading	9 249	1. Penjaringan	6 927
5. Cilincing	10 991	16. Kemayoran	35 522	23. Pesanggrahan	19 051
42. Cipayung	10 398	10. Kembangan	12 851	34. Pulo Gadung	18 119
41. Ciracas	18 863	4. Koja	27 566	15. Sawah Besar	20 584
37. Duren Sawit	18 421	38. Kramat Jati	23 175	18. Senen	29 715
17. Gambir	12 532	39. Makasar	9 611	26. Setiabudi	12 136
12. Grogol Petamburan	23 829	28. Mampang Prapatan	18 810	7. Taman Sari	16 512
32. Jagakarsa	15 416	33. Matraman	36 017	13. Tambora	49 841
36. Jatinegara	29 901	22. Menteng	13 500	21. Tanah Abang	18 863
19. Johar Baru	58 774	2. Pademangan	13 758	3. Tanjung Priok	17 916
8. Kalideres	15 210	14. Palmerah	31 147	27. Tebet	24 496

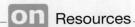 Resources

📄 **Digital doc** Jakarta district map (doc-42477)

4.6.2 Affordable housing in the megacity

Informal settlements are generally less common in megacities in more economically developed countries. While many residents in these megacities may not end up in an informal settlement, rental prices, the size of accommodation and commute times can often still present challenges for them. The closer to the centre of the city, the higher the land value. This pushes up property prices, in particular housing costs, and results in many people having to choose smaller, more affordable properties in the city, or look for accommodation further out in the suburbs. The further away from the centre they live, the cheaper and larger the house; however, with that comes increased travel costs if their place of work is in the middle of the city.

EXAMPLE: Tokyo's micro apartments

For those choosing to live in the middle of the megacities, space is at a premium. This can result in very small living spaces at high prices. In Tokyo, for example, some people rent micro apartments. These contain a toilet, small living space that doubles as the lounge and kitchen, and a loft style bed. This provides residents with somewhere that has the basic facilities at a lower cost than a larger apartment.

FIGURE 4.49 Making the most of small spaces is important in Tokyo.

Micro apartments are designed to maximise every inch of space. However, the lack of personal space and storage can be challenging. The rising popularity of these tiny homes underscores the pressing issue of housing affordability in densely populated urban centres.

With many jobs being based in the centre of the cities, and space to build new housing for the expanding population being on the edge, people often travel in vast numbers daily from the suburban fringes to reach their place of work. The map in figure 4.50 shows the average travel time to work for people in New York City. In the United Kingdom, many towns outside of Greater London are commuter towns, where people live in these towns but travel into the capital every day for work (see figure 4.51). This can result in hours spent commuting, impacting on time spent at home.

Ultimately, megacities are continuing to grow, and the issue of affordable housing is one that will be a top priority for countries such as India. It is also an issue that needs to be addressed worldwide, even in the most developed countries such as Australia.

FIGURE 4.50 Average travel time to work in minutes, New York City, 2019

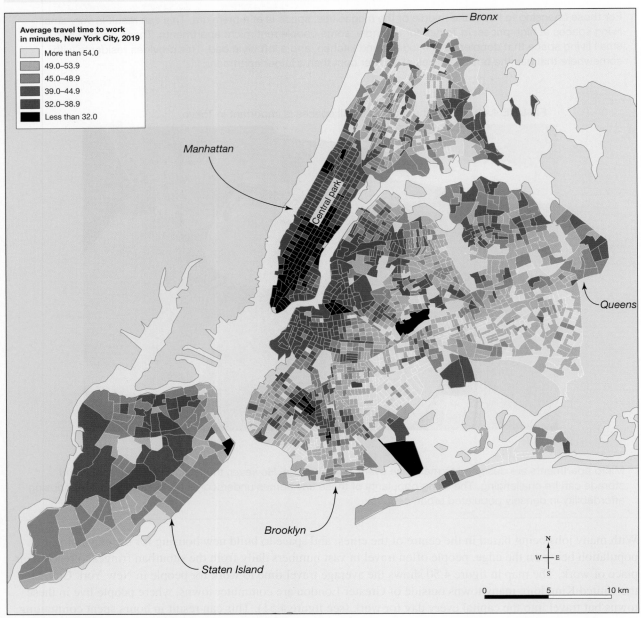

Average travel time to work
in minutes, New York City, 2019

- More than 54.0
- 49.0–53.9
- 45.0–48.9
- 39.0–44.9
- 32.0–38.9
- Less than 32.0

Manhattan

Central park

Bronx

Queens

Brooklyn

Staten Island

0   5   10 km

Source: U.S. Department of Transportation, Bureau of Transportation Statistics, National Transportation Atlas Database; U.S. Census Bureau, American Community Survey 2019 – Table B08135: Aggregate Travel Time to Work; OpenStreetMap. Map redrawn by Spatial Vision.

FIGURE 4.51 Percentage of workers in each Middle Super Output Area (MSOA) who commute to a MSOA in Greater London

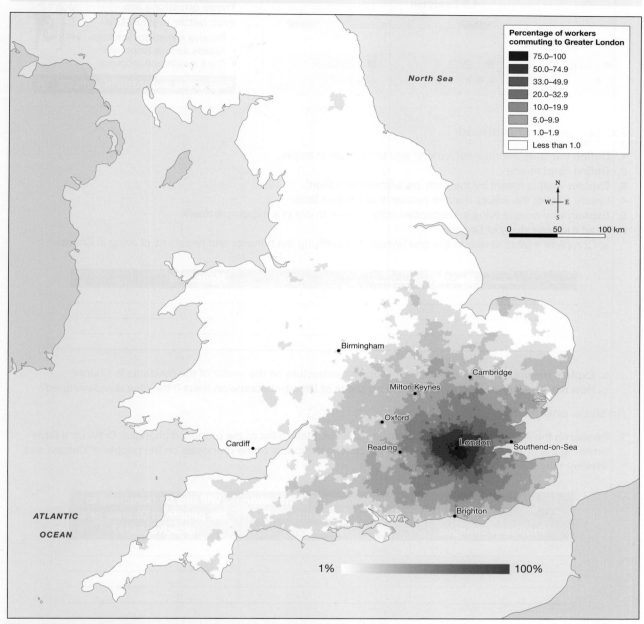

Source: Data from Office for National Statistics. Map drawn by Spatial Vision.

4.6 Exercise

Explain and comprehend

1. **Define** the term 'informal settlement' and provide an example.
2. **Define** 'land tenure'.
3. **Explain** what is meant by the term 'pseudo-urbanisation'.
4. Create a list of the issues that the residents of Dharavi face.
5. **Explain** why people living in megacities may choose to live in a micro apartment.
6. Read the case study of Dharavi.
   a. Complete a table similar to the one provided, identifying the benefits and negatives of living in Dharavi.

Benefits	Negatives

   b. **Explain** the impacts of the lack of sanitation infrastructure on the health of the residents in Dharavi.
   c. How may the lack of land tenure by the residents of Dharavi become an issue if the area is redeveloped?

Analyse and apply

7. **Research** the Adani redevelopment plans for Dharavi. Make a list of the proposed changes. Draw up a table similar to the one provided and decide on the potential impacts of the proposals on the people of Dharavi and whether they will be beneficial or not.

Proposed changes	Environmental, social and economic impacts on the people	Will this be beneficial to the people of Dharavi or a disadvantage?

8. **Discuss** the benefits and disadvantages of living further away from your work place.

Sample responses are available in your digital formats.

LESSON
4.7 Social opportunities and challenges in megacities

People are attracted to megacities for the numerous social and economic opportunities they offer, such as employment, education and health care. However, cities in the developing world lack the infrastructure and capacity to deal with a sustained and large influx of people. This results in many challenges, and in particular the development of shanty towns or slums on the outskirts of major cities. This lesson focuses on some of the social opportunities and challenges people find in megacities. Goal 11 of the UN's Sustainable Development Goals aims to make cities and human settlements inclusive, safe, resilient and sustainable. Figure 4.52 highlights some of the issues related to this goal.

4.7.1 Health challenges of living in the city

The urban poor generally experience poorer health due to a lack of access to clean water, fresh healthy food, and access to or the income to pay for health care and medicine when they fall ill. Crowded living conditions with inefficient or non-existent rubbish collection and sanitation also provide the perfect conditions for the spread of diseases, such as cholera, and parasitic infections, such as worms and diarrhoea. (Figure 4.53 shows some of the health conditions present in Kallyanpur slum in Dhaka, Bangladesh, for example.) These diseases are treatable, but without access to health care or the money to pay for medicine they can be fatal.

FIGURE 4.52 Sustainable Development Goal 11

11 SUSTAINABLE CITIES AND COMMUNITIES

MAKE CITIES AND HUMAN SETTLEMENTS INCLUSIVE, SAFE, RESILIENT AND SUSTAINABLE

SLUMS ON THE RISE

1.1 BILLION URBAN RESIDENTS ARE LIVING IN SLUMS (2020)

2 BILLION MORE ARE EXPECTED IN THE NEXT 30 YEARS

GLOBALLY, ONLY **ONE IN TWO** URBAN RESIDENTS HAVE CONVENIENT ACCESS TO **PUBLIC TRANSPORT** (2022)

AIR POLLUTION IS NO LONGER AN EXCLUSIVELY URBAN PROBLEM

TOWNS EXPERIENCE **POORER AIR QUALITY** THAN CITIES IN EASTERN AND SOUTH-EASTERN ASIA (2019)

GLOBALLY, **3 IN 4 CITIES** HAVE **LESS THAN 20%** OF THEIR AREA DEDICATED TO **PUBLIC SPACES AND STREETS** MUCH LOWER THAN THE TARGET OF 45-50% (2020)

IN THE DEVELOPING WORLD **1 BILLION PEOPLE LACK ACCESS** TO **ALL-WEATHER ROADS** (2022)

THE SUSTAINABLE DEVELOPMENT GOALS REPORT 2023: SPECIAL EDITION- UNSTATS.UN.ORG/SDGS/REPORT/2023/

Source: United Nations, Department of Economic and Social Affairs, Population Division (2019). World Urbanization Prospects: The 2018 Revision (ST/ESA/SER.A/420). New York: United Nations.

Some examples of health issues faced in megacities include the following:
- Air quality in megacities is often low, resulting in an increase in respiratory diseases such as asthma (see figure 4.54).
- The urban heat island effect can add to heat-related illnesses.
- Airborne viruses such as influenza and COVID-19 are also more easily spread in crowded urban areas (refer to lesson 4.5).

- Recent studies by the African Population and Health Research Centre found that people who live in urban slums also have higher child mortality and undernutrition rates, meaning babies and children growing up in these environments are more susceptible to other illnesses and are more likely to have stunted growth and development.

However, many people choose to move to the city in the knowledge they will likely have better access to, and possibly more affordable, health care.

FIGURE 4.53 Survey of health conditions in Kallyanpur slum, Dhaka

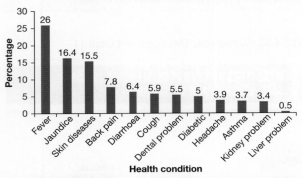

Source: Based on data from Latif, Mahmuda & Irin, Anjuman & Ferdaus, Jannatul. (2016). Socio-economic and health status of slum dwellers of the Kalyanpur slum in Dhaka city. Bangladesh Journal of Scientific Research. 29. 73. 10.3329/bjsr.v29i1.29760

FIGURE 4.54 Top 10 cities of air-pollution-related asthma

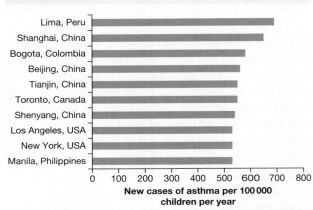

Source: Data based on Pattanun Achakulwisut, Michael Brauer, Perry Hystad, Susan C Anenberg. Global, national, and urban burdens of paediatric asthma incidence attributable to ambient NO_2 pollution: estimates from global datasets. Lancet Planet Health 2019; 3: e166–78 Published Online April 10, 2019 http://dx.doi.org/10.1016/ S2542-5196(19)30046-4.

4.7.2 Education in megacities

Education is another pull factor of megacities. Areas closer to the city centres, with their higher land values and well-established services and infrastructure, often attract residents on higher incomes. Those on lower incomes tend to live further out, perhaps in slum areas or the suburbs. This often creates social segregation between low and middle–high income earners within cities. This, in turn, can be reflected in the services provided to each area, with some areas better able to gather more revenue from taxes and council rates than others. In countries such as the United States, in which a significant proportion of the funding for public education comes from local taxes or funding, school districts with a high proportion of high-income families collect more revenue. This has a direct impact on standards of educational facilities.

FIGURE 4.55 The amount of funding schools receive is often unequal and based on location.

Rural areas with lower populations often have fewer educational opportunities, particularly in developing countries where less government expenditure goes to school facilities. In developed countries, many young people move from rural areas to large cities for tertiary education that is unavailable in the rural areas. Figure 4.56 shows that between 2000 and 2010 the overall education levels of the Chinese population improved; however, it also highlights the inequalities between rural and urban education levels.

FIGURE 4.56 Composition of residents' qualifications in rural and urban areas, China, 2000 and 2010

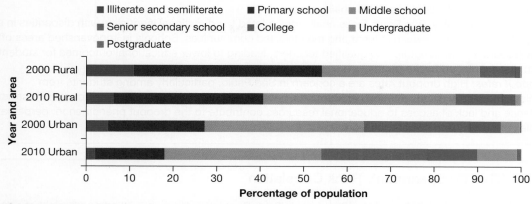

Legend:
- ■ Illiterate and semiliterate
- ■ Primary school
- ■ Middle school
- ■ Senior secondary school
- ■ College
- ■ Undergraduate
- ■ Postgraduate

Y-axis (Year and area): 2000 Rural, 2010 Rural, 2000 Urban, 2010 Urban

X-axis (Percentage of population): 0, 10, 20, 30, 40, 50, 60, 70, 80, 90, 100

Source: Xiang, L., Stillwell, J. Rural–Urban Educational Inequalities and Their Spatial Variations in China. Appl. Spatial Analysis 16, 873–896 (2023). https://doi.org/10.1007/s12061-023-09506-1

CASE STUDY: Health and education in São Paulo, Brazil

Quick facts: São Paulo
- **Location** Brazil
- **Population** 22 million (metro)
- **GDP per capita** US$16 000

São Paulo is one of two megacities in Brazil and has experienced significant urbanisation and population growth over the past century, evolving from a regional centre to a global megacity. With over 12 million inhabitants within the city limits and more than 21 million in the metropolitan area, São Paulo faces numerous challenges related to food, health and education.

FIGURE 4.57 São Paulo, Brazil

Food security

Situated in the tropics, São Paulo is in a region with high agricultural output and should be able to ensure food security for all its people. However, in 2021 experts estimated that 36 per cent of Brazilians were at risk of being unable to afford food for themselves or their families, mostly as a result of food price increases.

Health challenges

São Paulo faces the following challenges relating to health care:

- *Health care access:* While São Paulo boasts world-class medical facilities, access to health care is unequal across the city. Residents of impoverished areas often face barriers such as long distances to healthcare facilities, financial constraints and inadequate infrastructure. This disparity in access contributes to disparities in health outcomes and exacerbates existing social inequalities.
- *Air pollution:* São Paulo struggles with severe air pollution, primarily due to high levels of vehicle emissions and industrial activity. Poor air quality poses significant health risks, contributing to respiratory diseases, cardiovascular problems, and other health issues among residents. Vulnerable populations, such as children and the elderly, are particularly affected by the adverse effects of air pollution.

FIGURE 4.58 A thick layer of air pollution over the city of Sao Paulo, Brazil

Education challenges

São Paulo faces the following education challenges:

- *Educational inequality:* São Paulo faces challenges related to educational inequality, with disparities in access to quality education between affluent and marginalised communities. Schools in impoverished areas often lack resources, infrastructure and qualified teachers, leading to lower educational outcomes for students from disadvantaged backgrounds.
- *Dropout rates:* High dropout rates are a concern in São Paulo, particularly among students from socioeconomically disadvantaged backgrounds. Factors such as poverty, inadequate school infrastructure, violence and lack of access to educational resources contribute to the dropout phenomenon, perpetuating cycles of poverty and social exclusion.

SKILLS ACTIVITY: Spearman's Rank Correlation

1. Using data from table 4.8, complete a Spearman's Rank Correlation to examine the hypothesis that a link exists between urbanisation rate and number of people with a university education. See the Resources tab for a guide on completing a Spearman's Rank Correlation.

TABLE 4.8 Urbanisation rate and population with university education, China

Region	Estimated urbanisation rate %, 2020	Population with university education per 100 000 persons by region, 2021
Shanghai	87	33 872
Beijing	85	41 980
Tianjin	81	26 940
Guangdong	76	15 699
Zhejiang	71	16 990
Jiangsu	70	18 663
Fujian	68	14 148
Liaoning	64	18 216
Chongqing	63	15 412
Inner Mongolia	60	18 688
Shandong	58	14 384
Hainan	56	13 919
Heilongjiang	55	14 793
Shanxi	55	17 358
Hubei	54	15 502
Jilin	53	16 738
Hunan	52	12 239
Shaanxi	52	18 397
Jiangxi	50	11 897
Anhui	50	13 280
Ningxia	50	17 340
Hebei	49	12 418
Guangxi	48	10 806
Sichuan	48	13 267
Henan	46	11 744
Qinghai	46	14 880
Yunnan	42	11 601

(continued)

TABLE 4.8 Urbanisation rate and population with university education, China (continued)

Xinjiang	39	16 536
Guizhou	38	10 952
Gansu	38	14 506
Tibet	26	11 019

Sources: https://www.researchgate.net/publication/317258928_New-type_urbanization_in_China_Predicted_trends_and_investment_demand_for_2015-2030
https://www.stats.gov.cn/english/PressRelease/202105/t20210510_1817191.html#:~:text=Among%20the%20national%20population%5B2,and%20349658828%20persons%20with%20primary

4.7.3 Food security in megacities

Globalisation of food production has revolutionised the food industry but is only part of the solution to the problem of providing food security for all the world's urban populations. As countries become increasingly urbanised, their land use patterns, infrastructure and economies change. Each of these factors then affect the production and availability of food. This challenge presents in different ways in developed and developing nations.

TABLE 4.9 Food security challenges for developing and developed countries

Challenges	FOOD SECURITY IN MEGACITIES	Opportunities
Less economically developed countries		**Less economically developed countries**
Developing countries face two challenges in particular: 1. providing adequate infrastructure to connect producers with consumers efficiently 2. providing food at reasonable prices. People living in slum areas often have very limited access to fresh fruit and vegetables. Consequently, their food supply mainly consists of cheap foods with low nutritional value. Where fresh fruit and vegetables are available, they are often unaffordable, with the poorer urban households spending more than half their income on food. Therefore, they are highly vulnerable to increases in food prices.	**FIGURE 4.59** Street vendor in India **FIGURE 4.60** Hanoi suburbs, Vietnam 	Urban agriculture has the capacity to overcome these challenges by providing a secure source of nutritional food for the urban poor. Its role in developing countries in particular is important. For example, in Hanoi, Vietnam, 80 per cent of fresh vegetables and more than a third of the city's egg supply is produced by urban and peri-urban (between built-up suburbs and rural landscapes) farms.

(continued)

TABLE 4.9 Food security challenges for developing and developed countries *(continued)*

Challenges	FOOD SECURITY IN MEGACITIES	Opportunities
More economically developed countries		**More economically developed countries**
In Australia, the food bowls adjacent to both Melbourne and Sydney are threatened by urban sprawl. The loss of adjacent farmland to accommodate Melbourne's expansion, for example, means that the food bowl's ability to meet the city's food requirements is likely to drop below 20 per cent by 2051. This will increase the cost of perishable food, because having food production and processing close to market has significant cost savings for both producers and consumers. The reliance on imported food became an issue during the COVID-19 pandemic as transport routes were disrupted, leading to food shortages in many places across the world.	**FIGURE 4.61** Supermarket shelves in London during the COVID-19 pandemic  **FIGURE 4.62** Hydroponic vertical farm 	Developed countries can approach this issue in different ways. In 1999, the idea of a 'vertical farm' as a way for densely populated areas to feed themselves was launched in the United States. The idea was that fruit and vegetables could be supplied fresher, and also that these vertical farms would have a smaller footprint compared to the current system of importing vegetables into cities.

 Resources

📄 **Digital doc** How to complete a Spearman's Rank Correlation (doc-42476)

EXAMPLE: Metro farms in Seoul

While many countries are investing in vertical farms in urban areas, in Seoul, South Korea, they are also farming underground. A company called Farm8 rents space in subway stations in the city's metro and has set up vertical farm systems to provide 30 kilograms of fresh vegetables a day to the city above.

The proximity of these farms to consumers reduces the carbon footprint associated with food transportation, contributing to a more sustainable urban food system. Additionally, this method of farming helps to improve food security and resilience in the face of climate change and urbanisation. These farms highlight the potential for creative solutions to urban food production challenges, particularly in cities that can afford such solutions.

FIGURE 4.63 Metro farm, Seoul

4.7 Exercise

Explain and comprehend

1. **Identify** the challenges that living in a megacity may present to residents' health.
2. **Explain** how urban sprawl could lead to food insecurity in megacities.
3. **Explain** how vertical farming could help to reduce food insecurity in urban areas.
4. With reference to figure 4.56, **explain** how living in a megacity in China could give you better education opportunities.
5. Using data from figure 4.64, **explain** why people may choose to live in an urban area if they have a medical condition.

FIGURE 4.64 Number of practising physicians per thousand people, China

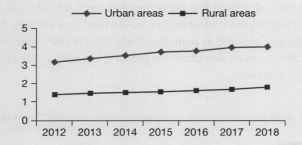

6. Using table 4.10, **describe** the pattern of farming in Melbourne's food bowl.

TABLE 4.10 Food types produced in Melbourne's food bowl

Food type	Proportion of Victoria's production occurring in Melbourne's food bowl
Dairy	12%
Sugar	0%
Fruit	8%
Oil crops	7%
Cereal grains	3%
Vegetables	47%
Red meat	15%
Chicken meat	81%
Rice	0%
Legumes	1%
Eggs	67%

Source: Sheridan, J., Larsen, K. and Carey, R. 2015 Melbourne's food bowl: Now and at seven million. Victorian Eco-Innovation Lab, The University of Melbourne

Sample responses are available in your digital formats.

LESSON
4.8 Economic opportunities and challenges in megacities

4.8.1 Jobs in the city

Many people who choose to move to cities do so in the hope of better employment opportunities. Jobs in the city tend to be more varied and offer higher wages than those in rural areas. Highly skilled jobs are also more often located in industries in urbanised areas (see figure 4.65). However, the jobs that migrants are eventually employed in often depends on the education and skills they have and the jobs that are on offer. As seen in figure 4.66, London and the areas around it offer higher wages than less urbanised areas in England. For some large cities such as London, people are able to commute from outside the wider metropolitan area due to public transport options. This eases the strain on resources for the city, but adds to the congestion of travellers, and is reliant on a sustainable transport network.

FIGURE 4.65 Distribution of skills jobs in London compared to the whole of Great Britain

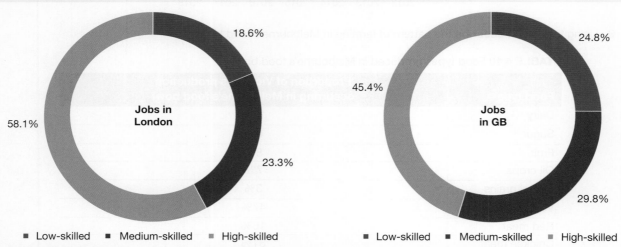

Source: The World Factbook. Central Intelligence Agency. Retrieved from https://www.cia.gov/the-world-factbook/field/gdp-composition-by-sector-of-origin/

FIGURE 4.66 Median weekly earnings across England and Wales

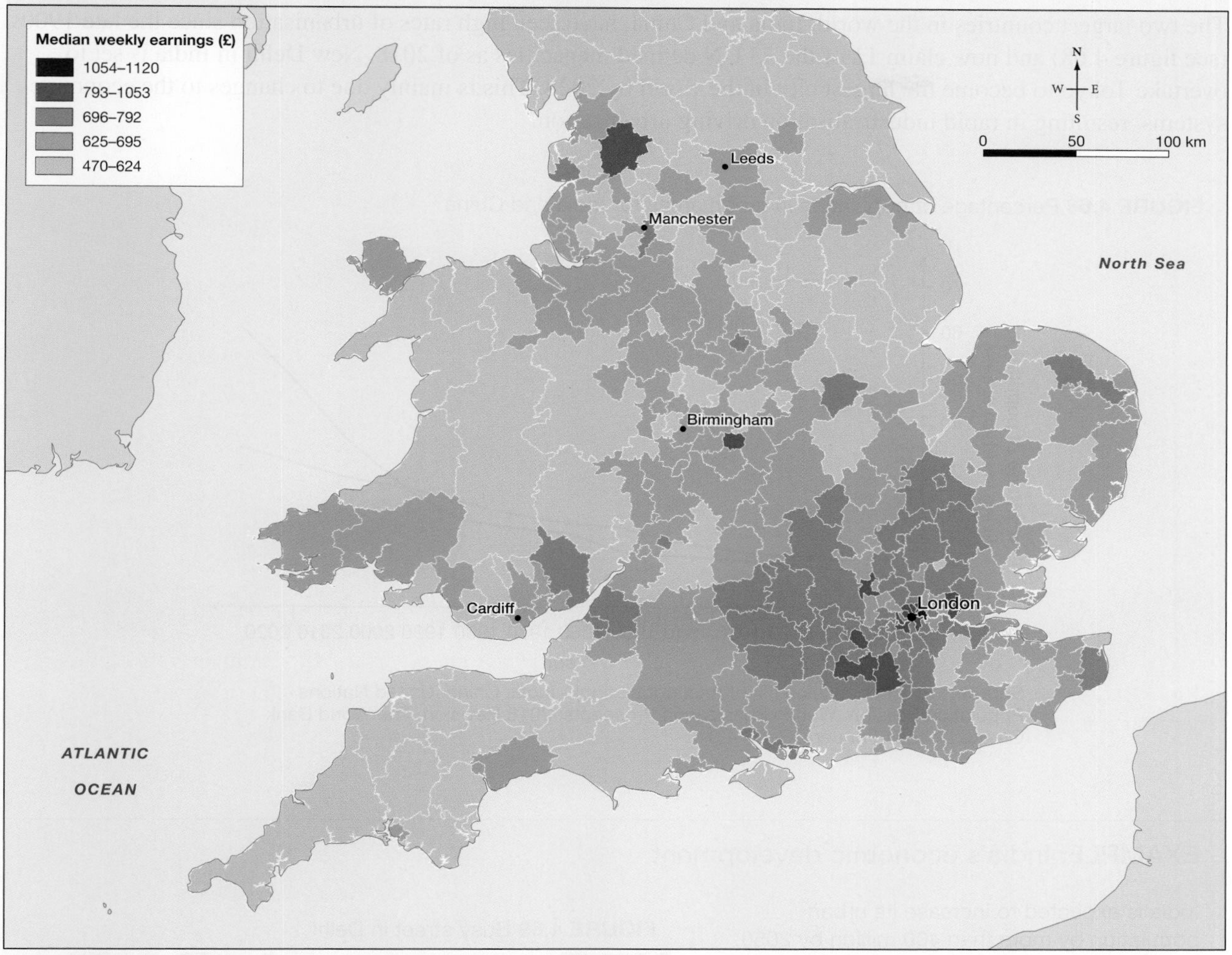

Median weekly earnings (£)
- 1054–1120
- 793–1053
- 696–792
- 625–695
- 470–624

North Sea

ATLANTIC OCEAN

Leeds

Manchester

Birmingham

Cardiff

London

0   50   100 km

Source: Annual Survey of Hours and Earnings from the Office for National Statistics (ONS), 2023. Map redrawn by Spatial Vision.

People in the slum areas of megacities often find employment in the informal sector. A study of Kallyanpur slum in Dhaka, for example, showed that nearly 25 per cent of the residents were involved in the garment industry, and nearly 20 per cent employed as rickshaw pullers (see figure 4.67). This work may be unregulated, and at times inconsistent, but it usually pays more than farming in more rural locations.

FIGURE 4.67 Nature of work for residents in the Kallyanpur slum, Dhaka

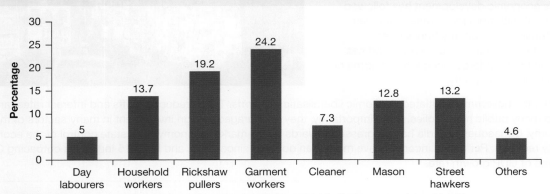

Source: Latif, Mahmuda & Irin, Anjuman & Ferdaus, Jannatul. (2016). Socio-economic and health status of slum dwellers of the Kalyanpur slum in Dhaka city. Bangladesh Journal of Scientific Research. 29. 73. 10.3329/bjsr.v29i1.29760

4.8.2 Urban growth and economic change

The two largest countries in the world, India and China, have seen high rates of urbanisation since the late 1900s (see figure 4.68) and now claim 11 of the 33 UN defined megacities as of 2018. New Delhi in India is set to overtake Tokyo to become the largest city in the world by 2028. This is mainly due to changes to the economic systems, resulting in rapid industrialisation driving urbanisation.

FIGURE 4.68 Percentage of population in urban areas for India and China

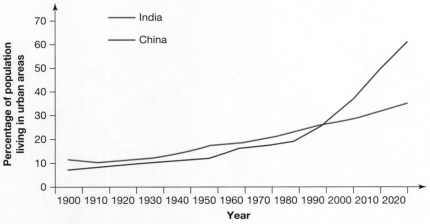

Source: Urban population (% of total population) - India, China. United Nations Population Division. World Urbanization Prospects: 2018 Revision. The World Bank Group. Licensed under CC-BY 4.0

EXAMPLE: India's economic development

India is expected to increase its urban population by more than 400 million by 2050. The annual growth in the urban population in India between 2010 and 2015 was the highest among the major economies. As in China, migrants from the rural areas make up a sizeable proportion of the urban population. The last census revealed that over one-third of the urban population comprises migrants from India's rural areas.

India's economic development has followed a different path from many other countries. Structurally, the economy transitioned quickly from agriculture to services and has become the third-largest economy in terms of purchasing power parity.

FIGURE 4.69 Busy street in Delhi

In 1991, the government initiated economic liberalisation reforms. These reduced tariffs and interest rates, and ended many public monopolies. More importantly, they encouraged foreign investment in many sectors of the economy. Consequently, India has progressed towards a free-market economy, with state control of the economy greatly reduced. Per capita incomes have more than doubled since 2005, and by 2015 India was outpacing China in terms of GDP growth rate.

CASE STUDY: China's economic development

Quick facts: China
- **Capital city** Beijing
- **Population** 1.4 billion
- **GDP per capita** US$12 720
- **Number of megacities** 17

The late 1970s marked a major turning point in the direction of China's economic and subsequent urban development. Some of the main changes implemented include the following.

- *1978:* Deng Xiaoping became premier and immediately launched a series of reforms aimed at transforming China into a more market-based economy. These reforms included creating Special Economic Zones (SEZ) in south-eastern coastal China.

- *1980:* Four SEZs were designated around four small cities in the Guangdong and Fujian provinces. One of these cities, Shenzhen, was a small fishing town but is now a hugely successful trade hub and manufacturing centre with a population of more than 12 million.

FIGURE 4.70 Shanghai, with a population of 26.8 million, is the largest city in China and the third biggest in the world.

- *1984:* Economic and Technological Development Zones (ETDZs) were created as foreign trade areas in cities such as Shanghai. These zones had a focus on specific industries, such as high-tech research and development.

- *2013:* Shanghai was one of four cities given Free Trade Zone (FTZ) status, meaning it benefited from faster foreign investments approvals and more relaxed trade regulations.

- *2017:* Seven more FTZ zones were approved but, unlike the previous ones, were located in underdeveloped areas in China's interior to boost regional economies.

  These economic initiatives inevitably had a huge impact on China's rate of urbanisation, resulting in the migration of 500 million rural Chinese people into the cities since 1980. With so many people, and therefore potential consumers, now living in urban areas, a significant shift has now occurred towards a consumption-driven economy.

FIGURE 4.71 The growth in GDP of eight Chinese cities

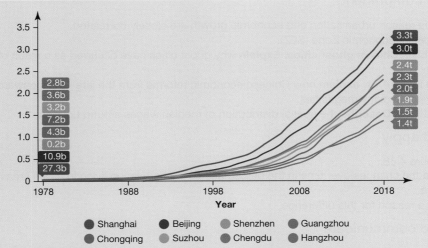

Shanghai
Beijing
Shenzhen
Guangzhou
Chongqing
Suzhou
Chengdu
Hangzhou

Source: © StatistaCharts. Data sourced from Channel Wu (Wu Xiaobo), CEIC via Caixin, Open Environmental Journal (Research Gate). Licensed under CC-BY
Note: b = billions; t = trillions.

On Resources

Digital docs GDP of provinces in China table template (doc-30358)
Blank map of China (doc-29173)

Weblinks China's ghost cities
20-Year Urbanization of Shenzhen, China

SKILLS ACTIVITY: Using spatial technologies

Visit the **ArcGIS, Rio de Janeiro: A Story Map weblink**. After reading the story and viewing the maps, answer the following questions.
1. Compare the distribution of elderly residents to young residents of Rio de Janeiro.
2. Explain why certain parts of the city are more vulnerable to the threat of landslides.
3. Outline a reason why areas of Rio that are vulnerable to sea-level rise are not necessarily the same as those that are susceptible to landslides.
4. Identify the likelihood of dengue, chikungunya and zika viruses spreading in Rio.
5. Explain what might happen to this likelihood of virus spread if Rio de Janeiro was located further south.

On Resources

Weblink ArcGIS, Rio de Janeiro: A Story Map

4.8 Exercise

4.8 Exercise

Learning pathways

■ LEVEL 1	■ LEVEL 2	■ LEVEL 3
2, 6	1, 5, 7	3, 4

These questions are even better in jacPLUS!
- Receive immediate feedback
- Access sample responses
- Track results and progress

Find all this and MORE in jacPLUS ▶

Explain and comprehend

1. **Suggest** one reason urbanisation and economic growth are closely correlated.
2. **Define** a Special Economic Zone (SEZ).
3. Visit the weblink **China's ghost cities. Explain** why ghost cities have occurred as a result of China's urbanisation.
4. Referring to figure 4.71, **explain** how Chinese economic reforms from the late 1970s onwards have led to the country having so many megacities.
5. Study figure 4.66. **Describe** the spatial distribution of median wages around London.

Analyse and apply

6. Examine figure 4.68.
   a. **Compare** the changes over time in the percentage of the population living in urban areas for India and China.
   b. **Propose** a reason for this difference.

Propose and communicate

7. **Hypothesise** what is likely to happen to India's urbanisation rate and economic growth by 2050. Provide reasons for this hypothesis.

Sample responses are available in your digital formats.

LESSON
4.9 Sustainable development in megacities

LEARNING INTENTION

By the end of this lesson you should be able to:
- explain how urban planning can be used in the development of resilient cities to mitigate and adapt to the future impacts of urbanisation
- explain the impacts of urbanisation and the growth of megacities on human wellbeing and environments using various forms of data and information and spatial technologies, considering risks and opportunities posed by changing land.

Source: Geography General Senior Syllabus 2024 © State of Queensland (QCAA) 2024; licensed under CC BY 4.0.

4.9.1 Challenges and opportunities for developed countries

Waste management and sanitation

Large populations produce large amounts of waste that need to be processed and disposed of somewhere, but cities don't have the extensive tracts of land to do this. So where does the waste go? Some developed countries, such as Australia, export some types of waste to other countries. China and India, for example, have at different times received Australian recycling for processing. This may be a 'sustainable' option for the developed nation's environment, but not for the developing nations, which may not have the means of processing the waste in a sustainable or safe way.

CASE STUDY: Waste disposal in New York City

Quick facts: New York City
- **Location** United States
- **Population** 20 million (metro)
- **Annual waste disposal budget** US$1.5 billion

The population of New York City produces around 12 000 tonnes of rubbish a day (with only about 2000 of that recyclables). The annual budget for waste removal in New York is approximately US$1.5 billion. In addition to this, city authorities need to manage street sweeping, and general cleaning and maintenance of public facilities and streets. Rubbish collected in New York is managed in a number of ways:
- shipped or trucked about 1000 kilometres to one of the state's landfills (80 per cent of mixed residential waste)
- incinerated for fuel (20 per cent of mixed residential waste)
- recycled, either locally (mostly paper and metals) or overseas (mostly plastics).

FIGURE 4.72 Rubbish on the side of the road in New York City

New York has implemented policies to help manage this volume of waste in a more sustainable way. In 1989, they began a compulsory recycling program, which runs as part of a broader plan that sees waste separated and managed accordingly. Other initiatives implemented as part of the city's waste reduction plan include:

- reduce the percentage of recyclables that are sent to landfill (the approximate capture rate for common recyclables, such as paper and plastic, is currently about half)
- develop household and community composting and separate organic matter collection
- reduce or eliminate non-recyclable material, such as single-use plastic bags and styrofoam food containers
- implement separate disposal programs for electronics that have reusable and recyclable component parts
- encourage recycling and reuse of clothing and consumer goods
- provide incentives for businesses to improve their waste management processes
- implement school programs to change waste disposal habits of the next generation, including a 'zero waste' scheme.

 Resources

 Weblinks Waste in New York City
Waste in New York State

Infrastructure

Poor infrastructure also negatively affects the sustainability and liveability of cities. In some instances, land on the rural-urban fringe of developed nations' megacities is sold off to create new housing and industrial estates. As cities move outward, they often expand at a faster rate than infrastructure. These factors can have a flow-on social effects, combining to create social segregation between low- and middle-high income earners.

Figure 4.74 shows some of the impacts of living in closer to the CBD or living in the suburbs.

FIGURE 4.73 Low-density housing on the outer limits of Melbourne.

FIGURE 4.74 Social and economics impacts of living closer to or further from the CBD

Advantages

Close to the city

Access to a wide variety of services:
- medical facilities
- higher education
- specialist shops
- public transport
Short commute

Suburbs

More green space
Better air quality

Better sense of community

More affordable housing, larger houses

Disadvantages

More expensive housing

Traffic congestion
Higher crime rates

Long commute times to jobs, education or medical care; increased car reliance and air pollution

Urban sprawl; loss of agricultural land

Possible socio-spatial segregation due to different income levels

Services yet to be provided:
- medical facilities
- public transport
- schools
- shopping complexes

Distance from the CBD

The development of digital communications — for example, the rise of e-commerce — also presents opportunities and challenges for large urban populations as the traditional ways of earning an income, such as manufacturing, trade and service industries, change. During the COVID-19 pandemic (refer to lesson 4.5) many people had to work from home, which was possible due to ICTs, and this has changed the perception of the need to travel daily to work in offices in the city centre.

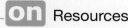 Resources

Weblinks Health impacts of suburban sprawl
Japan's CO_2 emissions hit second-highest on record

Density

When carefully planned and well-managed, high-density urban areas can improve social, environmental and economic sustainability when compared to low-density areas. Firstly, in densely populated cities, the amount of space required per person is significantly less than in low-density suburban or rural areas. Apartment or town house living allows more people to live on less land. This efficient use of space minimises urban sprawl and reduces the need for land clearance, preserving natural habitats and agricultural land on the urban fringes.

Additionally, high-density urban environments allow for greater accessibility to existing public transportation networks. With more people living in close proximity to each other, public transit systems become more economically sustainable and extensive, providing convenient alternatives to private car travel. Increased usage of public transport helps relieve traffic congestion and reduces greenhouse gas emissions associated with car use, contributing to improved air quality and reducing causes of climate change. Moreover, the compact nature of densely populated cities promotes walkability and cycling infrastructure. Shorter distances between destinations, combined with well-designed pedestrian pathways and dedicated cycling lanes, encourage active modes of transportation. Walking and cycling not only reduce reliance on fossil fuel-powered vehicles but also promote physical activity and improve public health.

FIGURE 4.75 High-density residential buildings in Singapore

CASE STUDY: Opportunities and challenges in Tokyo

Quick facts: Tokyo
- **Location** Japan
- **Population** 40 million (metro)
- **GDP per capita** US$51 000

As one of the world's largest megacities, Tokyo grapples with the challenges and opportunities inherent in rapid urban growth. Tokyo's urban planning framework includes thorough zoning regulations, infrastructure investments and disaster preparedness strategies. For instance, its comprehensive public transportation system reduces reliance on cars, alleviating congestion and carbon emissions. Moreover, Tokyo's green space initiatives, such as rooftop gardens and urban parks, enhance biodiversity, reduce heat island effects and promote community wellbeing.

Tokyo's population density is over 6000 people per square kilometre. This allows public transport systems to be efficient and makes the city more walkable, but also strains

FIGURE 4.76 Shibuya, Tokyo

infrastructure and resources, which can affect residents' quality of life. However, innovative urban planning measures, such as mixed-use development and efficient waste management systems, mitigate these pressures.

Tokyo's limited land availability encourages vertical growth, or increasing density, rather than spreading outwards. This can lead to challenges in preserving green spaces and ecosystems. Yet, Tokyo's commitment to sustainability with initiatives such as the Tokyo 2040 Plan, which promotes compact city development and energy-efficient infrastructure, balances urban growth with environmental conservation.

4.9.2 Challenges and opportunities in developing countries

Population predictions indicate that over the next 25 to 30 years most urban growth will occur in Africa and Asia — historically the least urbanised continents in the world. It is also in these places, where birth rates remain high, that urban growth will continue. It is estimated that:

- Asia will have 30 megacities by 2025 (see figure 4.77).
- Africa will have only three, but will have the fastest rate of population increase.
- Of the world's population growth, 90 per cent will take place in the cities of the developing world.
- The world's largest cities will be in Africa by 2100.
- By 2030, more than 2 billion people worldwide will be living in overcrowded slums.

Many countries in the developing world experience extreme poverty, civil unrest and famine. These push factors tend to drive people to abandon their rural homes and move to cities, lured by the prospect of safety, employment, shelter and an improved standard of living. The issues faced in these nations mean that the quality of life of their many inhabitants is seriously threatened.

FIGURE 4.77 Predicted megacities by 2030

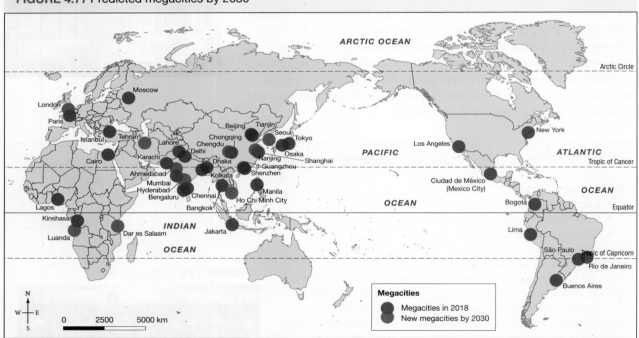

Source: United Nations, Department of Economic and Social Affairs, Population Division (2018). The World's Cities in 2018—Data Booklet (ST/ESA/ SER.A/417).

Moving to an urban environment does, however, offer opportunities that are not available in small rural or regional centres in many developing countries. Access to health care and education might be sporadic or far more difficult in remote areas, and opportunities for work might be rare. In times of drought or civil unrest, aid agencies are more likely to be accessible in urban centres.

Accommodating rapid population growth

Cities in the developing world lack the infrastructure and capacity to deal with a sustained and large influx of people. The result is the development of shanty towns or slums on the outskirts of major cities, where crime rates are high due to a lack of resources and employment.

FIGURE 4.78 A view of Maharashtra, India, one of the largest slums in the world

The rapid rise of shanty towns also presents an issue for health, sanitation, waste disposal and air pollution (refer to lesson 4.5). Without access to power grids or clean fuels, low grade fuels that produce high levels of smoke and toxic gasses are used for cooking, heat and light sources. In 2021, the World Health Organization estimated that in developing countries about 29 per cent of the population rely on solid fuels for cooking.

CASE STUDY: Opportunities and challenges in Mumbai

Quick facts: Mumbai
- **Location** India
- **Population** 18 million (metro)
- **GDP per capita** US$11 000

As one of the world's largest megacities, Mumbai, India, grapples with both challenges and opportunities as it continues to grow. The city's explosive population growth and rapid urbanisation pose significant challenges to its infrastructure, economy and society. Its informal settlements, or slums, are signs of the housing crisis and socioeconomic disparities that impact many urban dwellers. Overcrowding and inadequate access to basic services, such as sanitation and health care, exacerbate health risks and social inequalities. Moreover, Mumbai's coastal location renders it vulnerable to climate change impacts, including sea-level rise and extreme weather events. Addressing these urbanisation challenges requires holistic approaches that integrate urban planning, social welfare programs and environmental resilience strategies.

FIGURE 4.79 Mumbai, India

As well as its challenges, Mumbai also presents opportunities for sustainable development and growth. The city's vibrant economy fuels innovation and investment in various sectors, including technology, finance and entertainment. Furthermore, initiatives such as the Mumbai Metropolitan Region Development Authority (MMRDA) aim to improve transportation infrastructure, enhance connectivity and promote equitable urban development.

Impact on the natural environment

The expansion of a city's urban footprint also makes significant changes to the natural environment. Clearing vegetation can reduce biodiversity and natural habitats. Natural drainage is impacted as green spaces are lost and waterways disrupted, increasing flood risk. Extracting water from aquifers or draining land can result in subsidence.

As urban areas grow, food will need to travel increasingly vast distances from farms and factories to feed the population of a megacity, adding to each individual's carbon footprint (the total greenhouse gas emissions produced as a result of their activities). The cost of transporting this food also creates issues for the urban poor as the cost of transport is added to the cost of the produce at the market.

Expanding cities and populations in developed nations also have greater energy needs to maintain their lifestyles. Generating high levels of electricity also presents challenges for the natural environment, particularly in countries that rely on coal-fired power stations, which emit CO_2.

As shown in figure 4.81, while cities worldwide only occupy 2 per cent of the total land area, they contribute greatly to not only economic activity but also consumption, emissions and waste production.

As urban populations increase, the environmental quality tends to worsen, particularly in developing countries that lack the resources to adequately tackle the problem of pollution. The traditional view of the relationship between economic development and the biophysical environment is that economic growth leads to environmental decline. This means that, as a country develops economically, its natural environment will suffer the effects of land degradation, air and water pollution, resource depletion and loss of biodiversity.

However, in recent years, this traditional view of economic development and the environment has been challenged by some economists, who have argued that once a country reaches a certain level of income, the quality of its environment will start to improve rather than continue to decline.

FIGURE 4.80 Coal-fired thermal power plants at dusk in Hohhot, Inner Mongolia, China

FIGURE 4.81 Global economic contributions and environmental impacts of cities

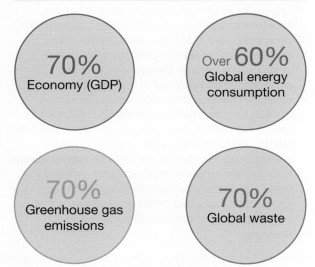

70% Economy (GDP)

Over **60%** Global energy consumption

70% Greenhouse gas emissions

70% Global waste

Source: From The New Urban Agenda by Habitat III, http://habitat3.org/

Stronger economies and better environmental quality

The **Environmental Kuznets Curve** (see figure 4.82) was created to demonstrate this view, based on the work of economist Simon Kuznets. Kuznets developed a hypothetical curve that graphs economic inequality against income per capita over the course of time. He suggested a country transitioning from being a primarily rural and agricultural economy to an industrialised urban economy would initially experience increasing levels of inequality, but these levels would eventually peak and then decline. Using various indicators of environmental degradation, the Environmental Kuznets Curve suggests an environment will become increasingly degraded as per capita income increases until a certain point at which the community has a strong enough economy and access to the required technology to implement strategies to better manage and protect the quality of their environment.

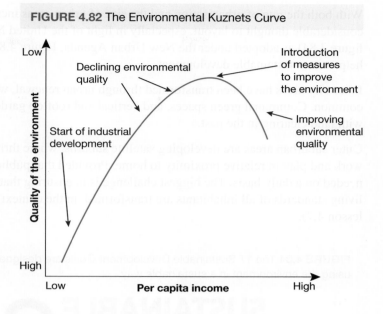

FIGURE 4.82 The Environmental Kuznets Curve

Figure 4.83 shows the sulphur dioxide emissions over the past 170 years. It could be interpreted that sulphur dioxide emissions have followed the Environmental Kuznets Curve theory. For example, Europe's emissions increased rapidly from the mid-1940s until peaking in 1980 and then starting to decline, as environmental legislation was brought in to reduce the levels of sulphur dioxide emissions. Similar patterns can be seen in other developed regions, whereas developing regions are still at the peak of emissions and yet to start to decline.

Environmental Kuznets Curve
a hypothetical curve that graphs environmental damage against income per capita over the course of time

FIGURE 4.83 Sulphur dioxide emissions by world region (tonnes per year)

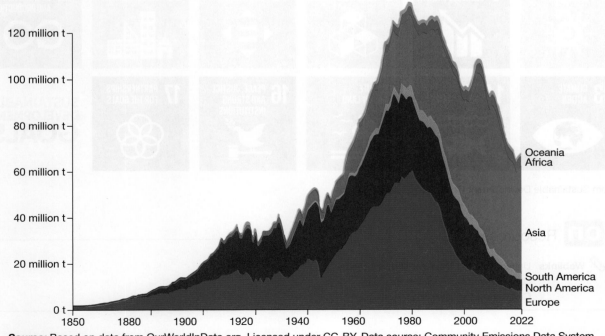

Source: Based on data from OurWorldInData.org. Licensed under CC-BY. Data source: Community Emissions Data System (CEDS) based on Hoesly et al. (2018)

4.9.3 Sustainable megacities

With both the rate of urbanisation and the size of urban areas increasing, it is important that town planners give considerable thought to layout, especially in light of the United Nations Sustainable Development Goals (see figure 4.84) as adopted under the New Urban Agenda. Figure 4.85 highlights some of the ways city planning can help ensure sustainable development.

Inner-city areas have been transformed through urban renewal, where medium- to high-density housing is more common. Communal green spaces, and vertical and rooftop gardens help address some of the issues associated with land clearing in the past.

Outer suburban areas are developing satellite suburbs, where thriving communities enable individuals to live, work and play in relative proximity to home. Provided that public transport networks also keep pace, cars are not needed on a daily basis. The biggest challenge is in ensuring that as cities expand in developing countries, the living standards of all inhabitants are transformed, in the context of Sustainable Development Goal 11 (refer to lesson 4.7).

FIGURE 4.84 The 17 Sustainable Development Goals are designed to ensure a prosperous future for all while using the environment in a sustainable way.

From Sustainable Development Goals, 2015 United Nations

---

on Resources

🔗 **Weblinks** United Nations: The New Urban Agenda
UN Sustainable Development Goals
Cities and urbanisation

FIGURE 4.85 Cities play an important role in ensuring sustainable development.

ROLE OF CITIES IN SUSTAINABLE DEVELOPMENT

Cities
play a central role in moving the sustainable energy agenda forward.

Current global share of renewable energy supply is

11%

The diversity of renewable energy resources is vast and research indicates a potential contribution of renewable energy reaching

60%
of total world energy supply.

Sustainable urban mobility
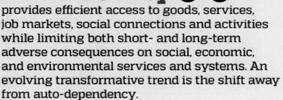
provides efficient access to goods, services, job markets, social connections and activities while limiting both short- and long-term adverse consequences on social, economic, and environmental services and systems. An evolving transformative trend is the shift away from auto-dependency.

Good governance
is crucial for developing, maintaining, and restoring sustainable and resilient services and social, institutional, and economic activity in cities. Many city governments are weakened due to limited power and responsibility over key public services, including planning, housing, roads and transit, water, land-use, drainage, waste management and building standards.

Source: Copyright © United Nations Human Settlements Programme, 2016, World Cities Report 2016

SKILLS ACTIVITY: Creating population pyramids

Nigeria is a low-income country that is home to one of the fastest growing megacities in the world. The metropolitan area of Lagos, Nigeria's largest city, is home to around 21 million people. Japan is a high-income country and its capital, Tokyo, is the world's most populated megacity, with around 40 million people in its metropolitan area. Both countries have very different population profiles: Japan's population is ageing and declining, while Nigeria's is young and growing. These demographics can lead to different challenges and opportunities in each country's megacities.

1. Use the data in table 4.11 to create population pyramids for Nigeria and Japan. Check the Resources tab for a guide to creating a population pyramid.

TABLE 4.11 Population data for Nigeria and Japan

Nigeria's population data			Japan's population data		
Age	Male	Female	Age	Male	Female
0–4	18 186 588	17 720 903	0–4	2 112 808	2 014 247
5–9	15 990 598	15 561 303	5–9	2 441 552	2 333 895
10–14	14 365 206	13 879 771	10–14	2 679 158	2 558 848
15–19	12 392 441	11 933 175	15–19	2 804 148	2 681 779

(continued)

TABLE 4.11 Population data for Nigeria and Japan *(continued)*

Nigeria's population data			Japan's population data		
20–24	10 266 501	9 906 705	20–24	3 042 548	2 906 363
25–29	8 408 160	8 169 115	25–29	3 125 979	2 994 763
30–34	7 010 193	6 840 646	30–34	3 180 970	3 033 986
35–39	6 023 469	5 916 015	35–39	3 474 773	3 347 631
40–44	5 255 903	5 167 756	40–44	3 819 190	3 714 855
45–49	4 267 099	4 210 892	45–49	4 608 558	4 492 935
50–54	3 267 954	3 263 587	50–54	4 695 396	4 621 492
55–59	2 543 462	2 586 131	55–59	4 074 823	4 051 348
60–64	1 955 318	2 033 555	60–64	3 681 109	3 729 616
65–69	1 408 470	1 502 281	65–69	3 545 202	3 716 034
70–74	911 885	992 013	70–74	4 206 950	4 671 066
75–79	548 175	597 344	75–79	3 458 583	4 136 654
80–84	244 564	277 905	80–84	2 589 919	3 520 935
85–89	731 57	91 343	85–89	1 547 605	2 607 840
90–94	12 603	18 955	90–94	640 440	1 528 257
95–99	1094	2206	95–99	166 364	595 814
100+	46	135	100+	17 646	122 423

2. Discuss the challenges that might be caused by each country's demographics, as shown by their population pyramids.
3. Evaluate the possible economic opportunities that might exist in the next 30 to 50 years for the following cities.
   a. Lagos
   b. Tokyo

4.9 Exercise

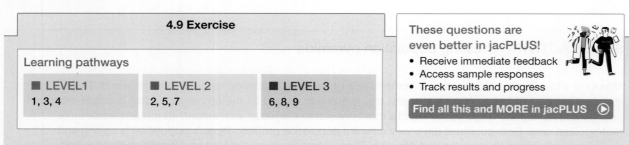

4.9 Exercise
Learning pathways

■ LEVEL 1	■ LEVEL 2	■ LEVEL 3
1, 3, 4	2, 5, 7	6, 8, 9

These questions are even better in jacPLUS!
- Receive immediate feedback
- Access sample responses
- Track results and progress

Find all this and MORE in jacPLUS ▶

Explain and comprehend

1. **Explain** the ways in which e-commerce and new technology might affect liveability in a megacity in a developed nation.
2. **Explain** the ways in which rapid population growth might affect liveability in a megacity in a developing nation.
3. **Summarise** the potential challenges and opportunities for individuals in a megacity in a developed nation. Consider factors including employment, housing, transport, sanitation and waste management, health and education, food and water security, energy needs and social factors such as community engagement or crime.
4. **Summarise** the potential challenges and opportunities for individuals in a megacity in a developing nation. Consider factors including employment, housing, transport, sanitation and waste management, health and education, food and water security, energy needs and social factors such as community engagement or crime.

Analyse and apply

5. Technological developments have led to changing job markets and economies in developed nations.
   a. **List** potential challenges that might arise in a city with high levels of unemployment.
   b. **Propose** and justify one strategy that authorities might employ to combat one of the challenges you identified.
6. Rapid increases in urban population present challenges for communities and for the individuals who relocate.
   a. **List** potential challenges that might be faced by a low-income family in a developing country moving from a rural area to an urban area.
   b. **Suggest** and justify one strategy that would minimise the negative impact of such a move for the family. In your answer, outline clearly who would need to implement the strategy: government authorities or other agencies such as non-government support agencies or community groups.
7. With reference to the Environmental Kuznets Curve, **explain** why urban areas in developing countries are more likely to have poor air quality than urban areas in developed countries.
8. Study figure 4.83.
   a. **Describe** the differences in sulphur dioxide emissions between Europe and Asia.
   b. **Suggest** a possible explanation for these differences.
9. Analyse figure 4.86 and suggest possible reasons for the difference in carbon footprint between rural and urban locations in India.

FIGURE 4.86 Contribution of various sources to carbon footprint in rural and urban areas, India

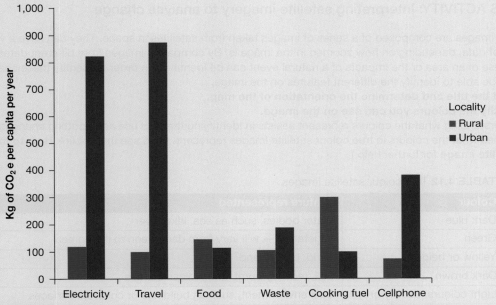

Source: © 2014 The Authors. Published by Elsevier Ltd. Selection and/or peer-review under responsibility of the Organizing Committee of ICTMS-2013. Open access under CC BY-NC-ND license.

Sample responses are available in your digital formats.

4.10 APPLY YOUR SKILLS — Interpreting satellite imagery

4.10.1 Interpreting satellite imagery

Over the past four decades, Dublin, the capital of Ireland, has experienced significant growth. The city has grown from having a metro population of around 900 000 in the 1980s to 1.4 million in the 2020s. As the city's population has grown, so too has its environmental footprint. Satellite imagery provides a lens to observe and analyse this change.

SKILLS ACTIVITY: Interpreting satellite imagery to analyse change

Satellite images are composed of a series of images taken from satellites in space. They can have a varying degree of data, depending on how zoomed in the image is. By comparing images from different dates, changes in land use of an area or the impacts of a natural event can be identified. In order to identify these changes, you need to be able to identify the different features on the image.

1. **Read the title and determine the orientation of the map.**
2. **Identify the colours you can see on the image.**
   Understanding what the colours represent assists in identifying the land use and spotting changes. Table 4.12 outlines what the colours in true colour satellite images represent. (Also see the weblink **How to interpret a satellite image** for further help.)

TABLE 4.12 True colour satellite images

Colour	Feature represented
Dark blue	Water bodies, such as sea, lake, river
Green	Vegetation — will vary from dark green to light green
Yellow or beige	Sand, barren land
Dark brown	Soil, recently burnt land, volcanic areas
Light colours, usually grey	Built environment, such as buildings or concrete surfaces
White (or light grey)	Snow or ice, or occasionally a building with a white roof or cloud

3. **Identify the shapes and patterns.**
   Studying the shape of features can assist you in understanding how the natural environment has been altered. For example:
   - The coastline can give an indication of whether it has been modified by humans — for example, a very straight coastline may have a sea wall, or you may be able to identify a harbour wall.
   - A straight river rather than a meandering one may have had its course altered to allow for new urban development or to mitigate flooding impacts.
   - Forest areas that show trees growing in straight rows may indicate a plantation rather than a natural forest.

When you study built areas, you can often identify urban planning methods and infrastructure growth. For example:

- Many newer urban areas, such as Melbourne's suburbs or New York City, have a carefully planned grid system of streets, whereas many older towns or cities, such as Dublin's city centre (see figure 4.87), appear to grow out from the CBD in a circular pattern.
- New road systems can be seen as the infrastructure develops to account for the growing urban population

4. **Estimate the growth of areas.**

   To estimate the expansion of an area, using a grid over the satellite image can assist in determining the percentage change. (Grids and lines have been added to the images in figure 4.87.) As well as identifying urban sprawl, an increase in building density can also be seen.

FIGURE 4.87 Dublin's urban area in (a) 1984 and (b) 2022, with grids and lines added

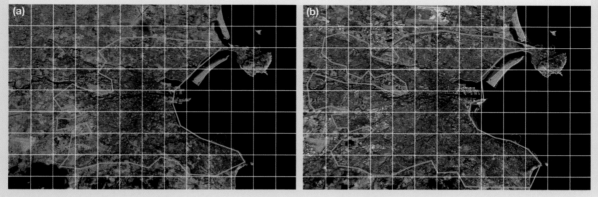

False-colour satellite images

Some satellite images have different colour patterns because they use infrared light. In these satellite images, vegetation will sometimes appear red and built areas will appear purple, depending on the infrared light used. Explore the **Why is that forest red and that cloud blue?** weblink to find out more.

Analyse the satellite images

Refer to the two images shown in figure 4.88.

FIGURE 4.88 Dublin's urban area in (a) 1984 and (b) 2022

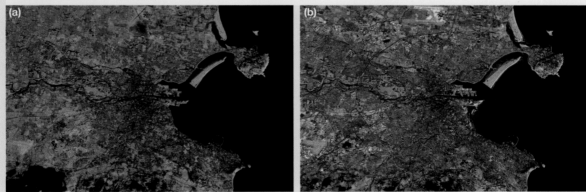

1. Describe the changes to Dublin's size over time.
2. Identify some of the infrastructure changes you can see between 1984 and 2022.
3. Explain why these infrastructure changes would be necessary.
4. Estimate the extent that the forested area around Dublin has declined.
5. Suggest reasons for the change in land use in the south west corner of the image.

6. Describe the change in the appearance of land use at the Dublin port. What does this suggest has happened?
7. Explain the possible impact of these changes on the natural environment of the area.
8. Create a plan for Dublin to cater for future population growth. Consider how city planners could maintain or improve social, environmental and economic sustainability.

on Resources

Weblinks   How to interpret a satellite image
Why is that forest red and that cloud blue?

LESSON
4.11 Review

4.11.1 Summary

4.2 Global patterns of urbanisation

- Megacities result from global population growth and urbanisation, rooted in early communities' ability to produce food surpluses and division of labor.
- The Industrial Revolution accelerated urban growth in Europe and North America, with cities such as New York and Chicago expanding rapidly.
- As economies shifted from agriculture to industry, urbanisation occurred.
- Urbanisation rates have slowed, but urban populations continue to increase globally, particularly in Africa and Asia.
- Urbanisation is closely linked to economic growth, with cities accounting for around 70 per cent of the world's GDP, and globalisation further intertwines with urbanisation through increased interconnection and trade.

4.3 Vulnerability to natural hazards in megacities

- Urbanisation concentrates wealth in large cities, making economies more vulnerable to disasters.
- Poorly managed urban growth in developing countries leads to illegal settlements in hazardous areas, exacerbating vulnerability to natural hazards.
- Climate change and rapid urban expansion increase flood risk in megacities worldwide.
- Impermeable urban surfaces and inadequate drainage systems contribute to urban flood risk.
- Climate change intensifies extreme weather events, such as heatwaves and heavy rainfall, increasing the vulnerability of megacities such as Tokyo and Jakarta.

4.4 Water and energy in megacities

- Access to clean water is a human right, but in megacities such as Jakarta and Manila, residents in urban slums often pay significantly more for water, highlighting issues of affordability and inequality.
- Urban water supplies must meet three criteria: ample and sustainable supply, good water quality, and an efficient distribution network.
- Rapid urbanisation often outpaces government capacity to manage water needs, leading to critical shortages, especially during droughts.
- Outdated and inefficient energy infrastructure in many megacities, particularly in the developing world, relies heavily on fossil fuels, contributing to air pollution, greenhouse gas emissions, and energy insecurity.
- Meeting the energy needs of megacities requires a comprehensive approach that considers urban planning, technological innovation and environmental sustainability.

4.5 Sanitation and waste management in megacities

- COVID-19 provides a recent example of how the higher density of urban areas can result in a quicker spread of disease.
- Poorly planned or rapid urbanisation exacerbates waste management challenges, with megacities such as Manila and Colombo facing environmental catastrophes due to inadequate infrastructure and poverty.
- Waste accumulation in urban areas contributes to the spread of diseases such as dengue, with densely populated and poorly managed areas being particularly vulnerable to outbreaks.

4.6 Housing people in megacities

- In developing countries, poor rural migrants often move to urban areas that are unprepared for their arrival. This can lead to informal settlements, or slums, developing.
- Some slums have become similar to functioning cities themselves, with vibrant communities.
- In more economically developed countries, slums are less common, but many urban residents face challenges such as high costs and long travel times.

4.7 Social opportunities and challenges in megacities

- Cities tend to have more health and educational services available, which can be important pull factors.
- A major challenge for cities is ensuring that food production can meet the needs of increasing populations.

4.8 Economic opportunities and challenges in megacities

- Urban growth and economic development are intertwined, with cities serving as hubs of innovation, productivity and opportunity.
- Cities concentrate human capital, industries and infrastructure, fostering economies of scale, specialisation and market demand.
- Urbanisation promotes social mobility, investment and innovation, driving economic dynamism and competitiveness.
- The influx of rural migrants into Chinese cities has fuelled urban growth and consumption-driven economic expansion.
- India has urbanised at a slower rate than China, but is expected to continue to see major urban growth and subsequently major economic growth.

4.9 Sustainable development in megacities

- Sustainable waste management strategies, such as recycling and composting, are crucial to reduce environmental impact and promote resource conservation.
- Implementation of renewable energy sources and energy-efficient practices can mitigate the environmental footprint of expanding urban areas.
- Planning for green spaces, vertical gardens and rooftop gardens can enhance biodiversity, improve air quality, and provide recreational areas for urban residents.
- Promotion of public transport, pedestrian-friendly infrastructure and cycling networks reduces reliance on cars, alleviates traffic congestion and lowers carbon emissions.
- Adoption of sustainable urban planning principles, including ecocity development and compact, high-density housing, prioritises environmental preservation and enhances the resilience of cities to climate change.

4.11.2 Key terms

anthropogenic change caused or influenced by people, either directly or indirectly

cumecs an abbreviation of 'cubic metres per second'

discharge the volume of water flowing through a river channel, measured in cumecs

Environmental Kuznets Curve a hypothetical curve that graphs environmental damage against income per capita over the course of time

evapotranspiration the process by which water moves into the atmosphere. This occurs through evaporation (from the ground or surfaces) and through the transpiration of plants.

globalisation the process by which the world has become increasingly inter-connected through freer movement of capital, goods and services. It is reflected in the value of cross-border world trade expressed as a percentage of total global GDP.

gross domestic product also known as GDP; the total value of all goods and services produced within a country, usually over a period of one year

lag time the time between the peak rainfall and the peak river discharge

megacities cities with 10 million or more inhabitants

pseudo-urbanisation a type of urbanisation without economic growth that results in large numbers of poverty-stricken residents living in informal settlements

pull factors factors that attract people to an area

tenure the occupancy or lease of an area of land

thermal mass capacity to absorb and retain heat energy

urban heat island microclimates created in an urban environment by the hard, dark surfaces that attract and retain heat, such as roads and buildings

urbanisation growth in the proportion of a population living in urban environments

KEY QUESTIONS REVISITED

1. What factors led to the growth of megacities?
2. What patterns or trends can be seen in where and how urban areas and megacities have developed?
3. What predictions can be made about future patterns of megacity development?
4. How do countries manage the challenges presented by urban development? Do responses differ between developed and developing countries?
5. Can megacities be planned in a way that promotes sustainability and the wellbeing of the people who live there?

4.11.3 Exam questions

4.11 Section I – Short answer question

▶ Question 1 (4 marks)

Study figure 4.89.

> **FIGURE 4.89** Share of urban population living in slums and the location of existing and predicted megacities

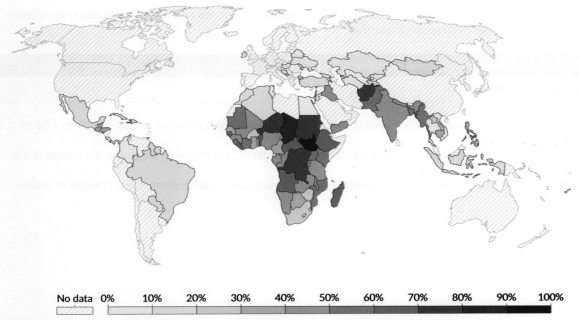

No data 0% 10% 20% 30% 40% 50% 60% 70% 80% 90% 100%

Source: OurWorldInData.org/urbanization. Licensed under CC BY. Data source: UN Human Settlements Programme

Many of the megacities in developing countries have slum areas. Explain why slums are more prevalent in the developing world.

▶ Question 2 (4 marks)

Study figure 4.90.

FIGURE 4.90 Exposure to PM2.5 air pollution versus GDP per capita, 2019

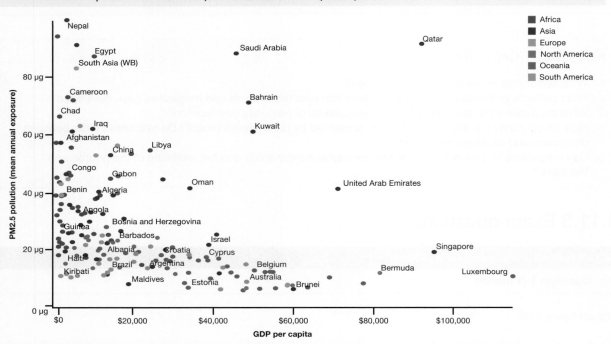

Source: OurWorldInData.org/air-pollution. Licensed under CC BY. Data source: Multiple sources compiled by World Bank (2024); World Bank (2023)

With reference to figure 4.90, explain why megacities in developing countries are more likely to have higher levels of air pollution than those in developed countries.

1.12 Section II – Extended response question

▶ Question 3 (16 marks)

In a written response of approximately 450–600 words, refer to the stimulus provided in figures 4.91 to 4.96 to respond to the following.
- Analyse the figures to identify the geographical challenges facing the megacity Kinshasa, the capital of the Democratic Republic of Congo.
- Apply geographical understanding to generalise about how these challenges can impact people or places.

FIGURE 4.91 Democratic Republic of the Congo, floods and landslides, December 2022

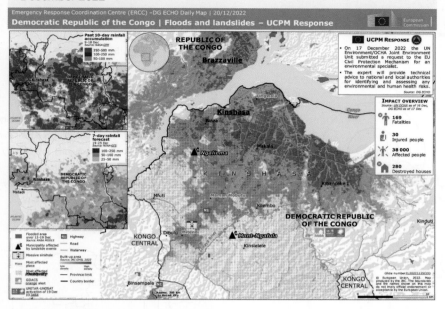

Source: ECHO (2022), Democratic Republic of the Congo | Floods and landslides - UCPM Response.

FIGURE 4.92 The main Kuluna gangs in Kinshasa, Democratic Republic of the Congo

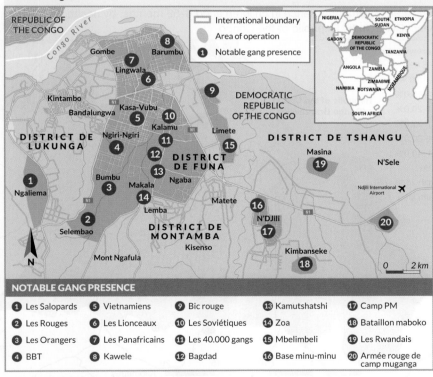

Source: © 'Global Initiative Against Transnational Organized Crime', Criminals or Vigilantes? The Kuluna Gangs of the Democratic Republic of Congo, May 2021, p. 5.

FIGURE 4.93 Precarious neighbourhoods in Kinshasa, Democratic Republic of the Congo

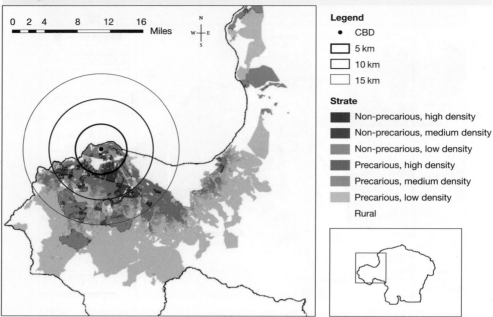

Source: The World Bank. Project Information Document, Kinshasa Multisector Development and Urban Resilience Project (P171141)
Note: Precarious neighbourhoods are those with low liveability and walkability, suffering from poor services, and with various environmental and disaster risks.

FIGURE 4.94 Malaria prevalence in children aged 6–59 months, by health area in Kinshasa, Democratic Republic of the Congo

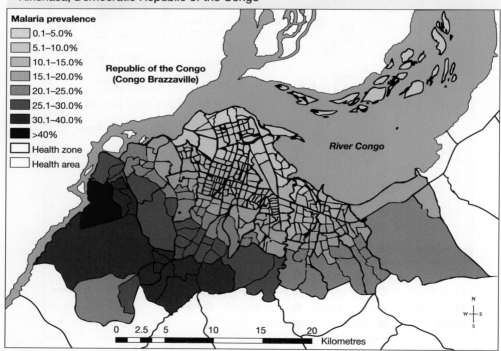

Source: Ferrari G, Ntuku HM, Schmidlin S, Diboulo E, Tshefu AK, Lengeler C. A malaria risk map of Kinshasa, Democratic Republic of Congo. Malar J. 2016 Jan;15 27. doi:10.1186/s12936-015-1074-8. PMID: 26762532; PMCID: PMC4712518.

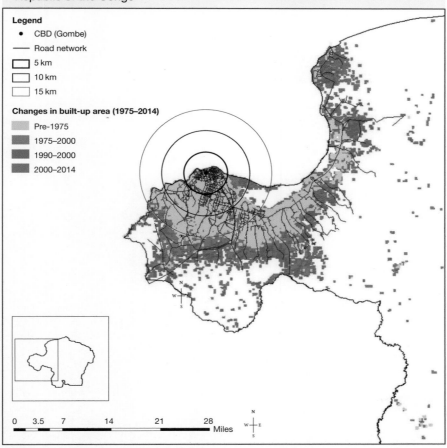

FIGURE 4.95 Spatial expansion of built-up areas, Kinshasa, Democratic Republic of the Congo

Legend

● CBD (Gombe)
— Road network
☐ 5 km
☐ 10 km
☐ 15 km

Changes in built-up area (1975–2014)

Pre-1975
1975–2000
1990–2000
2000–2014

0   3.5   7   14   21   28 Miles

Source: Yele Maweki Batana, et al (2021), Poverty and Equity Global Practice. Policy Research Working Paper 9857.

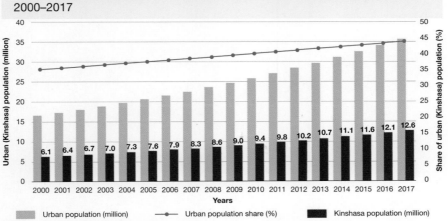

FIGURE 4.96 Urban population trends in Democratic Republic of the Congo, 2000–2017

Kinshasa population (million): 6.1, 6.4, 6.7, 7.0, 7.3, 7.6, 7.9, 8.3, 8.6, 9.0, 9.4, 9.8, 10.2, 10.7, 11.1, 11.6, 12.1, 12.6

Urban population (million) | Urban population share (%) | Kinshasa population (million)

Source: Yele Maweki Batana, et al (2021), Poverty and Equity Global Practice. Policy Research Working Paper 9857.

Sample responses are available in your digital formats.

GLOSSARY

acidification an increase in hydrogen ions, which lowers the pH of the water

animal invasions the introduction of a non-indigenous animal to an area that negatively impacts the environment

anthropogenic processes processes that involve human activity; for example, the burning of fossil fuels to produce electricity

anthropogenic change caused or influenced by people, either directly or indirectly

aquifers layers of rock that can hold large quantities of water in the pore spaces

asthenosphere the upper layer of the mantle, below the lithosphere, usually more than 100 km below the surface. It is where rock becomes molten and allows the solid tectonic plates to move over it.

atmospheric hazards a potentially damaging natural event generated in the troposphere, such as a severe storm, tropical cyclone (typhoons and hurricanes), tornado, blizzard and wind storm

bioaccumulation the process by which pollutants enter and concentrate through the food chain by passing from one food source to another by being eaten

biodegrade to break down through natural processes

biological processes that are vital for organisms to live, for example, plants require the process of photosynthesis to survive

biophysical environment both living (biotic) and non-living (abiotic) surroundings of an organism or population, made up of the elements of the atmosphere, hydrosphere, lithosphere and biosphere

climatological hazard a hazard that occurs due to the climatic conditions of an area, such as bushfires, droughts and heatwaves

counterurbanisation the migration of people from urban to rural areas

cumecs an abbreviation of 'cubic metres per second'

demographic profile a detailed description of a group of people, including information such as their ages, genders, ethnicities, incomes, education levels, jobs, family sizes and where they live

discharge the volume of water flowing through a river channel, measured in cumecs

disease a condition that causes harm to, or interferes with, the normal functioning of a living thing

ecological hazard an interaction between living organisms or between living organisms and their environment that could have a negative effect

endemic located in a specific area

environmental Kuznets curve a hypothetical curve that graphs environmental damage against income per capita over the course of time

epicentre the point on the Earth's surface directly above the focus when an earthquake has occurred

epidemic a disease outbreak that affects many people in a specific region

evapotranspiration the process by which water moves into the atmosphere. This occurs through evaporation (from the ground or surfaces) and through the transpiration of plants.

exposure the degree or likelihood of a place, person or thing being affected by a hazard, in terms of risk assessment

extreme weather event a weather event that is rare at a particular place and/or time of year, with unusual characteristics in terms of magnitude, location, timing or extent

faults large cracks in the Earth's crust, often associated with the boundaries of the Earth's tectonic plates

flow movement when rock, soil or sand mix with water and air and move downhill in a flow

focus where an earthquake rupture occurs in the crust or mantle. Seismic waves radiate away from the focus.

garden city area that contains proportionate areas of residences, industry and agriculture

gentrification the process of older areas, usually of lower socioeconomic status, being slowly bought out and renovated by wealthier people

geographic scale the spatial extent or level of detail at which a geographic area is analysed or represented, ranging from local to global

geological hazards a potentially damaging natural event occurring in the Earth's crust, such as a volcanic eruption, earthquake or tsunami

geomorphic hazards a potentially damaging event on the Earth's surface — such as an avalanche, landslide or mudslide — that is often caused by a combination of natural and human processes

geomorphic related to the formation of the Earth's surface and its changes

globalisation the process by which the world has become increasingly inter-connected through freer movement of capital, goods and services. It is reflected in the value of cross-border world trade expressed as a percentage of total global GDP.

green belt area of largely undeveloped, wild or agricultural land surrounding or neighbouring urban areas

gross domestic product also known as GDP; the total value of all goods and services produced within a country, usually over a period of one year

hazard zone an area that may be affected by a natural hazard; for example, areas vulnerable to flooding based on past events or areas likely to be affected by pyroclastic flows from a volcano

hazard something that has the potential to cause harm. It may be obvious (for example, a flooded section of road) or not obvious (for example, a damaged hidden electrical wire).

hydrological hazard an extreme event with a high-water component, such as flash flooding, cyclones, ice melt, storm surges and tsunamis

indigenous something that is native to or originaes from a specific place; for example, kangaroos are indigenous to Australia. Note that the word often has a different meaning when capitalised. The term 'Indigenous' refers to First Nations Peoples of Australia of Aboriginal or Torres Strait Islander descent.

infectious diseases contagious or communicable diseases that are spread by being passed from one person to another

infill development construction that occurs on vacant or underused land within an existing urban area

inter-tropical convergence zone (ITCZ) the zone near the Equator where trade winds of the northern and southern hemispheres meet. The intense heat, warm water and high humidity create what is an almost permanent band of low pressure. The monsoon trough seen on weather charts is part of the ITCZ.

invasive animal a non-indigenous animal species that has been introduced to a specific area by people (either intentionally or accidentally) and has multiplied to an extent that it threatens to damage or is damaging the economic, environmental or social value of a place

invasive plant a non-indigenous plant species that has been introduced to a specific area by people (either intentionally or accidentally) and has multiplied to an extent that it threatens to damage or is damaging the economic, environmental or social value of a place

lag time the time between the peak rainfall and the peak river discharge

landslide the large-scale movement of rock, debris and soil down a slope due to unstable conditions, which may be caused by heavy rainfall, an earthquake or a volcanic eruption. A landslide under the ocean can cause a tsunami.

liquefaction when saturated or partially saturated soil loses its firmness and displays the properties of a liquid, such as when an earthquake shakes and loosens wet soil in low-lying areas, and the soil loses rigidity and moves like fluid, covering things in its path

magma hot molten rock formed below or within the Earth's crust. It reaches the surface through volcanic or plate tectonic activity and becomes lava and eventually igneous rock.

magnitude a measure of size; for example, earthquakes are measured according to magnitude on the Richter Scale

megacities cities with 10 million or more inhabitants

metropolitan area a densely populated urban area with significant economic activity, cultural diversity and social interaction

monofunctional development developments that have a single function or purpose, such as a development that is solely for residential buildings

mudflow when large amounts of suspended silt and soil move quickly down a slope. They tend to occur mostly on steep slopes but can happen anywhere ground is unstable due to loss of vegetation.

natural disaster a large natural event, such as a cyclone, flood, earthquake or landslide, that causes considerable loss of life, damage to property and infrastructure, and/or destroys sections of the environment

natural hazard an extreme event occurring either in the lithosphere or in the atmosphere. It can be highly destructive and cause considerable harm to living things and property. Examples include tropical cyclones, tornadoes, earthquakes and volcanoes.

non-point sources broad areas a pollution hazard originates from; for example, run-off from city streets

P-waves also known as primary waves; high-frequency seismic waves that travel fastest and are measured first at a seismic station. P-waves can pass through solid rock and liquids.

pandemic a disease outbreak that is prevalent across a wide geographic area, including beyond the region in which the outbreak began

permafrost soil, rock or sediment that remains frozen for two or more years, commonly found in polar regions and high mountain areas

physical environment natural and built surroundings

plant invasions the introduction of a non-indigenous plant to an area that negatively impacts the environment

point sources a particular location from which a pollution hazard originates; for example, an industrial site or an oil tank

polar vortex a large area of low pressure and cold air that typically resides over the polar regions during the winter months, but can occasionally shift southward, bringing frigid temperatures and winter weather to lower latitudes

pollutants substances introduced into the environment that are potentially harmful to human health and the natural environment

pseudo-urbanisation a type of urbanisation without economic growth that results in large numbers of poverty-stricken residents living in informal settlements

pull factors factors that attract people to an area

push factors factors that encourage people to leave an area

pyroclastic clouds rapidly moving currents of hot air, gases and ash that run from the crater down the sides of a volcano. They are extremely lethal due to their high speed and lack of sound.

pyroclastic cones steep conical volcanic cones built by a combination of lava flows and cinder/ash from pyroclastic eruptions. They can form quickly and remain active for long periods.

qualitative data based on personal experience

quantitative data that can be measured as an exact quantity

risk management strategies and actions to reduce or mitigate risk based on the known consequences of encountering a hazard

risk the potential for something to go wrong. This is a subjective assessment about actions that may be predictable or unforeseen.

rural areas with low population densities, agricultural activities and other characteristics of the countryside

sea change relocation from an urban area to a coastal or seaside location for a lifestyle change

sea floor spreading the divergence of two oceanic crust plates

seismic waves waves of energy travelling away from an earthquake; are like huge vibrations and may travel through the Earth's mantle and crust or along the surface

slope failure when the pull of gravity causes a hill or mountain slope to collapse

social demography the statistical study of populations using measures such as age, sex and income

subduction zones the areas of the mantle in which convergent plates collide. Under the ocean, these areas are called trenches.

subduction a geological process where two tectonic plates collide at convergent boundaries

subsidence when part of the land sinks or collapses

suburbanisation the development and outward spread of new suburban areas

tectonic plates slow-moving plates that make up the Earth's crust. Volcanoes and earthquakes often occur at the edges of plates.

tenure the occupancy or lease of an area of land

thermal mass capacity to absorb and retain heat energy

toxicological related to the negative impacts of chemical substances

tree change relocation from a city or suburb to a rural or forested area for a lifestyle change

urban heat island microclimates created in an urban environment by the hard, dark surfaces that attract and retain heat, such as roads and buildings

urban sprawl process of outward expansion of urban areas from the CBD into the surrounding countryside and invading adjacent towns, regions and undeveloped land

urban areas with high population densities, human development and other characteristics typical of cities and towns

urbanisation growth in the proportion of a population living in urban environments

vector-borne diseases diseases that are carried by organisms, such as mosquitos or fleas, that are capable of transmitting disease-producing bacteria, viruses and parasites from one person to another

vulnerability the degree of risk faced by a place, person or thing, based on an approaching hazard's potential impact, the place's degree of preparedness, and the resources available to respond

vulnerability the degree of risk faced by a place, person or thing, based on an approaching hazard's potential impact, the place's degree of preparedness, and resources available to respond

wildfire an uncontrolled fire that spreads through vegetation

wind shear a sudden change in wind speed and/or direction over a relatively short distance in the atmosphere, generally due to altitude

INDEX